HIGH PRESSURE EFFECTS ON CELLULAR PROCESSES

CELL BIOLOGY: A Series of Monographs

G. M. Padilla, G. L. Whitson, and I. L. Cameron. THE CELL CYCLE: *Gene-Enzyme Interactions,* 1969

Arthur M. Zimmerman (editor). HIGH PRESSURE EFFECTS ON CELLULAR PROCESSES, 1970

HIGH PRESSURE EFFECTS ON CELLULAR PROCESSES

Edited by ARTHUR M. ZIMMERMAN

DEPARTMENT OF ZOOLOGY
UNIVERSITY OF TORONTO
TORONTO, CANADA

1970

AP

ACADEMIC PRESS New York and London

ACADEMIC PRESS, INC.
111 Fifth Avenue, New York, New York 10003

United Kingdom Edition published by
ACADEMIC PRESS, INC. (LONDON) LTD.
Berkeley Square House, London W1X 6BA

LIBRARY OF CONGRESS CATALOG CARD NUMBER: 70-114220

PRINTED IN THE UNITED STATES OF AMERICA

LIST OF CONTRIBUTORS

Numbers in parentheses indicate the pages on which the authors' contributions begin.

ROBERT R. BECKER (71), Department of Biochemistry and Biophysics, Oregon State University, Corvallis, Oregon

HENRY EYRING (1), Department of Chemistry, University of Utah, Salt Lake City, Utah

H. FLÜGEL (211), Institute of Marine Research, University of Kiel, Kiel, Germany

FRANK H. JOHNSON (1), Department of Biology, Princeton University, Princeton, New Jersey

J. A. KITCHING (155), School of Biological Sciences, University of East Anglia, Norwich, England

J. V. LANDAU (45), Department of Biology, Rensselaer Polytechnic Institute, Troy, New York

DOUGLAS MARSLAND (259), Research Professor Emeritus, New York University, New York, New York

RICHARD Y. MORITA (71), Department of Microbiology and Department of Oceanography, Oregon State University, Corvallis, Oregon

TETUHIDE H. MURAKAMI (131, 139), Department of Physiology, Okayama University Medical School, Okayama, Japan

C. SCHLIEPER (211), Institute of Marine Research, University of Kiel, Kiel, Germany

ARTHUR M. ZIMMERMAN (139, 179, 235), Department of Zoology, University of Toronto, Toronto, Canada

SELMA B. ZIMMERMAN (179), Glendon College, York University, Toronto, Canada

CLAUDE E. ZOBELL (85), Scripps Institution of Oceanography, University of California, San Diego, La Jolla, California

This book is dedicated to Professor Douglas Marsland who remains a constant source of enthusiasm, stimulation, and enlightenment to both his colleagues and his students.

PREFACE

This book reflects the current knowledge of investigators whose chief concern has been to understand the effects of high pressure on cellular processes. Since the "classic" monograph "The Kinetic Basis of Molecular Biology" by Johnson, Eyring, and Polissar appeared in 1954, there has not been any extensive review of hydrostatic pressure research. This volume bridges the information gap which has occurred over the past two decades. It includes reviews of hydrostatic pressure on life processes from the molecular level to the organismic level. It will serve as a comprehensive treatise for all workers in the field of high pressure and will aid students of biology to interpret oceanic life processes. It is hoped that this work will be of interest to biochemists, physiologists, microbiologists, and protozoologists, as well as to marine scientists, who are concerned with life processes in the depths of the ocean. The enthusiasm of the authors should act as a stimulus for new research.

The organization of the chapters and the treatment of the subject matter are intended to give the reader a basic understanding of the mechanisms of high pressure and its effects on organisms of increasing complexity in form and organization. Johnson and Eyring discuss the kinetic basis of hydrostatic pressure, which remains a foundation for the interpretation of physiological activity. Landau, ZoBell, Morita, and Becker discuss various facets concerning the effects of pressure on bacterial systems and emphasize the molecular basis for these effects. Kitching, Murakami, S. Zimmerman, and A. Zimmerman discuss pressure effects on protozoa, relating the physiological, structural, and biochemical interactions. Also included is a review of Japanese high pressure literature by Murakami. Zimmerman and Marsland in their contributions consider pressure effects on the cell cycle in marine eggs. Marsland further evaluates, in extensive detail, pressure effects in interpreting a mechanism for cell division. In the chapter by Flügel and Schlieper pressure effects on higher organisms are considered.

In reality, man arose from the ocean—the first synthetic systems had as their cradle, the stormy seas. Eventually life left the sea, but the mystery of the ocean remains a challenge to man. How are certain physiological processes able to function at great oceanic depths, whereas other life processes cannot withstand such harsh environment? More than 67% of the earth is covered by water and the volume of all land above sea level is only one-eighteenth of the volume of the ocean (*Littoral Lines* 4 [3], August, 1969). One of the most important factors for understanding life that exists below the surface is an understanding of the effects of hydrostatic pressure. It is hoped that this book will give the reader greater insight into the effects of high pressure on cellular processes.

ARTHUR M. ZIMMERMAN

Toronto, Canada
January, 1970

CONVERSION TABLE

psi[a]	atm[b]	kg/cm^2	N/m^2 [c]
14.696	1	1.033227	1.01325×10^5
1,000	68	70	68.9×10^5
2,000	139	141	137.8×10^5
3,000	204	211	206.7×10^5
4,000	272	281	275.6×10^5
5,000	340	352	344.5×10^5
6,000	408	422	413.4×10^5
7,000	476	492	482.3×10^5
8,000	544	562	551.2×10^5
9,000	612	633	620.1×10^5
10,000	680	703	689.0×10^5
11,000	749	773	758.9×10^5
12,000	817	844	827.8×10^5
13,000	885	914	896.7×10^5
14,000	953	984	965.6×10^5
15,000	1,021	1,055	$1{,}034.5 \times 10^5$
16,000	1,089	1,125	$1{,}103.4 \times 10^5$
20,000	1,361	1,406	$1{,}379.3 \times 10^5$
50,000	3,402	3,515	$3{,}447.1 \times 10^5$
100,000	6,805	7,031	$6{,}897.2 \times 10^5$
150,000	10,207	10,546	$10{,}342.2 \times 10^5$
300,000	20,414	21,092	$20{,}684.5 \times 10^5$

[a] Pounds per square inch.
[b] Atmosphere.
[c] Newtons/meter^2.

CONTENTS

3. Hydrostatic Pressure Effects on Selected Biological Systems

Richard Y. Morita and Robert R. Becker

4. Pressure Effects on Morphology and Life Processes of Bacteria

Claude E. ZoBell

5. Japanese Studies on Hydrostatic Pressure

Tetuhide H. Murakami

6. A Pressure Study of Galvanotaxis in *Tetrahymena*

Tetuhide H. Murakami and Arthur M. Zimmerman

10. High Pressure Studies on Synthesis in Marine Eggs

Arthur M. Zimmerman

11. Pressure–Temperature Studies on the Mechanisms of Cell Division

Douglas Marsland

CHAPTER 1

THE KINETIC BASIS OF PRESSURE EFFECTS IN BIOLOGY AND CHEMISTRY

Frank H. Johnson and Henry Eyring

I. Introduction

The purpose of this chapter is to give a brief resume of the background of experimental studies on the biological effects of increased hydrostatic pressures, together with the development of a rational, even though only partial, understanding of the observed phenomena from the point of view of physical chemical theory. It would be neither feasible nor appropriate to undertake a detailed treatment of either the early background or the later biological investigations which have been considered at some length by Johnson *et al.* (1954) in a book which includes several chapters giving the bare essentials of modern reaction rate theory. While the general theoretical basis can be hardly more than alluded to in this chapter, and application of the theory can only be briefly illustrated, the remaining chapters by the different authors include competent and thorough expositions of various aspects of the biological problems, with indications of their present status in the fields of research involved.

A. Early Experiments

The first laboratory research on the biological effects of high hydrostatic pressures was inspired by the discovery of organisms living in the depths of the sea, as far down as 6000 m below the surface, where the pressure amounts to some 600 atm (8820 psi). Prominent among the expeditions which pioneered dredging such depths was that of the *Talisman* in the latter part of the nineteenth century (1882–1883), and prominent among the investigators who soon thereafter began to experiment with high pressures and biological processes were Regnard and Certes. Starting in 1884 their papers were published in the *Comptes Rendus* of the Académie des Sciences and of the Société de Biologie of Paris; a full monograph on the subject by Regnard appeared in 1891.

During the period of these early studies, as well as of later ones, certain circumstances had an important influence on the trend of the research as well as the results obtained. From the present perspective, the following are particularly noteworthy.

At first, the equipment used for producing high pressures in biological laboratories limited the amount of increased pressures to less than those which are now known to occur in the deepest parts of the sea, i.e., of the order of 1000 atm or roughly 15,000 psi, where various types of animal and bacterial life have been found to exist (Bruun, 1951, 1956, 1957; Wolff, 1960; ZoBell, 1952; ZoBell and Morita, 1957, 1959). In the early studies, the maximum pressure generally did not exceed 600 atm, which was well within the limits of those which occur in natural habitats. As later investigations showed, this range of pressures is perhaps most interesting of all from a physiological point of view, inasmuch as the changes they cause are often immediate, quantitatively great, and, if not maintained for too long a time, are largely reversible on decompression.

Within the above range of from 1 to 600 atm, it was soon discovered that some biological processes were profoundly retarded under increased pressure, whereas others were accelerated; in some instances accelerated by moderately increased pressure then retarded at somewhat higher pressures. In still other instances, increased pressure appeared to have practically no effect. In the published accounts of these experiments, temperature was usually not mentioned. There was no reason to expect or even suspect that the effect of pressure at one temperature would be significantly different from what it might be at another temperature. It was only natural, therefore, that the temperature involved was the one most convenient for the experiments, or the one which seemed most favorable for the organism or the process. The fact that a given biological process could be influenced by a given pressure in quite an opposite manner at different temperatures, under otherwise the same conditions, was not observed until some decades later in studies on muscle and bacterial luminescence, as discussed below (pp. 20–32). It seems a bit strange now that such a phenomenon was not even accidentally stumbled upon during the intervening years.

During the years which intervened between the early studies and those just mentioned in regard to muscle and bioluminescence, two developments of fundamental importance to later work took place. The first, very largely empirical, consisted of the rather gradual accumulation of further information concerning the effects of pressure on biological systems with increasing application of higher and higher pressures as equipment became more generally available for attaining greater pressures in the laboratory. Thus it was soon discovered that, at room temperature, pressures of the order of 5000 to 15,000 atm coagulated egg albumin (Bridgman, 1914) and carboxyhemoglobin Bridgman and Conant, 1929), inactivated enzymes, viruses, and toxins, as well as killed bacteria and yeasts (Basset and collaborators, 1932, 1933a,b,c, 1935a,b,c, 1938; Giddings *et al.*, 1929; Larson *et al*, 1918; Lauffer and Dow, 1941; Luyet, 1937a,b; Macheboeuf and collaborators 1933, 1934). As a consequence of such studies, an idea that "high hydrostatic pressure denatures proteins" developed and became a widely accepted generalization, so much so in fact that the opposite effect, *viz.*, a retardation and even reversal of protein denaturation under the influence of somewhat less pressure and somewhat higher temperatures (depending on the specific system involved), seemed an almost incredible phenomenon at the time when it was gradually beginning to gain acceptance, in the early 1940's, through studies on various processes (cf. Johnson *et al.*, 1954).

The second of the two developments referred to above was of much more general, fundamental significance, *viz.*, the advent of the modern theory of precisely how temperature and pressure affect the rates of chemical reactions; a theory which, with respect to temperature, may be considered to have begun with Arrhenius (1889) and to have culminated in the modern theory of absolute reaction rates (Eyring, 1935a,b; Wynne-Jones and Eyring, 1935; Polanyi and Evans, 1935; Glasstone *et al.*, 1941). While it is manifestly impossible in any brief space to do justice to the hard-won intellectual victories of the many dedicated, brilliant efforts which led to the present theory, and while it is similarly impossible to give an adequate account even of the full conceptual basis of the theory, it seems useful nevertheless to point out certain relationships with particular reference to the action of pressure. For, the same theory which provides a rational understanding of the influence of temperature on chemical reaction rates does likewise for the influence of pressure, and in a manner closely akin, in part, to well-understood formulations of thermodynamic equilibria. Although the latter formulations do not depend on a knowledge of the structure of atoms and molecules, modern reaction rate theory is strongly so dependent and could not have been arrived at prior to the quantum theory, or without the aid also of classical and statistical mechanics, as well as the principles of thermodynamics.

B. The Arrhenius Theory

In studying the influence of temperature upon the rate of hydrolysis of sucrose by acids, Arrhenius (1889) found that the increase in rate with rise in temperature, about 12% per degree, was much too great to be accounted for in terms of the effect of temperature upon the kinetic energy of the molecules, about 1/6 percent per degree, or upon the amount of dissociation of the acids. In order to account for this large effect of temperature, Arrhenius introduced the idea of active molecules M_a, molecules with sufficient energy that reaction could take place, as opposed to inactive molecules M_i having less than sufficient energy for reaction. He assumed that at any temperature the number of active molecules would be proportional to the number of inactive molecules, and that

$$M_a = k \cdot M_i \tag{1}$$

the proportion of active molecules increased rapidly with a rise in temperature. Data from experiments showed that the velocity of reaction Q_{t_0} at one temperature t_0 and the velocity Q_{t_1} at a different temperature t_1 is described by the equation:

$$Q_{t_1} = Q_{t_0} \cdot e^{A\cdot(T_1 - T_0):T_0 T_1} \tag{2}$$

in which A is a constant, e is the base of natural logarithms, and T_0 and T_1 are the absolute temperatures corresponding to t_0 and t_1, respectively. Arrhenius considered Eq. (2) in relation to van't Hoff's theoretically derived expression (1884) for the temperature dependence of an equilibrium constant K in a reversible reaction between A and B to give C and D, according to specific rate constants $k_{\prime}$ for the forward and $k_{\prime\prime}$ for the backward reaction. Letting C represent the concentration of the reactant denoted by the subscript,

$$\begin{aligned} k_{\prime} C_A C_B &= k_{\prime\prime} C_E C_D \\ k_{\prime\prime}/k_{\prime} = K &= C_E C_D / C_A C_B \end{aligned} \tag{3}$$

The temperature relation of the k's is given by

$$d \ln k_{\prime}/dT - d \ln k_{\prime\prime}/dT = q/2T^2 \tag{4}$$

where q is the heat in calories liberated per gram molecule when A and B are converted to C and D. It was pointed out that although Eq. (4) does not give the relation between temperature and a single reaction rate constant k, this relation has the form

$$(d \ln k)/dT = (A/T^2) + B \tag{5}$$

in which A and B are constants.

In accordance with the above ideas, Arrhenius wrote

$$(d \ln k)/dT = q/2T^2 \tag{6}$$

in which the constant B is omitted as required by data from experiments. On integration, Eq. (6) becomes

$$kT_1 = kT_0 e^{q(T_1 - T_0)/2T_0 T_1} \tag{7}$$

which, aside from the difference in notation, is Eq. (2), $q/2$ being equal to A.

In later work, especially in application to biological processes of various degrees of complexity, the q of Eq. (7) was changed to the symbol μ and referred to as activation energy, or sometimes as temperature characteristic or critical thermal increment (reviewed in Johnson *et al.*, 1954, Chapter 8). The more familiar form of the Arrhenius equation, with v denoting velocity and A an empirical constant, both derived from experiments, is

$$v = Ae^{-\mu/RT} \tag{8}$$

from which we derive Eq. (9) which has often been conveniently applied to biological data from experiments for determining a value for μ:

$$\mu = 2 \times 2.3 \left[\frac{\log_{10} k_{T_1} - \log_{10} k_{T_0}}{(1/T_0) - (1/T_1)} \right] \tag{9}$$

where k_{T_0} represents the observed velocity or rate constant at a temperature T_0, k_{T_1} the observed velocity or rate constant at a different, usually higher, temperature, and R, the gas constant, is given the approximate value of 2 instead of a more exact value which would in most instances represent a false accuracy; the factor 2.3 is included to change from natural logarithms to logarithms to the base 10.

The Arrhenius Eq. (8) proved quantitatively satisfactory to account for a remarkable number of chemical reactions; a linear or nearly linear relation was found, over a very considerable range of temperatures, between the logarithm of the velocity of reaction and the reciprocal of the absolute temperature, the slope of the line being $\mu/2$, expressed as activation energy in calories. As a rule, the higher the activation energy, the higher the temperature required for a measurable reaction rate, but this was not invariably so. Moreover, two different reactions with numerically the same activation energies were sometimes found to proceed at very different velocities at a given temperature, i.e., the slope of the line in a plot of log velocity vs the reciprocal of the absolute temperature was the same but the intercept was different, a difference that was numerically accounted for by a difference in the value of the coefficient A pertaining to the different reactions.

C. Interpretation of the Arrhenius Equation

The value of μ and of A had to be determined by experiment; neither one had a precise meaning, although the general idea of what is now called normal and activated states was of utmost importance and is embodied in modern reaction rate theory, albeit with certain conceptual differences, as referred to in later paragraphs. With reference to the meaning of μ, it is evident that the number of molecules N_a which in a population N_t acquires by absorbing radiant energy or by successive collisions with other molecules sufficient energy to equal or to exceed the minimum energy for reaction to occur is given by the Boltzman-like expression:

$$N_a/N_t = (\Sigma\omega_a/\Sigma\omega_t)e^{-(E/kT)} \tag{10}$$

where $\Sigma\omega_a$ and $\Sigma\omega_t$ are the weighted number of separate states corresponding to the respective energies. Thus, raising the temperature has an exponential effect in increasing the number of molecules with sufficient activation energy E, and, if it is assumed that the velocity of reaction is proportional to the ratio N_a/N_t, it follows that the velocity will be given by an equation of the form of the Arrhenius Eq. (8).

Prior to the understanding of the energies of molecules which became possible through the quantum theory and other advances, efforts were made to find a rational basis for the constant A of Eq. (8). The kinetic theory of gases sufficed to calculate the actual number Z of collisions between two species of molecules, A and B, from the concentration n, the mean collision radii r_A and r_B, and the mass m_A and m_B of A and B, respectively, in accordance with Eq. (11):

$$Z = n_A n_B (r_A + r_B)^2 \left[8\pi kT \left(\frac{1}{m_A} + \frac{1}{m_B} \right) \right]^{1/2} \tag{11}$$

where k represents the Boltzmann constant, i.e., the gas constant divided by Avogadro's number, or (R/N), and T is the absolute temperature. In certain instances, practically every collision resulted in reaction, whereas in others it appeared that only one in many powers of 10 resulted in reaction, and even with an activation energy known from measurement in experiments, it became necessary to multiply the number Z by a "probability" or "steric" factor P, so that the Arrhenius Eq. (8) was thus modified in Eq. (12):

$$v = PZe^{-E/RT} \tag{12}$$

From experiments it was found necessary to assign values of P ranging, for "slow" reactions, from 10^{-1} to 10^{-8}, whereas in other instances the value of P had to be much greater than unity. The meaning of A in the Arrhenius equation was obviously not satisfactorily interpretable in terms of P and Z.

D. The Arrhenius Relation in Biological Processes

Had the Arrhenius equation contained an explicit expression for the influence of pressure on chemical reaction rates, as it did for temperature, it is reasonable to believe that many more studies on the effects of pressure would have been undertaken than actually were. As it was, Arrhenius himself applied his temperature theory to biological processes, as did numerous others, while the field of pressure studies remained separate, seemingly quite distinct, attracting far fewer investigators. Temperature studies, however, often revealed that in biology the measured process conformed rather closely to the Arrhenius relation, through at least a limited range in temperature, although in going from a relatively low to a relatively high temperature, usually from somewhat above 0° to 40° or 50°C, the process always went through a maximum, generally referred to rather vaguely as an "optimum." As a result, the value of μ computed from a seemingly straight-line portion of the curve relating log v to $1/T$ in a low temperature range was generally higher than a value of μ computed from data pertaining to a higher temperature range. The fact that, in a very complex biological process such as respiration, growth, mental activity, etc., any significant portion of the "temperature activity curve" conformed to the Arrhenius equation could be logically interpreted as indicating that the rate of the total process was limited by one slowest reaction. The change in apparent activation energy, which in some instances appeared to occur rather abruptly, became the subject of a vast amount of discussion and controversy during the third and fourth decades of this century. In any event, whether the change was unmistakenly abrupt or, more commonly, by subtle, gradual and continuous changes, considering the overall process as one reaction, the value of μ was bound to decrease in nearing the optimum from the low temperature side, then become zero at the optimum and become negative with further rise in temperature. At temperatures higher than the so-called optimum, it was generally assumed that the decrease in rate resulted from a thermal destruction of an essential catalyst, a presumably irreversible process which in itself might easily conform to the Arrhenius equation or might turn out to be more complicated, according to the particular system and the conditions involved. The easily reversible changes in velocity with changes in temperature in the neighborhood of the optimum, especially with reference to a single enzyme *in vitro*, remained obscure until the fact was gradually established that enzymes and other proteins were capable of undergoing reversible changes of various sorts which, in enzymes, affected the catalytic activity through a reaction which exhibited certain characteristics of protein denaturation. The role of such reactions in biological temperature–activity curves is discussed briefly in later paragraphs.

II. The Theory of Absolute Reaction Rates

The modern theory of absolute reaction rates makes it possible *in principle* to predict the absolute rate of a chemical reaction on the basis of physical properties of the reactants. The available mathematical tools are not precise enough, however, to do so accurately without recourse to certain empirically determined data, especially the activation energy, since an error of 1.3 kcal in the predicted value of the activation energy would make a tenfold difference in the predicted rate of reaction at room temperature. Although it is difficult to give even an inkling of the theory in a short space, it seems worth while to mention a few of the salient features with reference to their bearing on the Arrhenius theory, as well as on the mechanism through which pressure influences reaction rates, and application of the theory to biological processes in general.

A. The Activated Complex or Transition State

In essence, the theory of absolute reaction rates states that any chemical reaction involves the formation of an unstable intermediate compound between the reactants. The intermediate compound, designated either as the activated complex or the transition state, has a lifetime of less than one period of vibration between two atoms in any molecule, i.e., a lifetime of the order of 10^{-13} sec. Once formed, the activated complex decomposes at a temperature–dependent frequency which is the same for all reactions and is given by the expression kT/h, where k is the Boltzmann constant, i.e., R/N, T is the absolute temperature, and h is Planck's constant which has the measured value of 6.62×10^{-27} ergs/sec. It follows that, provided the activated complex decomposes in the direction of reaction products rather than in the direction of reconstitution of the reactants, the rate of the reaction is governed by the probability of formation of the activated complex. The chances of decomposition to products is designated by the letter kappa ($\varkappa$), the numerical value of which for many, if not most, reactions is unity or very nearly unity. The probability of formation of the activated complex is given by $e^{-(\Delta G^{\ddagger}/RT)}$, where the double dagger ($\ddagger$) indicates that the symbol refers to the activated complex, and $\Delta G^{\ddagger}$ represents the Gibbs free energy change in going from the normal to the activated state, or more properly, from the normal states to the activated states, inasmuch as the distribution of energy among the molecules in either of these states is not uniform.

B. Energy Barriers and Potential Energy Surfaces

The formation of the activated complex requires that the reactant molecules first acquire a certain minimum energy by absorption of radiation or by collision with other molecules before successful combination of the reactants into the unstable complex can occur. There is thus an "energy barrier" in a reaction, and the chief problem of calculating the rate of a reaction from first principles consists in calculating the height of this barrier, which may be pictured in the simplest case as a sort of saddle-shaped elevation or pass between two mountain valleys, with the reactants on one side and the products on the other. The height of the barrier is obtained from a potential energy surface, very few of which have been calculated, and only for the simplest sort of reactions.

A three-dimensional model of a potential energy surface as a whole is pictured diagrammatically in Fig. 1A. Each valley represents the potential energy of a diatomic molecule, in this particular instance for H_2, as a function of the distance separating the atoms vibrating in nearly simple harmonic motion with a period of about 10^{12} per sec in the different quantum levels of energy. Each of the two valleys, at right angles to each other, has, on the side facing away from the central plateau, a steep wall indicating the rapid rise in potential energy as the very slightly compressible atoms come close together; the less steep walls directly opposite represent the potential energy at distances indicated in angstrom units numbered at the base on the right-hand side, the dissociation energy of 102 kcal being indicated as a plateau where each H is uncombined with another.

By way of an analogy, though not literally applicable, one can imagine a ball rolling about in one of the two valleys depicted in Fig. 1A, with potential energies varying with the velocity, pathway, and position of the ball in the valley. Only if the ball proceeds in the right way and with sufficient energy will it cross the barrier into the adjacent valley. When this happens, it will usually be by the easiest route over the barrier, but in some instances it will be along a higher route with excess energy, which, in this analogy, would correspond to the higher than minimum quantum level of energy of the activated complex. At high temperatures, the ball might acquire sufficient energy to roll up on the plateau and disappear forever from these valleys. This could be considered analogous to dissociation of the reactant, but dissociation of no diatomic molecule

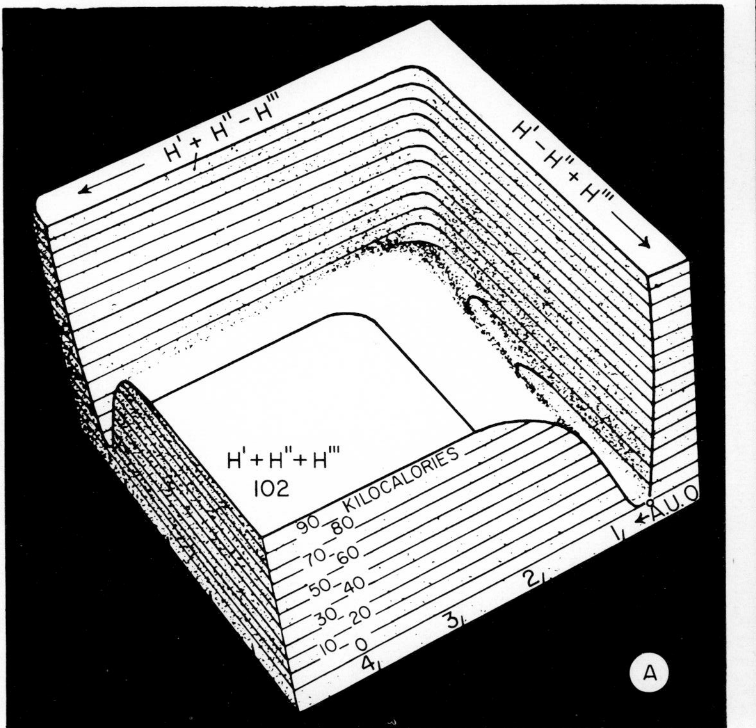

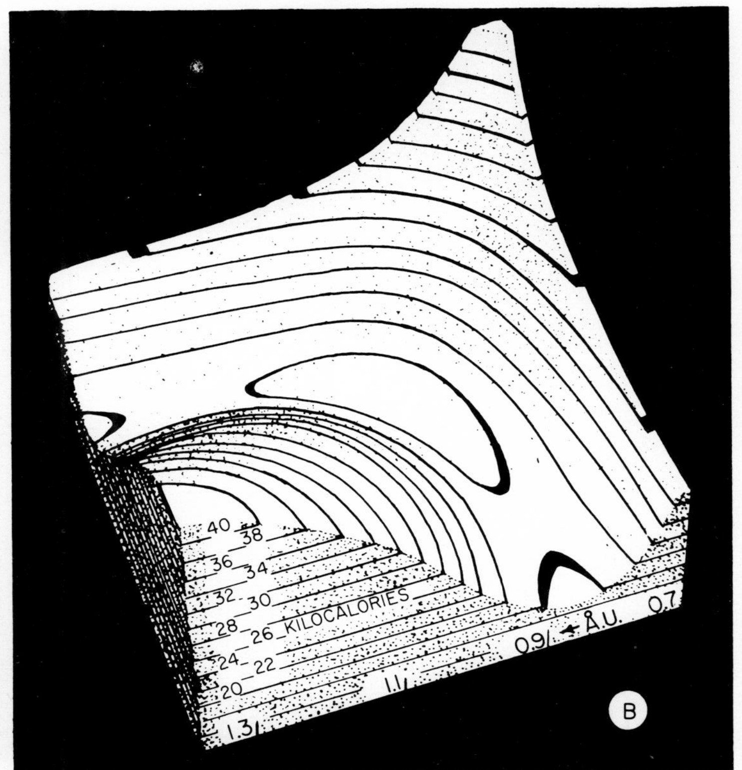

FIG. 1. (A) Drawing made from a photograph of a model of the potential energy surface for the reaction $H + H_2 \leftrightharpoons H_2 + H$, constructed by Goodeve (1934) according to the contour lines calculated by Eyring and Polyani (1931), this model taking into account only the resonance forces. (B) Drawing from the same source showing a close-up of the saddle region of a model in which the coulombic as well as resonance forces have been taken into account.

ever occurs at room temperature, except in solution, where it may occur immeasureably fast to form ions.

C. The Morse and London Equations

Referring to Fig. 1A, the cross section of the valley at either of the open ends has very nearly the shape of a parabola near the bottom, but no equation has been derived which describes the whole curve exactly. An equation accurate enough for most purposes, however, has been proposed by Morse (1929):

$$E = D' \left[e^{-2a(r-r_0)} - 2e^{-a(r-r_0)}\right] \tag{13}$$

where E is the potential energy, D' is the heat of dissociation of the molecule plus the zero point energy, i.e., plus the lowest half-quantum level of energy, r_0 is the equilibrium distance separating the atoms, r is any other distance apart, and a is equal to $\pi\nu(2\mu D')^{1/2}$, where μ, the "reduced mass" equals $m_1m_2/(m_1 + m_2)$, the masses of the atoms being represented by m_1 and m_2. With a knowledge of the masses of the atoms and with spectroscopic data, Eq. (13) provides a means of calculating potential energies as a function of the distance separating the atoms, and for the purposes involved it is usually sufficient to do so for distances of about 0.5 to 4.0 Å, the equilibrium distance being never very far from 1 Å.

The problem of calculating the height of the energy barrier involves calculating the energies for all possible configurations of the reactants, and this requires as many dimensions as are required to fix the potential energy. London (1928, 1929) derived an equation for the potential energy E for a three-atom and for a four-atom system; for a three-atom system the London equation is

$$E = Q - \{(1/2)\,[\alpha - \beta)^2 + (\beta - \gamma)^2 + (\gamma - \alpha)^2]\}^{1/2} \tag{14}$$

where Q represents the sum of the Coulombic energies of the three atoms A, B, and C, while α, β, and γ represent the corresponding exchange energies. Eyring (1930) suggested that the Morse Eq. (13) could be used to compute the potential energies of different pairs of atoms for use in the London Eq. (14), and Eyring and Polanyi (1930, 1931) calculated a potential energy surface for the reaction $H + H_2 \rightarrow H_2 + H$, making the simplifying assumption that the Coulombic energy is a constant fraction of the total binding energy, as had been suggested by the Heitler and London (1927) and Sugiura (1927) calculations for H_2.

D. Statistical Calculation of Reaction Rates

Although the approximate method afforded by the London equation does not yield sufficiently accurate potential energy surfaces to obtain a value for the activation energy which can be relied upon in a priori calculations of absolute rates of reaction, the calculated surfaces become satisfactory if the percent Coulombic energy is chosen such that E_0, the activation energy at absolute zero, agrees with the value obtained from experiments. With a satisfactory value for the activation energy E, the problem of predicting the rate of reaction is that of calculating the rate at which mass points move across the energy barrier separating the products from the reactants. The reaction occurs through a continuous change of coordinates in passing from the normal states of reactants in one valley, through the activated complex at the top of the barrier, to the products in the other valley. At equilibrium, half of the activated complexes will be proceeding toward the right and half toward the left. Thus, considering a length δ at the top of the valley, and taking the average velocity v of traversing this length, the reaction rate is given by the number C^* of activated complexes in unit volume along the length $\delta \times \frac{1}{2} \times$ the frequency v/δ of crossing $\times$ the quantum mechanical chance $\varkappa$ that the system having crossed the barrier will continue through to reaction products:

$$\text{Rate of reaction} = C^*(\tfrac{1}{2})(v/\delta)\varkappa \tag{15}$$

In accordance with the theory under discussion, the concentration C^* of activated complexes in a length δ at the top of the barrier is given by Eq. (16):

$$C^* = K^\ddagger \frac{(2\pi m^\ddagger kT)^{1/2}\delta}{h} C_{\mathrm{AB}} C_{\mathrm{C}} \tag{16}$$

where C_{AB} and C_{C} represent the concentration of reactants AB and C, respectively, in the reaction $\mathrm{AB + C \rightarrow A + BC}$, and the other symbols are as follows: $K^\ddagger$ is the quasi-equilibrium constant for the normal and activated molecules, $m^\ddagger$ is the effective mass of the activated complex for motion along the reaction coordinate, k is the Boltzman constant, h is Planck's constant, and T is the absolute temperature. The average velocity $\bar{v}$ in the forward direction along the length δ is

$$\bar{v} = \frac{\int_0^\infty p/m^\ddagger) e^{-p^2/2m^\ddagger kT}\, dp}{\int_0^\infty e^{-p^2/2m^\ddagger kT}\, dp} = \left(\frac{2kT}{\pi m^\ddagger}\right)^{1/2} \tag{17}$$

where p is the momentum along the reaction coordinate, and the equation for velocity is the one familiar in statistical mechanics. Substituting in Eq. (15) from Eqs. (16) and (17),

$$\text{Rate of reaction} = K^{\ddagger} \frac{(2\pi m^{\ddagger} kT)^{1/2}\,\delta}{h} \left(\frac{2kT}{\pi m^{\ddagger}}\right)^{1/2} \frac{1}{\delta} \frac{1}{2} \varkappa\, C_{AB} C_{C} \tag{18}$$
$$= K^{\ddagger} \varkappa (kT/h) C_{AB} C_{C}$$

Equation (18) is a general equation, applicable to any type of reaction involving the crossing of an energy barrier—in oxidation, hydrolysis, diffusion, etc. In this equation the length δ along the barrier has been taken such that

$$\delta = h/(2\pi m^{\ddagger} \mathrm{k}\mathrm{T})^{1/2} \tag{19}$$

This length usually lies between 0.1 and 1.0 Å. The precise value of δ does not matter, since it cancels out in the final expression.

E. Meaning of $K^{\ddagger}$

The quasi-equilibrium constant $K^{\ddagger}$ is analogous to the equilibrium constant K of a thermodynamic equilibrium, but is conceptually different in that the latter is given by the ratio between the specific rates of a reaction in the forward and backward direction, k_f and k_b, respectively, each of which is given by an equation having the form of Eq. (18), whereas the quasi-equilibrium denoted $K^{\ddagger}$ does not represent a ratio between the rate of formation and rate of decomposition of the activated complex. Rather, $K^{\ddagger}$ may be said to represent the probability that the reactants will get into the activated state and then decompose with the universal frequency kT/h in the forward direction according to the value of $\varkappa$, or be reflected backward $(1 - \varkappa)$ of the time. The activated complex itself is the same, whether it is formed according to the specific rate constant for the forward reaction, i.e., from reactants in the initial state of the equilibrium, or from the products in the final state of the equilibrium, and it is a basic premise of the theory that the number of activated complexes at any instant at the top of the barrier has no influence on the rate of their formation from either direction. Moreover, while thermodynamic equilibria can be predicted from first principles, there is no way to predict, short of the rate theory under discussion, how long it will take for equilibrium to be reached.

For all practical purposes the quasi-equilibrium constant $K^{\ddagger}$ behaves in a manner similar to a true equilibrium constant K, which in terms of the changes in free energy (ΔG), enthalpy (ΔH), entropy (ΔS), and volume (ΔV), is given by Eq. (20):

$$K = exp\left(-\frac{\Delta G}{RT}\right) = exp\left(-\frac{\Delta H}{RT}\right) exp\left(\frac{\Delta S}{R}\right)$$

$$= exp\left(-\frac{\Delta G_1 + \int_{p=1}^{p} \Delta V\,dp}{RT}\right)$$

$$= exp\left(-\frac{\Delta G_1}{RT}\right) exp\left[-\frac{\overline{\Delta V}\,(p-1)}{RT}\right] \simeq$$

$$exp\left(-\frac{\Delta G_1}{RT}\right) exp\left(-\frac{\Delta V p}{RT}\right) \quad (20)$$

Similarly, for rate processes, the quasi-equilibrium constant $K^{\ddagger}$ is given by Eq. (21):

$$K^{\ddagger} = exp\left(-\frac{\Delta G^{\ddagger}}{RT}\right) = exp\left(-\frac{\Delta H^{\ddagger}}{RT}\right) exp\left(\frac{\Delta S^{\ddagger}}{R}\right) =$$

$$exp\left(-\frac{\Delta G_1{}^{\ddagger} + \int_{p=1}^{p} \Delta V^{\ddagger}\,dp}{RT}\right)$$

$$= exp\left[-\frac{\Delta G_1{}^{\ddagger} + \overline{\Delta V^{\ddagger}}\,(p-1)}{RT}\right]$$

$$\simeq exp\left(-\frac{\Delta G_1{}^{\ddagger}}{RT}\right) exp\left(-\frac{\overline{\Delta V^{\ddagger}} p}{RT}\right) =$$

$$exp\left(-\frac{\Delta H_1{}^{\ddagger}}{RT}\right) exp\left(\frac{\Delta S_1{}^{\ddagger}}{R}\right) exp\left(-\frac{\overline{\Delta V^{\ddagger}} p}{RT}\right) \quad (21)$$

From Eqs. (18) and (21), the specific reaction rate k' can be written:

$$k' = \varkappa(kT/h)\, e^{-\frac{\Delta G^{\ddagger}}{RT}} = \varkappa(kT/h)\, e^{-\frac{\Delta G_1{}^{\ddagger}}{RT}}\, e^{-\frac{p\Delta V^{\ddagger}}{RT}} \quad (22)$$

Here $\Delta V = \partial \Delta G/\partial p$, and if ΔV varies appreciably with pressure, an average value must be used. The same argument applies to $\Delta V^{\ddagger}$. The subscripts to ΔG_1 and $\Delta G^{\ddagger}{}_1$ mean these are the values at 1 atm.

Parenthetically, it is worth noting at this point that for any reaction to proceed at a measurable rate at ordinary temperatures, the Gibbs free energy of activation $\Delta G^\ddagger$ must amount to between roughly 20,000 and 30,000 cal. In those instances, exemplified by thermal denaturation of proteins characterized by high energies of activation $\Delta H^\ddagger$, a high entropy of activation $\Delta S^\ddagger$ brings the value of $\Delta G^\ddagger$ into the above range.

F. Partition Functions

In further reference to the theoretical prediction of absolute rates of reaction: if the activation energy is known it remains necessary only to obtain a satisfactorily accurate value for the entropy of activation, $\Delta S^\ddagger$, to put into Eqs. (21) and (22), provided of course that the conditions under consideration exclude from importance, for the time being, the term pertaining to pressure and the volume change of activation. A numerical value of $\Delta S^\ddagger$ can be obtained through calculation of partition functions that for any atom or molecule, including the activated complex, have the value $F = \Sigma e^{-\epsilon_i/kT}$, where ε_i is the energy of the ith state of the molecule and the summation is over all states. Since the equilibrium constant for any system can be expressed in terms of the partition functions of the molecules involved, the equation for the specific rate constant k' can be written:

$$k' = \varkappa \frac{kT}{h} K^\ddagger = \varkappa \frac{kT}{h} e^{-\frac{\Delta H^\ddagger}{RT}} e^{\frac{\Delta S^\ddagger}{R}} = \varkappa \frac{kT}{h} \left(\frac{F^\ddagger}{F_A F_B} \right) e^{-\frac{E_0}{RT}} \quad (23)$$

where the notation is the same as in Eqs. (21) and (22), with the addition of $F^\ddagger$ the partition function per unit volume of the activated complex, F_A, F_B the partition functions of the reactants, and E_0 the activation energy at absolute zero. In deriving the expression for reaction rates in Eqs. (15)–(18), the partition function for translation for a length δ at the top of the barrier was considered, since this was involved in the expression for the probability of the occurrence of the activated complex at the top of the barrier. However, every atom or molecule has three translational degrees of freedom for motion as a whole in the x,y, and z directions. The partition function f_{tr} for three translational degrees of freedom is given by Eq. (24), where V is the volume of the container and the other symbols have the same meaning as above.

$$f_{tr} = [(2\pi mkT)^{3/2}/h^3)]\, V \quad (24)$$

Although a single atom has only these three degrees of freedom, any molecule composed of two or more atoms has $3n$ degrees of freedom, n representing the number of constituent atoms; for a diatomic molecule there are three translational, two rotational, and one vibrational degrees of freedom. The appropriate expressions pertaining thereto are derived from quantum and statistical mechanical theory. The physical data necessary for computing the various energies of a particular system can be obtained from spectroscopy (for nuclear, electronic, vibrational, and rotational energies) and from mass and velocity (for translational energies). While further details are beyond the scope of the present discussion, it is perhaps useful to illustrate the complexity of even a very "simple" reaction, such as that of hydrogen and iodine to give hydrogen iodide, by giving Eq. (25), which has been used with success to calculate the reaction rate (see Glasstone *et al.*, 1941). In Eq. (25), the subscripts 1, 2, and 3 refer to hydrogen, iodine, and the activated complex H_2I_2, respectively. Thus the velocity v of the reaction:

$$H_2 + I_2 \rightarrow \begin{matrix} H\text{——}H \\ / \quad\quad \backslash \\ I\text{————}I \end{matrix} \rightarrow 2HI$$

is given by

$$v = \frac{[H_2][I_2]\dfrac{(2\pi m_3 kT)^{3/2}}{h^3}\dfrac{8\pi^2(8\pi^3 A_3B_3C_3)^{1/2}(kT)^{3/2}}{2h^3}\prod_{i=1}^{5}\dfrac{1}{1-e^{-h\nu_i/kT}}\,e^{-E_a/kT}}{\dfrac{(2\pi m_1 kT)^{3/2}}{h^3}\dfrac{8\pi^2 I_1 kT}{2h^2}\dfrac{1}{1-e^{-h\nu_1/kT}}\dfrac{(2\pi m_2 kT)^{3/2}}{h^3}\dfrac{8\pi^2 I_2 kT}{2h^2}\dfrac{1}{1-e^{-h\nu_2/kT}}} \times \varkappa\frac{kT}{h} \quad (25)$$

If the activated complex is linear, with the two iodine atoms on the outside, as is probable, the rotational partition function must be changed to $\frac{8\pi^2 IkT}{\sigma h^2}$ and one more vibrational partition function must be added. Here I is the moment of inertia of the activated complex and σ is the symmetry number which for a symmetrical linear activated complex is 2.

G. Meaning of *A* in the Arrhenius Equation

With reference to the Arrhenius Eq. (8), Eq. (22) provides a rational and precise interpretation, inasmuch as the empirical constant A of Eq. (8) has the approximate equivalence:

$$A \cong \varkappa(kT/h)e^{\Delta S^\ddagger/R} \quad (26)$$

It will be noted that small deviations in the linearity of data plotted according to the Arrhenius equation would be expected because of the different values of T in kT/h at different temperatures, and also because of variations in $\Delta S^{\ddagger}$. The deviations, however, would not be expected to be quantitatively very important. Deviations have been found in many chemical reactions because of the factors just mentioned and also because of factors such as changes in the mechanism of the reaction, which can exert a much greater influence; the latter influences can be especially important in biological processes, as will be mentioned below.

H. Volume Changes of Activation $\Delta V^{\ddagger}$ and of Reaction ΔV

Equation (22) also provides a rational, quantitative basis for describing the influence of hydrostatic pressure on chemical reactions in condensed phases, such as the aqueous solutions which are of ubiquitous importance in biological processes. From Eq. (22) it follows, that at constant temperature:

$$\Delta V^{\ddagger} = 2.3RT \left[\frac{\log_{10} k_{p_1} - \log_{10} k_{p_2}}{p_2 - p_1} \right] \tag{27}$$

where $\Delta V^{\ddagger}$ is the volume change of activation, i.e., the ratio between the volume of the activated complex and that of the reactants, the gas constant R has the value 82.07 cm^3/mole, k_{p_1} is the rate constant for the reaction at pressure p_1 atm and k_{p_2} is the rate constant of the reaction at increased pressure, p_2 atm.

Equation (27) for the relation between pressure and rate of reaction is obviously of the same form as Eq. (9) for the relation between temperature and rate of reaction. Thus by plotting log $_{10}$ of the rate constant as a function of pressure in atmospheres, a linear relation should result if the reaction behaves in the manner of a single chemical reaction, and if $\Delta V^{\ddagger}$ is approximately independent of pressure the value of the volume change of activation $\Delta V^{\ddagger}$ can then be easily computed from the slope of the line.

Equation (27) is exactly analogous to Eq. (28), the equation for the volume change of reaction ΔV in thermodynamic equilibria:

$$\Delta V = 2.3RT \left(\frac{\log_{10} K_{p_1} - \log_{10} K_{p_0}}{p_1 - p_0} \right) \tag{28}$$

Thus a value for the volume change of reaction ΔV in going from the initial to the final states can be calculated in the same manner, from the appropriate data, as the volume change of activation in going from the normal to the activated states. In a practical sense, however, there is one important difference, namely, that a value for ΔV can be determined directly, e.g., by means of a dilatometer, from physical measurements of the volume of a reaction system at the start and at the end of the reaction, i.e., in the initial and final states, whereas $\Delta V^{\ddagger}$ *must* be determined indirectly and *only* through measurements of the rate of reaction at different pressures under otherwise identical conditions. The reason, of course, is simply that the fraction of a population of molecules which, in a reacting system, exists at any moment in the form of the activated complex is so small and the lifetime of about 10^{-13} sec is so short that no known physical method can measure directly the change in volume of the activated as compared with the normal states of the reactants.

The volume change of activation in ordinary chemical reactions is usually small, of the order of a few cubic centimeters per mole at most. Although no explicit, necessarily empirical or at best semirational, basis for describing the effects of pressure on chemical reaction rates was included in the Arrhenius equation, the theory of absolute reaction rates makes it possible to make certain a priori predictions that are in satisfactory agreement with the results of experiments with simple organic reactions. For example, if the activated state is more ionized than the reactants, $\Delta V^{\ddagger}$ will be negative because of an increase in electrostriction; increased pressure will therefore accelerate the reaction. If the activated state is less ionized, the converse will be true. Other mechanisms may also be involved. Calculated and observed values for a number of reactions, together with discussion of the theory, are given by Stearn and Eyring (1941).

III. Biological Processes

A. Volume Changes in Biological Reactions

Especially in regard to biological systems, factors in addition to the volume changes associated with ionization can be involved in the action of pressure, e.g., rearrangements which bring to the molecular surface groups that bond in varying degrees with the solvent. The stability and reactivity of enzymes and other proteins may be influenced by pressure through the above mechanisms, and perhaps also by changes in the structure of the solvent (Eisenberg and Kauzmann, 1969), and in certain instances by the degree of ionization of a commonly used buffer system, such as phosphate, which has the unusually large volume change of

$\Delta V = -24$ cm³/mole accompanying the dissociation of $H_2PO_4^-$ (Linderstrøm-Lang and Jacobsen, 1941). With this buffer, a pH near neutrality at atmospheric pressure would be expected to become about 0.4 of a pH unit more acid under 680 atm pressure, and a shift of this magnitude could result of itself in drastic changes in the rate or state of any biological system that happened to be especially sensitive to the concentration of hydrogen ions, although the net result would depend on the net effects of pressure on not only the buffer but on all other systems significantly involved in the total process.

Because of the large size and complex structures of key biological molecules, such as enzymes or other types of proteins and nucleic acids, the molecular volume changes of activation as well as of reaction are sometimes quite large, if not seemingly enormous. With respect to the latter, i.e., volume change of reaction, the splitting of between ten and fifteen peptide bonds per mole of β-lactoglobulin at 0° with trypsin as a catalyst is characterized by a volume decrease of more than 800 cm^3/mole of protein (Linderstrøm-Lang, 1950). This is understandable on the basis of a summation of smaller volume decreases (in some instances increases) pertaining to reactions of amino acids and simple peptides, e.g., the volume changes accompanying the formation of an ion pair, or a tetra- or longer peptide, or of an amino acid, di- or tripeptide dipole (Linderstrøm-Lang and Jacobsen, 1941). Although a change of this magnitude 800 cm^3/mole, in volume of the activated complex over that of the reactants is unlikely, it is not unusual to encounter volume changes of activation of the order of 100 cm^3/mole in certain reactions of biological molecules such as the rate of thermal denaturation of proteins (see Johnson *et al.*, 1954).

B. Absolute Rate Theory and Cellular Processes

Application of the theory of absolute reaction rates to processes taking place within intact living cells was first made in regard to bacterial luminescence (Brown *et al.*, 1942; Eyring and Magee, 1942; Johnson *et al.*, 1942a,b,c, 1945, 1954; McElroy, 1943). It was through these studies that a rational and quantitative, albeit obviously and admittedly oversimplified, interpretation of the action of hydrostatic pressure was made in regard to certain fundamental aspects that had previously remained unknown, unrecognized, or unexplained on the basis of physical chemical principles. These aspects included: the action of hydrostatic pressure on living biological reaction rates in general; the role of reversible thermal denaturation of enzymes in controlling the overall rate of a physiological process; the reversal by increased hydrostatic pressure of the shift in this equilibrium toward the denatured form of the enzyme at temperatures above the "optimum" for the process under the conditions involved; the retardation by increased hydrostatic pressure of the rate of irreversible denaturation of the limiting enzyme(s); the relation of the observed effects of increased hydrostatic pressure to the specific temperature activity curve of the process under study, whereby pressure accelerated, had little effect, or retarded the overall rate at temperatures above, close to, or below the optimum, respectively; and a relation between hydrostatic pressure and the action of typical narcotic agents such as urethan, ethyl alcohol, and chloroform. The seemingly separate fields of investigation pertaining to the biological action of pressure, temperature, and narcotics were thus brought together and a common basis in physical chemical theory was provided for an understanding of the phenomena encountered in nature as well as in the laboratory. All these advances were made possible by some unusual circumstances, including several which were responsible for the time's being "ripe," e.g., the increasing application of the then recently introduced theory of absolute reaction rates, the growing instances of profitable collaboration in research by specialists in different fields of investigation, the volume of accumulated data awaiting a rational interpretation, etc., and, in addition, some extraordinary advantages in using bacterial luminiscence as a type-reaction for the studies involved. Because this process is so favorable both for investigating fundamental mechanisms controlling the kinetics of physiological processes and for illustrating the interrelationships in the action of temperature, pressure, and chemical agents on such processes, the following paragraphs are pertinent.

With reference first to the advantages of bioluminescence in studying reaction rates in cells, the light-emitting enzyme reaction is practically unique in having a visible, accurately and instantaneously measurable natural indicator of its own reaction velocity. Thus, under given conditions, the brightness of the light provides an index to the overall reaction velocity of the system, the evidence for this being based on both the time course and on the influence of relative concentrations of enzyme and substrate on the intensity of light in cell-free, partially purified systems.

C. Bacterial Luminescence; Pressure–Temperature Relations

In a nonreproducing population of luminous bacterial cells under favorable physiological conditions, the intensity remains conveniently constant, in a steady state, for a period of minutes or longer. Sudden changes in intensity, owing for example to the action of quick changes in pressure applied to the system, can be virtually as quickly recorded. The following examples illustrate the manner in which the kinetics can be analyzed and interpreted on a general theoretical basis, which has proved sufficiently correct to anticipate with some success the results obtained

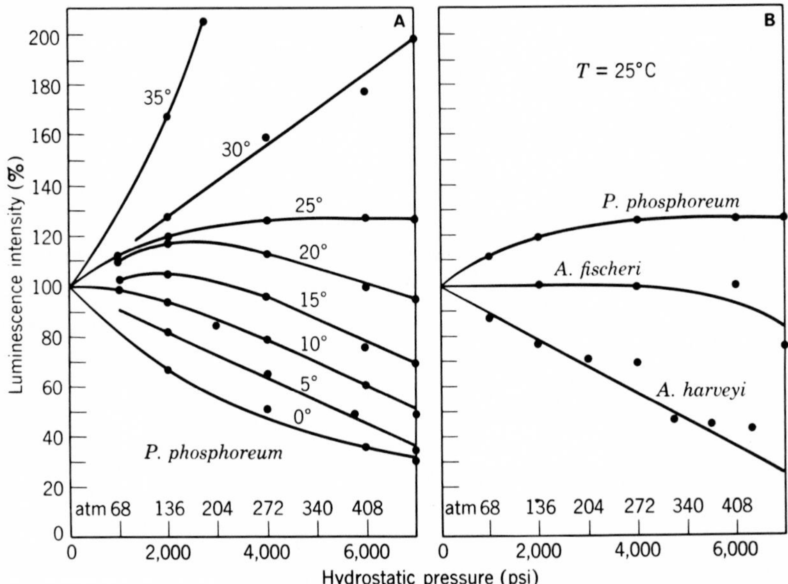

FIG. 2. (A) Pressure–temperature relations of luminescence intensity of a suspension of *Photobacterium phosphoreum* cells in isotonic, buffered salt solution. The observed intensity at normal pressure has been arbitrarily taken to equal 100 in order to show clearly the change in luminescence intensity with change in pressure at different temperatures. (B) Influence of increased pressure, at one temperature, on three species of luminous bacteria having naturally different optimum temperatures for luminescence intensity, at approximately 21°, 26°, and 32°C in *P. phosphoreum, Achromobacter fischeri,* and *A. harveyi,* respectively, each suspended in corresponding, buffered salt solution. (Data of Brown *et al.,* 1942.)

later in reference to seemingly remotely related phenomena, such as the activity of single nerve fibers or locomotion of drunken tadpoles, as will be mentioned presently.

When various pressures are applied to a suspension of luminous bacteria, the light intensity changes in a manner that depends on both the temperature of the experiment and the temperature activity curve of luminescence in the particular species of bacteria. In Fig. 2A, the change in intensity as a function of pressure at different temperatures is plotted in terms of percent, taking the intensity arbitrarily as 100 at normal pressure at each temperature. At the lowest temperature, 0°C, pressure decreases the intensity; with rise in pressure the decrease is exponential and the slope of the straight line resulting from plotting the log of intensity against hydrostatic pressure indicates a large volume change, amounting to about 50 cm^3/mole. At the highest temperature, 35°C, pressure increases the intensity; the rise in intensity with rise in pressure occurs with a steeper slope, indicating an even larger volume change. At intermediate temperatures, both types of change occur, i.e., first an increase then a decrease in intensity with rise in pressure. It will be recalled that in earlier work, the different types of effects were thought to be associated with the process involved, in the absence of evidence concerning an influence of temperature, rather than with one process at different temperatures.

D. Temperature and Pressure Optimums

The "optimum temperature" for brightness of luminescence in a suspension of the bacteria used for the foregoing experiments (Fig. 2A) was about 21°C. These bacteria were a strain of *Photobacterium phosphoreum*, a psychrophilic species that produced the brightest cultures when cultivated at 15° to 18°C, and generally failed to grow at temperatures as high as 25°C. Two other species, *Achromobacter fischeri* and *A. harveyi*, had a maximum growth temperature of 28°–30° and 37°–40°C, respectively, and optimum temperatures for brightness of luminescence in cell suspensions at around 26° and 32°C, respectively. These temperature relations may be considered genetically determined, characteristics of the species and strain, inasmuch as they pertain to growth and luminescence under the same experimental conditions. The important point in the present context is that in all three species the effect of pressure on the intensity of luminescence in cell suspensions is to decrease it at temperatures below the optimum, to have relatively little effect at the optimum, and to increase it at temperatures above the optimum. Thus, the direction of the observed pressure effect—causing a decrease, little effect, or an increase —could be changed by a suitable change in the species at a single temperature (Fig. 2B), as well as by a suitable change in temperature with a single species (Fig. 2A).

E. Pressure–Temperature Relations of Muscle Tension

Some relationships between the observed effects of pressure and temperature had been noted earlier in experiments on muscle contraction (Cattell and Edwards, 1928, 1930, 1932), and unpublished data on pressure–temperature relations of muscle tension (Fig. 3), remarkably similar to the data on luminescence intensity (Fig. 2A), had been obtained by Brown. Such relationships were indicated in abstracts (Brown, 1934, 1934-1935) together with suggestive evidence that the pressure–temperature relations were related to the specific temperature range to which the

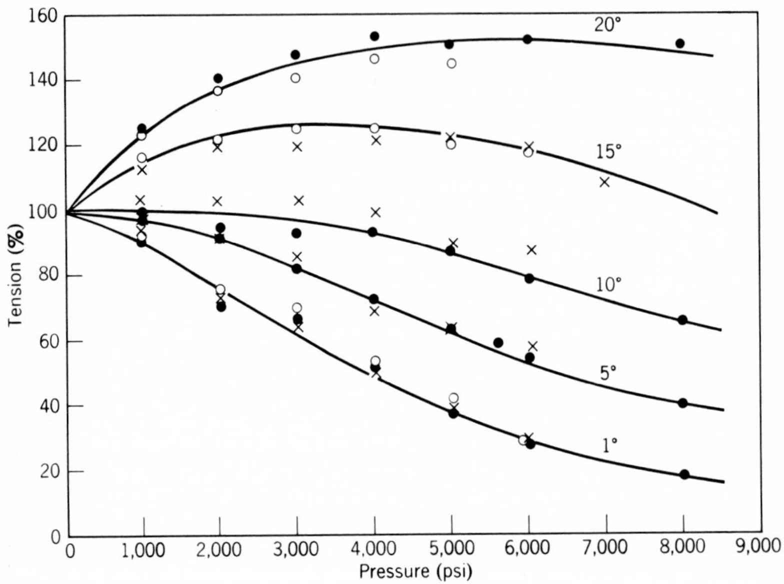

FIG. 3. Influence of increased pressure, at different temperatures, on the tension developed in auricular muscle tissue of the turt'e. The data, furnished by courtesy of D. E. S. Brown, are plotted in the manner of Fig. 2A, and are reproduced from Fig. 9.9 of Johnson *et al.* (1954).

organism involved was adapted or acclimitized. Publication of the extensive data of Brown and his collaborators on this subject was withheld until such data had become sufficient to justify definitive conclusions (Brown, 1957; Brown *et al.*, 1958; Guthe, 1957; Guthe and Brown, 1958), aided in part by development of the theoretical basis of the kinetics.

F. Reversibility of Pressure and Temperature Effects

A phenomenon of key importance to an understanding of the data of Fig. 2A was the reversibility of the effects of temperature, as well as those of pressure. Rapid application of pressure, at any temperature, was accompanied by an almost equally rapid change in intensity of luminescence, and the same effect occurred in the opposite direction when the pressure was suddenly released. Similarly, when a suspension of cells at atmospheric pressure was transferred from an ice bath to a hot water bath, the intensity increased with rise in temperature, went through a maximum or "optimum," then very rapidly decreased with further rise in temperature until the light became quite dim. If the rise in temperature was rapid and if the cells were maintained only momentarily at the highest temperature and were then transferred back to an ice bath, the light grew brighter with the process of cooling, again went through a maximum or optimum, then decreased again as the temperature continued to fall, finally returning to practically the same intensity as at the start, before heating.

For purposes of the present discussion, some incidental phenomena need not be considered in detail, e.g., the momentary "blackout" following a sudden decompression at a relatively high temperature. It should be noted and emphasized, however, that in accordance with general expectations, a rate process of irreversible, thermal destruction of the luminescence system occurred at temperatures above the optimum, the rate of destruction increasing very rapidly with rise in temperature. It is primarily because of the unique indicator of reaction velocity provided by the brightness of luminescence that the full temperature activity curve, from temperatures much below to temperatures much above the optimum, can be so easily observed and quickly measured. The ease of observing a quantitative reversibility of the effects of the high temperatures, however, depends also on certain differences in the temperature relations of the immediate effect through a shift in a mobile equilibrium and on a rate process of destruction at these temperatures, each being dependent on the system involved and the conditions of the experiments.

G. Quantitative Interpretation of the Pressure–Temperature Relations of Luminescence in Cells

On plotting the data of Fig. 2A as luminescence intensity on the ordinate against the reciprocal of the absolute temperature on the abscissa for three pressures, namely 1, 272, and 476 atm, the relations shown in Fig. 4 are seen. In this figure the points represent data from experiments and the smooth curves the result of calculations, in accordance with the following interpretations.

From the slope of the nearly linear portion of the curves in the low temperature range (Fig. 4), a numerical value for activation energy of the luminescence process as a whole at normal pressure can be computed and is found to be about 17,000 cal, which is within the range typical of enzyme reactions and of more complex processes limited by enzyme activity. Thus we may tentatively assume that in this range of temperature, the process of bacterial luminescence is limited by the activity of some enzyme, which, pending further evidence, we may refer to as bac-

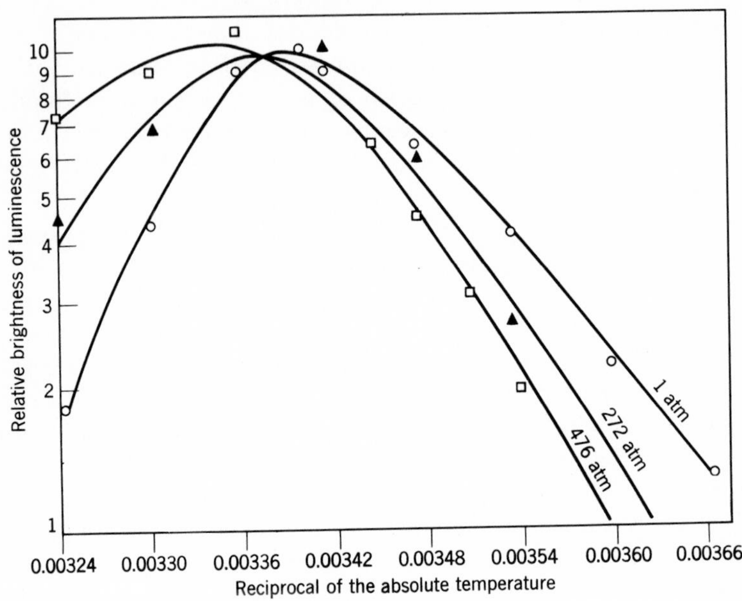

FIG. 4. The data of Fig. 2 plotted as a function of luminescence intensity on the logarithmic scale of the ordinate against the reciprocal of absolute temperature on the abscissa for three different pressures of 1, 272, and 476 atm. The smooth curves were calculated by Eyring and Magee (1942) in accordance with Eq. (30) as described in the text.

terial "luciferase," that catalyzes a light-emitting oxidation of a substrate, bacterial "luciferin." These terms, used here for convenience, are general terms for more specific, and not necessarily chemically related, enzymes and substrates responsible for luminescence in various types of luminous organisms (Harvey, 1952; Johnson, 1967).

In the high temperature range (relative to the optimum; Fig. 4), the slope of the line may be taken as a first approximation of the numerical value of heat of reaction ΔH_1 in a reversible denaturation of the luciferase whereby the state of the enzyme changes from active to inactive as the temperature is raised; by appropriate calculation, a numerical value for the entropy of reaction ΔS_1 can also be obtained. The assumption is merely that the intensity of luminescence decreases as the temperature rises toward and beyond the optimum, because even though the specific reaction rate k_1 of the luciferin–luciferase reaction increases with rise in temperature, the equilibrium constant K_1 likewise increases but much faster, quickly overriding the increase in k_1 at the optimum and higher temperatures. Thus, at temperatures above optimum, the velocity of the luciferin–luciferase reaction is greater than at the lower temperatures, but due to the reversible denaturation equilibrium, there is much less luciferase in the catalytically active state, with the net result that the light intensity undergoes a decrease rather than an increase with rise in temperature. Experimentally, the intensity as a function of time remains essentially constant for a sufficient interval to obtain fairly accurate data, except at relatively high temperatures, where some hysteresis is inevitable due to the irreversible destruction of the enzyme.

In regard to the influence of pressure, the data suggest that the limiting enzyme reaction proceeds with a volume increase of activation $\Delta V_1^{\ddagger}$, a numerical value for which may be obtained from data pertaining to low temperatures, where K_1 is negligibly small, with the aid of Eq. (27). Similarly, from data in the high temperature range, an approximate value for the volume change of reaction ΔV_1 can be obtained by means of Eq. (28).

On the basis of the foregoing evidence and assumptions, wherein the instantaneous intensity I of the luminescence is taken to be proportional to the velocity v of the luciferin–luciferase reaction, in a steady state condition whereby the concentration of luciferin LH_2 in effect remains constant and the total amount of luciferase, A_o, i.e., in the active A_n plus the reversibly inactivated A_d states, also remains constant. Eq. (29) provides the quantitative basis for the variation in intensity at different temperatures and pressures. While this equation is obviously oversimplified, it has proved adequate for the smooth curves calculated by Eyring and Magee (1942) shown in Fig. 4.

$$I \alpha v = \frac{bk_1(\mathrm{LH_2})(\mathrm{A_0})}{1 + K_1} = \frac{b\varkappa(kT/h)K_1^{\ddagger}(\mathrm{LH_2})(\mathrm{A_0})}{1 + K_1} \qquad (29)$$

In Eq. (29), b is a proportionality constant to allow for the units of measurement. For practical application to rates of intracellular processes, such as in the present instance (where the actual concentration of reactants LH_2 and A_o is unknown and the entropy of activation $\Delta S_1^{\ddagger}$ pertaining to k_1 cannot be readily obtained, but assuming that they are essentially constant under the experimental conditions involved), these different constants can be incorporated along with $\varkappa$ (k/h) into one unknown constant c, whereupon Eq. (29) becomes Eq. (30):

$$I\alpha v = \frac{cTe^{-(\Delta H_1\ddagger/RT)}}{1 + e^{-(\Delta H_1/RT)}\, e^{(\Delta S_1/R)}} = \frac{cTe^{-(\Delta E_1\ddagger/RT)}\, e^{-(p\Delta V_1\ddagger/RT)}}{1 + e^{-(\Delta E_1/RT)}\, e^{-(p\Delta V_1/RT)}\, e^{(\Delta S_1/R)}} \tag{30}$$

The numerical values used by Eyring and Magee (1942) in this equation for calculation of the curves in Fig. 4 are as follows (a value for the constant c being arbitrarily chosen to make the maximum of the curve for 1 atm equal to 10 on the logarithmic scale of the ordinate):

$$\Delta E_1\ddagger = 17{,}220 \quad \Delta V_1\ddagger = 546.4 - 1.813T$$
$$\Delta E_1 = 55{,}260 \quad \Delta V_1 = -922.8 = 3.206T \quad \Delta S_1 = 184 \text{ eu}$$

The large values for heat and entropy of reaction are indicative of large molecules such as are involved in denaturation reactions of proteins (Eyring and Stearn, 1939). The closeness of the agreement between the theoretical curves and the points from experiments (Fig. 4) is aided somewhat by including in the equation a temperature dependence in the values for ΔV_1^* and ΔV_1.

This temperature dependence of the volume changes might be real. On the other hand, it might reflect an oversimplification of the theoretical treatment, especially in regard to reactions taking place in the complex setting of a living organism. In 1942 it was not possible to determine whether or not similar results would be obtained with the luminescence system *in vitro*, apart from the intricate metabolic machinery and structural organization of an intact bacterial cell, because no cell-free luminescent extracts had been successfully obtained. Some years later such extracts were obtained, however, and it was shown that the minimal components for a brightly luminiscent reaction *in vitro* consisted of reduced flavine mononucleotide ($FMNH_2$), a long-chain aliphatic aldehyde (R-CHO) of not yet established function, and a bacterial enzyme (luciferase) (Strehler, 1953, 1955; McElroy *et al.*, 1953; Cormier and Strehler, 1953); crystalline preparations of the enzyme have been recently obtained (Kuwabara *et al.*, 1965) as well as evidence for the formation of several intermediary compounds in the overall light-emitting process catalyzed by the purified enzyme *in vitro* (Hastings *et al.*, 1966). Addition of reduced diphosphopyridine nucleotide (DPNH) to an aerated, aqueous solution of FMN plus decaldehyde results in a steady state luminescence of potentially long duration, with fundamentally the same characteristics as that which is emitted by intact cells with either pure or crude luciferase *in vitro*.

H. Pressure–Temperature Relations of Bacterial Luminescence *In Vitro*

With a partially purified enzyme preparation it was shown (Strehler and Johnson, 1954) that fundamentally the same pressure–temperature

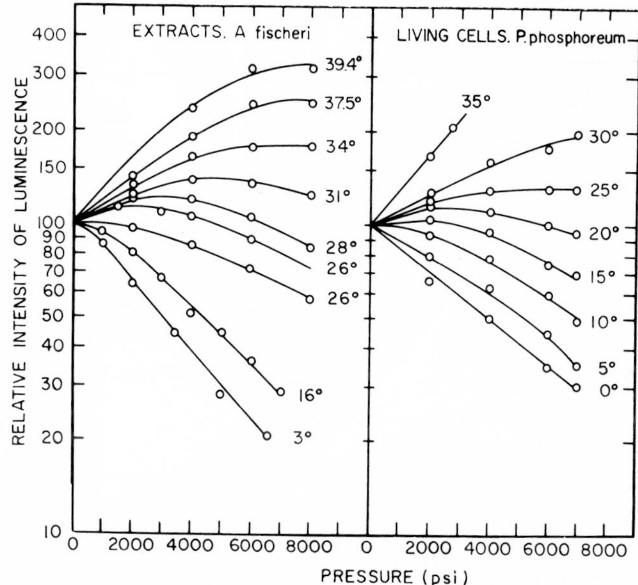

Fig. 5. Steady state levels of luminescence intensity in cell-free extracts of the luminous bacterium *Achromobacter fischeri* (left) and in living cells of *Photobacterium phosphoreum* (right) (from Strehler and Johnson, 1954). The data for the graph on the right are replotted from Brown *et al.* (1942). As in Fig. 2, the changes in intensity as a function of pressure at different temperatures are shown with reference in each case to the control as arbitrarily equal to 100 at normal pressure at the temperature involved. In the curves for luminescence of extracts, it should be mentioned that the change from one steady state level of intensity to another level, following sudden application or release of pressure, occurred much more slowly than in living cells. Allowance was made, therefore, wherever the decay in intensity of luminescence in extracts occurred at a significant rate. The lower of the two curves for 26°C in the graph on the left was obtained with an enzyme solution that had stood for several days at room temperature.

relations of luminiscence intensity occur *in vitro* (Fig. 5) as *in vivo* (Fig. 2A), although the change from one steady state level to another, following rapid application or release of pressure, was slower in extracts than in living cells. In any event, the data of Fig. 5 provide evidence that the luminescence system itself is directly affected by changes in temperature and pressure in the manner that has been discussed rather than through some obscure, unknown mechanism operative only amidst the complexities of a living organism. With the aid of highly purified components now available, it should be interesting to investigate the reaction step(s) most susceptible to influence of pressure.

I. General Theoretical Interpretation of Biological Temperature–Pressure Relations

The foregoing kinetic analysis of the pressure–temperature relations of luminescence intensity was undertaken through the simplest and most convenient approach, in this instance involving chiefly two reactions, designated by the specific rate constant k_1 and denaturation equilibrium K_1, controlling the net rate of the process. Numerical values for heats, entropies, and volume changes cannot be predicted a priori for any biological process, which usually lacks a detailed understanding of the reaction mechanism and in general is subject to modification by such widely recognized factors as templet mechanisms leading to asymmetric syntheses, differential diffusion across membranes, spatial organization of catalysts in the subcellular particulate form of mitochondria or other structures, and enzyme induction, and feedback inhibitions. Yet the basic theory for each elementary reaction is the same, and in many instances a complex biological phenomenon as a whole conforms, within limits of course, to the same quantitative laws governing the simplest reactions. Thus, by first measuring the appropriate rate and equilibrium constants under a few conditions, remarkably accurate predictions can sometimes, as in luminescence, be made for a variety of other conditions pertaining to the same variables.

Essentially all biological processes have an "optimum temperature" which varies with the specific organism, the specific process, and the specific conditions, including the recent past history of the organism within which time mechanisms of temporary adaptation may have been active. Very likely, essentially all biological processes also have an "optimum pressure," subject no doubt to variation according to the same factors, and it will be noted that reversible, slight alterations in the temperature of maximum luminescence intensity occur at different hydrostatic pressures (Fig. 4). Much greater reversible changes sometimes occur under the influence of chemical agents, such as narcotics, as will be discussed.

At this point, it should perhaps be emphasized that, although the foregoing method of analysis has the advantage of convenience and relative simplicity, the whole situation can be viewed, in terms of the theory of absolute reaction rates, in a different way. Thus in biological rates at the optimum or other temperatures the rate measures the concentration of the activated complex in the limiting reaction. The activated complex can be viewed as in equilibrium with its environment, and it is in fact composed of a species whose properties, heat content in particular, change with temperature. So long as the activated complex represents a more energy-rich configuration for its constituents than the average value outside of the activated complex, a rise in temperature will increase the rate; at the point where this is no longer true, the rate process will go through a maximum, familiarly termed "optimum" in biology. At still higher temperatures, the constituents of the activated complex are in equilibrium with even more energy-rich configurations characterizing the reversible, thermally inactivated states of the enzyme. In the context of this view, the slopes of the curve for luminescence intensity vs $1/T$ (Fig. 4) are a measure of the activation energy, whether the slope is negative as at low temperatures, zero at the maximum, or positive as at higher temperatures. These considerations apply to any biological reaction whose net rate is limited by a single species of activated complex; the situation is more complicated, of course, when the net rate is significantly governed by different species of activated complex.

The same view holds with regard to pressure. So long as the volume of the activated complex exceeds the average volume of its constituents outside of the complex, increased pressure will retard the reaction. At the point where the volumes become equal there is no change in rate under pressure. When the volume of the activated complex is less than that of the reactants, pressure will accelerate the rate.

J. The Active Intermediate Configuration of an Enzyme

As yet, the actual configuration of the activated complex is known only for the simplest reactions, such as $H_2 + I_2 \rightarrow 2HI$. No firm rule can be stated in regard to the nature of changes in structure of a protein enzyme accompanying the catalytic activity in reacting with a substrate, or in undergoing reversible or irreversible denaturation. When a large volume increase of the order of 50 or 100 $cm^3/mole$ is associated with the reaction, however, it is reasonable to view such changes as arising from an "unfolding" of the protein, whereby hydrophobic groups that are normally bonded inside are brought to the surface of the molecule and thereby exposed to the aqueous solvent; because of less attraction between these than of hydrophilic groups and water, the molecule in effect acquires a larger volume, even though the percent change in volume may be extremely small. This interpretation is not unique, but it is still useful and plausible. We discussed this general interpretation some years ago and pointed out also that ". . . a more active intermediate state of the enzyme is important in various circumstances. For example, sulfanilamide and urethan antagonize each other's inhibition of bioluminescence. Similarly the active configuration of the enzyme is favored by the appropriate temperature and pressure. Thus, with particular reference to the luminescent enzyme (in luminous bacteria - *sic.*), there is evidence that hydrogen ions, low temperatures, hydrostatic pressure and sulfanilamide each stabilize the molecule in states characterized by relatively low heat content and small volume, whereas hydroxyl ions, high temperature, low pressure, and urethan each stabilize the molecule in states characterized by relatively high heat content and large volume, the catalytically active system being intermediate. Apart from quantitative differences among different enzymes, the same concept of an intermediate configuration necessary for activity applies in general and is the basis for the optima that are found under the influence of various factors" (Johnson *et al.*, 1954, p. 29). The same concept, though clothed in new phraseology and stemming from lines of evidence different from the pressure–temperature relations considered in the present discussions, is implicit in generally accepted, contemporary views regarding the mechanism of enzyme action; cf., for example, review by Koshland and Neet (1968, p. 401) who remark, "It seems clear conformational changes at the active site are an essential feature of enzymes. Chemical tests in solution, physical chemical criteria and X-ray crystallography all provide strong support that in some enzymes substrate-induced conformational changes play a vital role. It does not follow that all enzymes require such a change but the evidence is beginning to accumulate that perhaps they do. . . ." Despite the vastly clearer insight into the nature of protein structure and enzyme reactivity which has arisen within the past few years, a number of details remain to be clarified, e.g., as Koshland and Neet point out (1968, p. 386), "The manner in which the conformational change is induced is not established yet." (See also Morita and Becker, or Chapter 3).

K. Influence of Hydrostatic Pressure on the Action of Narcotics

With particular reference to the action of what are now sometimes called "ligands" in reference to low molecular weight substances, (including inhibitors, activators, and natural substrates) which combine with proteins by noncovalent bonds (Koshland and Neet, 1968), the effects of increased hydrostatic pressure remain potentially useful. Prior to the pressure–temperature studies on bacterial luminescence there was apparently no record of a relation between hydrostatic pressure and the

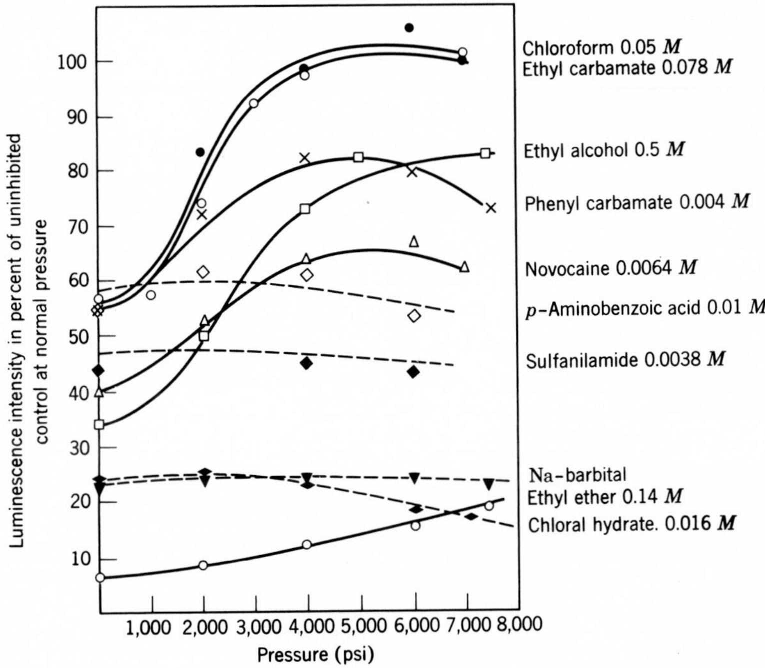

FIG. 6. Influence of increased hydrostatic pressures on the intensity of luminescence in cells of *P. phosphoreum* suspended in isotonic, phosphate buffered NaCl solution at 17° to 18°C, containing various narcotics and other drugs at concentrations indicated at the right of the corresponding curve in the figure. The changes in intensity in each instance are indicated in terms of the percent scale of the ordinate with respect to the intensity of the drug-free control at atmospheric pressure arbitrarily taken equal to 100. Solid lines are for curves pertaining to inhibitors whose effects at atmospheric pressure were partially or practically completely reversed by increased hydrostatic pressure. The broken lines are for inhibitors whose action was not significantly affected by increased pressure under the conditions involved. (Data of Johnson *et al.*, 1942b, as reproduced in Johnson *et al.*, 1954.)

action of a narcotic or other type of inhibitor. Relatively few investigations from this point of view have been carried out since. When such a relation was revealed (Fig. 6) (Johnson *et al.*, 1942b), the mechanism was obscure, but it gradually became somewhat understandable as further data concerning concentration and temperature, as well as hydrostatic pressure, were accumulated. In fact, from the earlier as well as present perspective, according to the concept of reversible "unfolding" of an enzyme molecule with a concomitant exposure of more hydrophobic groups to the solvent than in the normal, folded state, it would seem to follow that lipid-soluble ligands might easily further the unfolding process by combining with the hydrophobic groups. Since exposure of groups simply by raising the temperature results in a "denaturation" equilibrium shift accompanied by a large volume increase in the species of enzyme molecules involved, it would seem to follow also that the reversible effects of certain lipid soluble narcotic agents would be susceptible to change by variations in hydrostatic pressure, as was indeed found to be true. On the other hand, it would be reasonable to expect that certain other agents would act in a reversible manner with no appreciable volume change, and, consequently their effects would not be influenced by changes in pressure. Both types of effect are illustrated in Fig. 6.

In deriving explicit expressions for analyzing the two types of inhibitor action, the latter of the above two was designated as type I and the former as type II (Johnson *et al.*, 1942c, 1943, 1945, 1954). In general, type I typically involves an increase in apparent activation energy, a higher optimum temperature, no promotion of a reversible or irreversible protein denaturation, and practically no susceptibility to change by pressure. On the other hand, type II typically involves a decrease in apparent activation energy, a lower optimum temperature, promotion of an irreversible as well as reversible protein denaturation, and susceptibility to reversal and retardation of the denaturation by increased pressure. These phenomena have been already alluded to in a preceding paragraph with reference to the activated complex (p. 32). In the context of this chapter, it seems appropriate to mention the principle that has been found useful in arriving at explicit formulations without the details of their derivation and results of analyses, as follows.

In any enzyme reaction, under specified conditions of temperature, pressure, and chemical environment, the net rate of reaction depends on the fraction of active enzyme present in a total amount which may include inactive forms of various kinds. Whenever an appreciable fraction of the total exists in combination with the substrate involved in the catalytic process, this fraction must be taken into account, as in the Michaelis-Menten theory (1913), which, however, may lead to some uncertainties of interpretation even with a single enzyme system (cf. Kauzmann *et al.*, 1949). In a number of instances it is unnecessary to include the fraction of enzyme molecules combined with the substrate in order to analyze data for values of the heat, entropy, volume change, and ratio of ligand to enzyme molecules in the various equilibria which can be present. Where different types of ligands simultaneously present combine, in effect, with each other, the formulations become notably complicated. Otherwise, a "conservation equation" can be written to include the sum of the various species of the enzyme with their equilibrium constants. Assuming that one of these species is a thermally denatured, catalytically inactive form, on the basis that all enzymes are susceptible to inactivation by heat and that this reaction is likely to be at least partially reversible on cooling, the rate is given by Eq. (29), wherein the total enzyme, substrate, and specific reaction rate constant appear in the numerator, and the equilibrium constant for the reversible denaturation is in the denominator. When inhibitors are present, additional equilibrium constants times the concentration of the inhibitor (the concentration in each case raised to the power corresponding to the number of molecules of the inhibitor combining with one enzyme molecule) must be included in the denominator. On dividing Eq. (29) by the corresponding equation, which takes into account additional equilibria represented in the denominator, the numerators cancel out, and, on simplification, useful formulations are obtained in terms of ratios of velocities without to velocities with various concentrations of inhibitors at different temperatures and pressures. Formulations which can be applied to data from experiments to yield values for the heat, entropy, volume change, and ratio of combining molecules in the enzyme–inhibitor complex are different, of course, for the two types of inhibition mentioned above.

L. Retardation of Protein Denaturation by Pressure

The action of urethan or alcohol on the bacterial luminescence system cannot be satisfactorily analyzed without taking into account the ordinary, reversible thermal denaturation. In effect, such inhibitors lower the temperature required for significant amounts of the thermal denaturation. They are also apt to catalyze a rate process of irreversible denaturation at a rate which increases with concentration of inhibitor and with temperature but decreases with rise in pressure. An analogous phenomenon occurs in the precipitation of serum globulin (Johnson and Campbell, 1945, 1946) and tobacco mosaic virus (Johnson *et al.*, 1948), as well as inactivation of bacteriophage T5 (Foster *et al.*, 1949) and disinfection of bacterial spores (Johnson and ZoBell, 1949a,b). As a result of catalyzing an irreversible denaturation, the amount of inhibition of enzyme activity caused by urethan or alcohol tends to increase with time, whereas inhibitions of the other type, independent of the thermal denaturation, tend to remain relatively constant with time.

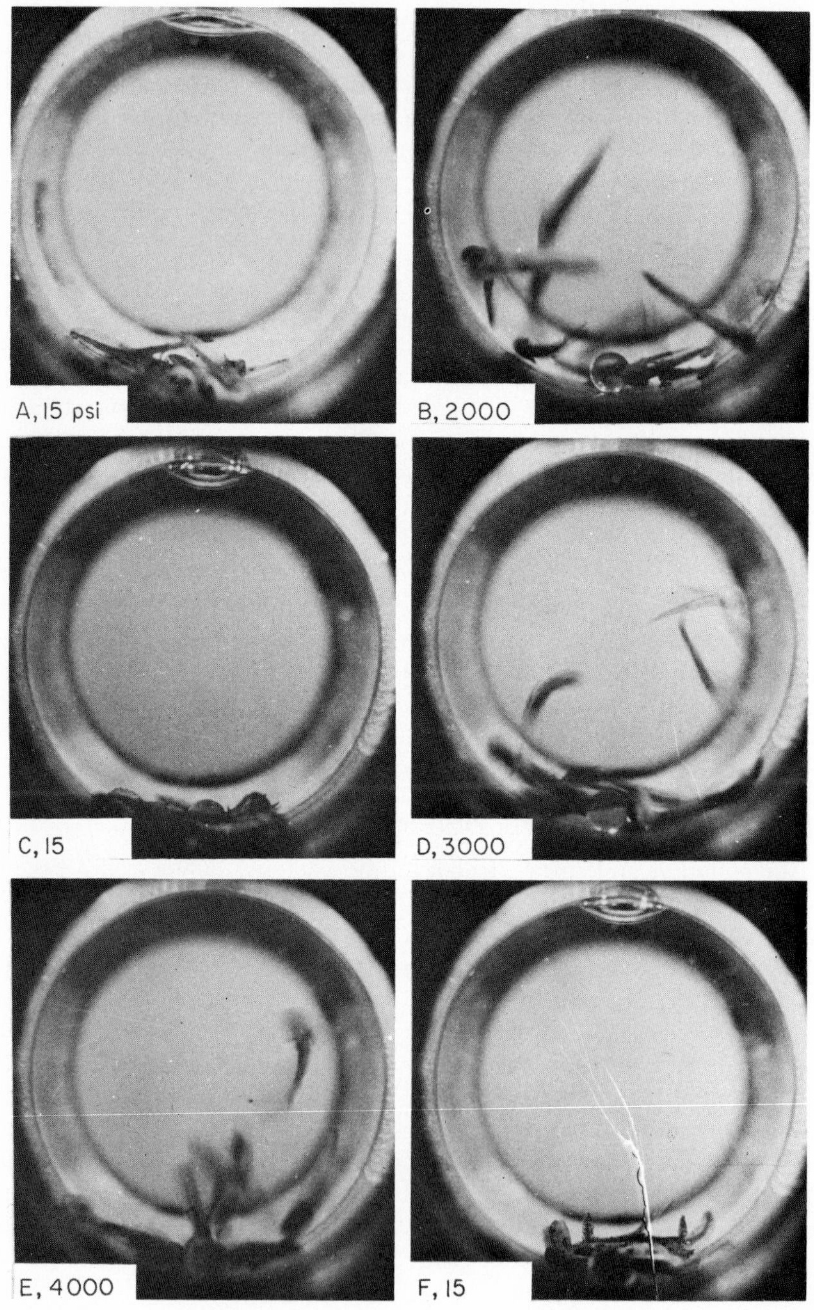
A, 15 psi
B, 2000
C, 15
D, 3000
E, 4000
F, 15

M. Pressure Reversal of Narcosis

The pressure effects illustrated in Fig. 6 served to predict the possibility of analogous phenomena in reference to a number of systems other than bacterial luminescence. Perhaps the most impressive is the pressure reversal of the alcohol narcosis of small aquatic animals such as the tadpoles of frogs and larvae of salamanders (Fig. 7) (Johnson and Flagler, 1951a, b). Unfortunately, it is not possible to conduct experiments of the sort involved with tadpoles on higher animals whose anatomy includes organs such as lungs which are filled with air; under increased hydrostatic pressure over the whole body these organs would be crushed by the pressures required for the effect in question. On the other hand, experiments with the isolated giant axon of the squid have revealed that the reduction in the action potential caused by immersing the nerve in sea water containing 3% ethanol was restored practically to normal through increasing the hydrostatic pressure or through lowering the temperature (Tasaki and Spyropoulos, 1957). These results pertaining to the activity of single nerve fibers lend credence to a belief that the phenomena observed in regard to narcosis of tadpoles take place through an effect of pressure on the nervous system, in a manner which, at the molecular level, is fundamentally similar to the results pertaining to the intensity of bacterial luminescence.

IV. Future Research

At the present time and in the near future it should be possible to specify in more detail and greater clarity the molecular mechanisms through which biological processes in living cells are modified by changes in hydrostatic pressure. The foregoing discussions, in addition to constituting an attempt to give an idea of the fundamental reaction rate theory which will inevitably be involved in the final analysis, should perhaps serve to emphasize the basic importance of temperature toward an understanding of the biological effects of pressure, a relationship which is amply illustrated elsewhere in this book.

Two general, basic problems that have been no more than touched

---

FIG. 7. Photographs through the window (2 inches in diameter) of a large pressure chamber showing the activity of *Amblystoma* in water containing 2.5% ethyl alcohol, with successive changes in hydrostatic pressure, over a total period of 4 minutes. The pressure pertaining to each picture is indicated as psi in the lower left corner of that picture. (From Johnson and Flagler, 1951b, and Johnson *et al.*, 1954.)

upon in this chapter but which future investigations may be expected to bring to an improved state of understanding are (1) the action of high pressures, usually in excess of 1000 atm, in denaturating proteins, as well as the relation of this phenomenon to that of the pressure retardation and reversal of protein denaturation, and to unfolding of macromolecules, and (2) the pressure–temperature relations of physiological processes in deep sea organisms more complex than bacteria, i.e., invertebrates and fishes in particular.

In regard to point (1), notable progress is evident in the interesting results of mostly recent and current research with both simple and complex systems. Although limitations of space prevent further discussion here, reference may be made, for example, to the following: Gill and Glogovsky (1965), Holcomb and Van Holde (1962), Horne *et al.*, (1968), Kettman *et al.* (1966), Murakami (1960, 1961, 1963), Nishikawa *et al.* (1968), Schneider (1963, 1966), Suzuki *et al.* (1968), Tongur (1949, 1951, 1952), Tongur and Kasatochkin (1950, 1952), Tongur and Kasatochkin (1954), and Kliman (1969).

In regard to the second of the above-mentioned basic problems, it is encouraging to note that efforts are indeed being undertaken in this direction, e.g., by Schlieper (1968) (see also Flügel and Schlieper, Chapter 9). It is remarkable that relatively little effort seems to have been made thus far to investigate the physiology of animals from the very deep sea. Unfortunately, forms of plant life, other than bacteria, appear to be excluded from this habitat because of its perpetual, practically complete darkness, which is relieved only by light of biological origin, emanating for example from luminous fishes and scarcely sufficient, so far as is known, to permit the occurrence of an indigenous type of photosynthetic organism. The hiatus of data on hadal physiology is no doubt attributable in large part to the difficulties of obtaining living specimens in the laboratory for the research needed. In principle, however, such difficulties are not insurmountable, and very likely would be quickly overcome if exploration of this last frontier of regions inhabited by life on earth were ever undertaken with anything resembling the immense scale and concerted effort now being exerted toward the exploration of extraterrestrial space. Yet no one can predict with certainty at the present time which one of these frontiers will ultimately yield the greater consequences to mankind.

ACKNOWLEDGMENTS

This has been aided in part by NSF grant GB 6836, and ONR Contract Nonr 4264(00). The authors are indebted also to John Wiley and Sons, Inc., New York, for permission to use figures and textual material from the book by Johnson, Eyring, and Polissar (1954), entitled "The Kinetic Basis of Molecular Biology."

REFERENCES

Arrhenius, S. (1889). Über die Reaktionsgeschwindigkeit bei der Inversion von Rohrzucker durch Säuren. *Z. Physik. Chem.* 4, 226–248.

Basset, J., and Macheboeuf, M. A. (1932). Etude sur les effets biologiques des ultrapressions: Résistance des bactéries, des diastases et des toxines aux pressions très élevées. *Compt. Rend.* **195**, 1431–1433.

Basset, J., Wollman, E., Macheboeuf, M. A., and Bardach, M. (1933a). Etudes sur les effets biologiques des ultra-pressions: Action des pressions très élevées sur les bactériophages et sur un virus invisible (virus vaccinal). *Compt. Rend.* **196**, 1138–1139.

Basset, J., Lisbonne, M., and Macheboeuf, M. A. (1933b). Action des ultra-pressions sur le suc pancréatique. *Compt. Rend.* **196**, 1540–1542.

Basset, J., Macheboeuf, M., and Sandor, G. (1933c). Etudes sur les effets biologiques des ultrapressions. Action des pressions très élevées sur les protéides. *Compt. Rend.* **197**, 796–798.

Basset, J., Wollman, E., Wollman, E., and Macheboeuf, M. A. (1935a). Etudes sur les effets biologiques des ultra-pressions; action des pressions très élevées sur les bactériophages des spores et sur les autolysines. *Compt. Rend.* **200**, 1072–1074.

Basset, J., Wollman, E., Macheboeuf, M. A., and Bardach, M. (1935b). Etudes sur les effets biologiques des ultra-pressions: Action des pressions élevées sur les tumeurs. *Compt. Rend.* **200**, 1247–1248.

Basset, J., Nicolau, S., and Macheboeuf, M. A. (1935c). L'action de L'ultrapression sur l'activité pathologique de quelques virus. *Compt. Rend.* **200**, 1882–1884.

Basset, J., Gratia, A., Macheboeuf, M., and Manil, P. (1938). Action of high pressure on plant viruses. *Proc. Soc. Exptl. Biol. Med.* **38**, 248–251.

Bridgman, P. W. (1914). The coagulation of albumen by pressure. *J. Biol. Chem.* **19**, 511–512.

Bridgman, P. W., and Conant, J. B. (1929). Irreversible transformations of organic compounds under high pressures. *Proc. Natl. Acad. Sci. U. S.* **15**, 680–683.

Brown, D. E. S. (1934). The pressure–tension–temperature relation in cardiac muscle. *Am. J. Physiol.* **109**, 16.

Brown, D. E. S. (1934–1935). "Cellular Reactions to High Hydrostatic Pressure," *Ann. Rep. Tortugas Lab.*, pp. 76 and 77. Carnegie Institution of Washington, Washington, D. C.

Brown, D. E. S. (1957). Temperature–pressure relation in muscular contraction. *In* "Influence of Temperature on Biological Systems," pp. 83–110. Am. Physiol. Soc., Washington, D. C.

Brown, D. E. S., Johnson, F. H., and Marsland, D. A. (1942). The pressure–temperature relations of bacterial luminescence. *J. Cellular Comp. Physiol.* **20**, 151–168

Brown, D. E. S., Guthe, K. F., Lawler, H. C., and Carpenter, M. P. (1958). The pressure, temperature, and ion relations of myosin and ATP–ase. *J. Cellular Comp. Physiol.* **52**, 59–77.

Bruun, A. F. (1951). The Philippine Trench and its bottom fauna. *Nature* **168**, 692–693.

Bruun, A. F. (1956). The abyssal fauna: Its ecology, distribution and origin. *Nature* **177**, 1105–1108.

Bruun, A. F. (1957). Deep sea and abyssal depths. *Mem. Geol. Soc. Am.* No. **67**, Vol. 1, 641–672.

Cattell, McK., and Edwards, D. J. (1928). The energy changes of skeletal muscle accompanying contraction under high pressure. *Am. J. Physiol.* **86**, 371–382.

Cattell, McK., and Edwards, D. J. (1930). The influence of hydrostatic pressure on the contraction of cardiac muscle in relation to temperature. *Am. J. Physiol.* **93**, 97–104.

Cattell, McK., and Edwards, D. J. (1932). Conditions modifying the influence of hydrostatic pressure on striated muscle, with special reference to the role of viscosity changes. *J. Cellular Comp. Physiol.* **1**, 11–36.

Cormier, M. J., and Strehler, B. L. (1953). The identification of kidney-cortex factor (KCF): Requirement of long-chain aldehydes for bacterial extract luminescence. *J. Am. Chem. Soc.* **75**, 4864–4865.

Eisenberg, D., and Kauzmann, W. (1969). "The Structure and Properties of Water." Oxford Univ. Press, London and New York.

Eyring, H. (1930). Verwendung optischer Daten zur Berechnung der Aktivierungswärme. *Naturwissenschaften* **18**, 915.

Eyring, H. (1935a). The activated complex in chemical reactions. *J. Chem. Phys.* **3**, 107–115.

Eyring, H. (1935b). The activated complex and the absolute rate of chemical reactions. *Chem. Rev.* **17**, 65–77.

Eyring, H., and Magee, J. L. (1942). Application of the theory of absolute reaction rates to bacterial luminescence. *J. Cellular Comp. Physiol.* **20**, 169–177.

Eyring, H., and Polanyi, M. (1930). Zur Berechnung der Aktivierungswärme. *Naturwissenschaften.* **18**, 914–915.

Eyring, H., and Polanyi, M. (1931). Ueber einfache Gasreaktionen. *Z. Physik Chem.* **B12**, 279–311.

Eyring, H., and Stearn, A. E. (1939). The application of the theory of absolute reaction rates to proteins. *Chem. Rev.* **24**, 253–270.

Foster, R. A. C., Johnson, F. H., and Miller, V. K. (1949). The influence of hydrostatic pressure and urethane on the thermal inactivation of bacteriophage. *J. Gen. Physiol.* **33**, 1–16.

Giddings, N. J., Allard, H. A., and Hite, B. H. (1929). Inactivation of the tobacco-mosaic virus by high pressure. *Phytopathology* **19**, 749–750.

Gill, S. J., and Glogovsky, R. L. (1965). Influence of pressure on the reversible unfolding of ribonuclease and poly-γ-benzyl-L-glutamate. *J. Phys. Chem.* **69**, 1515–1519.

Glasstone, S., Laidler, K. J., and Eyring, H. (1941). "The Theory of Rate Processes." McGraw-Hill, New York.

Goodeve, C. F. (1934). Three dimensional models of the potential energy of triatomic systems. *Trans. Faraday Soc.* **30**, 60–69.

Guthe, K. F. (1957). Myosin ATP-ase activity in relation to temperature and pressure. *In* "Influence of Temperature on Biological Systems," pp. 71–82. Am. Physiol. Soc., Washington, D. C.

Guthe, K. F., and Brown, D. E. S. (1958). Reversible denaturation in the myosin adenosine triphosphatase system. *J. Cellular Comp. Physiol.* **52**, 79–87.

Harvey, E. N. (1952). "Bioluminescence." Academic Press, New York.

Hastings, J. W., Gibson, Q. H., Friedland, J., and Spudich, J. (1966). Molecular mechanisms in bacterial luminescence: On energy storage intermediates and the role of aldehyde in the reaction. *In* "Bioluminescence in Progress" (F. H. Johnson and Y. Haneda, eds.), pp. 151–186. Princeton Univ. Press, Princeton, New Jersey.

Heitler, W., and London, F. (1927). Wechselwirkung neutraler Atome und homöopolare Bindung nach der Quantenmechanik. *Z. Physik* **44**, 455–472.

Holcomb, D. N., and Van Holde, K. E. (1962). Ultracentrifugal and viscometric studies of the reversible thermal denaturation of ribonuclease. *J. Phys. Chem.* **66**, 1999–2006.

Horne, R. A., Day, A. F., Young, R. P., and Yu, N. T. (1968). Interfacial water structure: The electrical conductivity under hydrostatic pressure of particulate solids permeated with aqueous electrolyte solution. *Electrochim. Acta* **13**, 397–406.

Johnson, F. H. (1967). Bioluminescence. *In* "Comprehensive Biochemistry" (M. Florkin and E. H. Stotz, eds.), Vol. 27, pp. 79–136. Elsevier, Amsterdam.

Johnson, F. H., and Campbell, D. H. (1945). The retardation of protein denaturation by hydrostatic pressure. *J. Cellular Comp. Physiol.* **26**, 43–46.

Johnson, F. H., and Campbell, D. H. (1946). Pressure and protein denaturation. *J. Biol. Chem.* **163**, 689–698.

Johnson, F. H., and Flagler, E. A. (1951a). Hydrostatic pressure reversal of narcosis in tadpoles. *Science* **112**, 91–92.

Johnson, F. H., and Flagler, E. A. (1951b). Activity of narcotized amphibian larvae under hydrostatic pressure. *J. Cellular Comp. Physiol.* **37**, 15–25.

Johnson, F. H., and ZoBell, C. E. (1949a). The retardation of thermal disinfection of *Bacillus subtilis* spores by hydrostatic pressure. *J. Bacteriol.* **57**, 353–358.

Johnson, F. H., and ZoBell, C. E. (1949b). The acceleration of spore disinfection by urethan and its retardation by hydrostatic pressure. *J. Bacteriol.* **57**, 359–362.

Johnson, F. H., Brown, D. E., and Marsland, D. A. (1942a). A basic mechanism in the biological effects of temperature, pressure and narcotics. *Science* **95**, 200–203.

Johnson, F. H., Brown, D. E., and Marsland, D. A. (1942b). Pressure reversal of the action of certain narcotics. *J. Cellular Comp. Physiol.* **20**, 269–276.

Johnson, F. H., Eyring, H., and Williams, R. W. (1942c). The nature of enzyme inhibitions in bacterial luminescence: Sulfanilamide, urethane, temperature and pressure. *J. Cellular Comp. Physiol.* **20**, 247–268.

Johnson, F. H., Eyring, H., and Kearns, W. (1943). A quantitative theory of synergism and antagonism among diverse inhibitors, with special reference to sulfanilamide and urethane. *Arch. Biochem.* **3**, 1–31.

Johnson, F. H., Eyring, H., Steblay, R., Chaplin, H., Huber, C., and Gherardi, G. (1945). The nature and control of reactions in bioluminescence. With special reference to the mechanism of reversible and irreversible inhibitions by hydrogen and hydroxyl ions, temperature, pressure, alcohol, urethane, and sulfanilamide in bacteria. *J. Gen. Physiol.* **28**, 463–537.

Johnson, F. H., Baylor, M. B., and Fraser, D. (1948). The thermal denaturation of tobacco mosaic virus in relation to hydrostatic pressure. *Arch. Biochem.* **19**, 237–245.

Johnson, F. H., Eyring H., and Polissar, M. J. (1954). "The Kinetic Basis of Molecular Biology." Wiley, New York.

Kauzmann, W., Chase, A. M., and Brigham, E. H. (1949). Studies on cell enzyme systems. III. Effects of temperature on the constants in the Michaelis-Menten relation for the luciferin-luciferase system. *Arch. Biochem.* **24**, 281–288.

Kettman, M. S., Nishikawa, A. H., Morita, R. Y., and Becker, R. R. (1966). Effect of hydrostatic pressure on the aggregation reaction of poly-l-valyl-ribonuclease. *Biochem. Biophys. Res. Commun.* **22**, 262–267.

Kliman, H. L. (1969). The solubility of 4-octanone in water, a model compound study of hydrophobic interactions at high pressure. Dissertation, Chemistry Dept., Princeton University, Princeton, N. J.

Koshland, D. E., and Neet, K. E. (1968). The catalytic and regulatory properties of enzymes. *Ann. Rev. Biochem.* **37**, 359–410.

Kuwabara, S., Cormier, M. J., Dure, L. S., Kreiss, P., and Pfuderer, P. (1965). Crystalline bacterial luciferase from *Photobacterium fischeri. Proc. Natl. Acad. Sci. U.S.* **33**, 342–345.

Larson, W. P., Hartzell, T. B., and Diehl, H. S. (1918). The effect of high pressures on bacteria. *J. Infect. Diseases* **22**, 271–279.

Lauffer, M. A., and Dow, R. B. (1941). The denaturation of tobacco mosaic virus at high pressures. *J. Biol. Chem.* **140**, 509–518.

Linderstrøm-Lang, K. (1950). Structure and enzymatic breakdown of proteins. *Cold Spring Harbor Symp. Quant. Biol.* **14**, 117–126.

Linderstrøm-Lang, K., and Jacobsen, C. F. (1941). The contraction accompanying enzymatic breakdown of proteins. *Compt. Rend. Trav. Lab. Carlsberg*, Ser. *Chim.* **24**, 1–46.

London, F. (1928). Über den Mechanismus der homöopolaren Bindung. *In* "Probleme der moderne Physik" (Sommerfeld Festschrift), pp. 104–113. Hirzel, Leipzig.

London, F. (1929). Quantenmechanische deutung Vorgangs der Aktivierung. *Z. Elektrochem.* **35**, 552–555.

Luyet, B. (1937a). Sur le méchanisme de la mort cellulaire par les hautes pressions: L'intensité et la durée des pressions léthales pour la levure. *Compt. Rend.* **204**, 1214–1215.

Luyet, B. (1937b). Sur le mécanisme de la mort cellulaire par les hautes pressions; modifications cytologiques accompagnant la mort chez la levure. *Compt. Rend.* **204**, 1506–1508.

McElroy, W. D. (1943). The application of the theory of absolute reaction rates to the action of narcotics. *J. Cellular Comp. Physiol.* **21**, 95–116.

McElroy, W. D., Hastings, J. W., Sonnenfeld, V., and Coulombre J. (1953). The requirement of riboflavin phosphate for bacterial luminescence. *Science* **118**, 385–386.

Macheboeuf, M. A., and Basset, J. (1934). Die Wirkung sehr hoher Drucke auf Enzyme. *Ergeb. Enzymforsch.* **3**, 303–308.

Macheboeuf, M. A., Basset, J., and Levy, G. (1933). Influence des pressions très élevées sur les diastases. *Ann. Physiol. Physicochim. Biol.* **9**, 713–722.

Matthews, J. E., Jr., Dow, R. B., and Anderson, A. K. (1940). The effects of high pressure on the activity of pepsin and rennin. *J. Biol. Chem.* **135**, 697–705.

Michaelis, L., and Menten, M. L. (1913). Der Kinetik der Invertinwirkung. *Biochem. Z.* **49**, 333–369.

Morse, P. M. (1929). Diatomic molecules according to the wave mechanics. II. Vibrational levels. *Phys. Rev.* **34**, 57–64.

Murakami, T. H. (1960). The effects of high hydrostatic pressure on cell division. *Symp. Cellular Chem.* **10**, 233–244 (in Japanese).

Murakami, T. H. (1961). The effects of high hydrostatic pressure on cell division: Synthesis of nuclei acids. *Symp. Cellular Chem.* **11**, 223–233 (in Japanese).

Murakami, T. H. (1963). Effect of high hydrostatic pressure on the permeability of plasma membrane under the various temperature. *Symp. Cellular Chem.* **13**, 147–156.

Nishikawa, A. H., Morita, R. Y., and Becker, R. R. (1968). Effects of solvent medium on polyvalylribonuclease aggregation. *Biochemistry* **7**, 1506–1513.

Polanyi, M., and Evans, M. G. (1935). Some applications of the transition state method to the calculation of reaction velocities, especially in solution. *Trans. Faraday Soc.* **31**, 875–894.

Regnard, P. (1891). "Recherches expérimentales sur les conditions physiques de la vie dans les eaux." Librairie de l'Academie de Médecine, Paris.

Schlieper, C. (1968). High pressure effects on marine invertebrates and fishes. *Marine Biol.* **2**, 5–12.

Schneider, G. (1963). Druckeinfluss auf die Entmischung flüssiger Systeme. II. Löslichkeit von H_20 und D_20 in Methylpyridinen und Methylpipyridinen. *Z. Physik. Chem. (Frankfurt)* [N.S.] **39**, 187–197.

Schneider, G. (1966). Phasengleichgewichte in flüssigen Systemen bei hohen Drucken. Z. Ber. Bunsengesel. Physikal. Chem. **70**, 497–613.

Stearn, A. E., and Eyring, H. (1941). Pressure and rate processes. *Chem. Rev.* **29**, 509–523.

Strehler, B. L. (1953). Luminescence in cell-free extracts of luminous bacteria and its activation by DPN. *J. Am. Chem. Soc.* **75**, 1264.

Strehler, B. L. (1955). Factors and biochemistry of bacterial luminescence. *In* "Luminescence of Biological Systems," pp. 209–255. Am. Assoc. Advance. Sci., Washington, D. C.

Strehler, B. L., and Johnson, F. H. (1954). The temperature–pressure-inhibitor relations of bacterial luminescence *in vitro*. *Proc. Natl. Acad. Sci. U.S.* **40**, 606–617.

Sugiura, Y. (1927). Über die Eigenschaften des Wasserstoffmoleküls im Grundstande. *Z. Physik* **45**, 484–492.

Suzuki, K., Miyosawa, Y., Tsuchiya, M., and Taniguchi, Y. (1968). Biopolymer solutions and model systems under high pressure. *Rev. Phys. Chem. Japan* **58**, 63–68.

Tasaki, I., and Spyropoulos, C. S. (1957). Influence of changes in temperature and pressure on the nerve fiber. *In* "Influence of Temperature on Biological Systems," pp. 201–220. Am. Physiol. Soc., Washington, D. C.

Tongur, V. S. (1949). Effect of pressure on denaturation of egg albumin. *Kolloidn. Zhur.* **11**, 274–279.

Tongur, V. S. (1951). Some problems of reversible denaturation in proteins. *Usp. Sovrem. Biol.* **31**, 391–412.

Tongur, V. S. (1952). Regeneration of egg albumins under pressure. *Biokhimiya* **17**, 495–503.

Tongur, V. S., and Kasatochkin, V. I. (1950). Reversibility of thermal denaturation of protein under pressure. *Dokl. Akad. Nauk SSSR* **74**, 553–556.

Tongur, V. S., and Kasatochkin, V. I. (1952). Effect of high pressures on thermal denaturation of proteins. *Khim. i Fiz.-Khim. Vysokomolekul. Soedin. Dokl. k Konf.*

Po Vysokomolekul. Soedin., 7-ya Konf., 1952 pp. 124–130. Izd. Akad. Nauk. S.S.S.R.

Tongur, V. S., and Kasatochkin, V. I. (1954). Protein regeneration under pressure. The kinetics and thermodynamics of the process. *Tr. Vses. Obshchestva Fiziologov, Biokhimikov i Farmakologov, Akad. Nauk SSSR* **2**, 166–176.

Tongur, V. S., and Kazmina, N. A. (1950). Effect of pressure on proteins. Renaturation of serum albumin under pressure. *Biokhimiya* **15**, 212–215.

Tongur, V. S., and Tongur, A. M. (1951). Regeneration of insulin under pressure. *Biokhimiya* **16**, 410–415.

van't Hoff, J. H. (1884). "Études de Dynamique Chimique." Frederik Muller & Co., Amsterdam.

Wolff, T. (1960). The hadal community, an introduction. *Deep-Sea Res.* **6**, 95–124.

Wynne-Jones, W. F. K., and Eyring, H. (1935). The absolute rate of reactions in condensed phases. *J. Chem. Phys.* **3**, 492–502.

ZoBell, C. E. (1952). Bacterial life at the bottom of the Philippine trench. *Science* **115**, 507–508.

ZoBell, C. E., and Morita, R. Y. (1957). Barophilic bacteria in some deep sea sediments. *J. Bacteriol.* **73**, 563–568.

ZoBell, C. E., and Morita, R. Y. (1959). Deep-sea bacteria. *In* "Galathea Report. Scientific Results of the Danish Deep-sea Round the World Expedition 1950–52," Vol. I, pp. 139–154. Copenhagen.

CHAPTER 2

HYDROSTATIC PRESSURE ON THE BIOSYNTHESIS OF MACROMOLECULES

J. V. Landau

I. Introduction

Over the past several decades the application of moderate hydrostatic pressures, i.e., 15–15,000 psi (pounds per square inch), to living systems has provided a wealth of information concerning the form and function of a host of biological species. This work, in the United States, has been largely the effect of a small group of pioneers and their students; Brown in instrumentation (1934) and muscle physiology (1936), Johnson in growth and function of microorganisms (Johnson & Lewin, 1946) and kinetic interpretation (Johnson *et al.*, 1954), Marsland in sol-gel phenomena in relation to cell division and ameboid movement (1956), and ZoBell in growth and function of marine microorganisms (Zobell and Oppenheimer, 1950).

As the interest in the oceanic portion of our planet increases, experimental work involving pressure, a fundamental environmental parameter of marine depths, should be markedly stimulated. Certainly there are profound questions as to the mechanisms of biological function at great ocean depths. However, the problems involved in the collection of deep sea organisms, their isolation, and their maintenance under laboratory conditions that would allow for valid experimentation are so numerous that, for the time being, more fruitful areas of investigation would seem to lie within the boundaries prescribed by the use of more readily accessible organisms. This approach has been taken with the understanding that initially at least a basis for the eventual comparison of the life processes of barophilic and nonbarophilic organisms can be established.

The relatively recent advances in the field of molecular biology have laid the groundwork for a more precise interpretation of the effects of pressure on macromolecular biosynthesis and aggregation. It is readily apparent that each of these processes requires the formation of specific molecular complexes which may be interpreted along the lines of Michaelis-Menten kinetics. If such a molecular interaction involves a volume increase, the effect of hydrostatic pressure application should be to inhibit the reaction. Similarly, a reaction accompanied by a volume decrease would be stimulated, and that which incurs no change in volume should be unaffected. Admittedly, this would seem a rather simplistic approach to the effects of an apparently nonspecific physical agent such as pressure when an entire organism with its myriad interrelated reactions is utilized as the experimental object. However, as an initial approach, it would seem to this author that a great deal of information can be obtained from experiments utilizing the whole cell. The problem, of course, lies in a reasonable interpretation of the experimental data. As will be shown, it has become evident that pressure has differential effects on some of the processes involved in the biosynthesis of protein. The judicious manipulation of pressure and temperature, the use of radioactive precursors, and, in some cases, the use of selective chemical inhibitors yield data which may be rationally approached on the basis of reaction rate theory. The techniques and theory of molecular biology allow for the subsequent analysis of pressure effects on cell-free synthesizing systems and the delineation of such effects on each of the known substeps of the system. A comparison of the present quantitative data on the whole cell with forthcoming data on the cell-free system should either justify the initial analytical approach or, at the very least, indicate new areas for consideration when dealing with the intact cell.

Although it is not the purpose of this author to present a comprehensive review of the literature, a brief discussion of those findings which have a direct bearing on the experiments to be presented in detail here is necessary in order to maintain scientific and historical perspective.

Johnson and Lewin (1946), while investigating disinfection of *E. coli* with quinine, reported that the application of hydrostatic pressure on the rate of growth of the organism could be stimulatory or inhibitory, depending on the magnitude of pressure and the temperature level. A justification for the application of absolute reaction rate theory on the basis of apparent first order kinetics of growth rates was made and the data interpreted in this manner. Johnson *et al.* (1954) reviewed pressure effects on a wide variety of biological systems and quantitatively interpreted them on this basis (see also, Chapter 1).

ZoBell and Oppenheimer (1950) demonstrated that cells of *Serratia marinorubra* grew into long filaments under pressure and temperature conditions that prevented reproduction. These filament-shaped cells reverted to normal-sized cells shortly after pressure was released by means of the establishment of cross walls along the length of the filament. The effect of pressure seemed to be significantly greater on cell wall formation than on the other processes involved in cell growth. Berger (1959) reported that *E. coli* and *Pseudomonas perfectomarinus* exhibited filamentous growth when pressurized and that this was a result of the inhibition of cross wall formation and was accompanied by an increase of cross wall precursor molecules in the culture medium. Pressures in the range of 750–2000 psi were found sufficient to cause growth into long filaments; longer cells frequently lysed, exhibiting a general deficiency of cell wall structure.

Morita and ZoBell (1956) demonstrated an inhibition of succinic dehydrogenase activity in *E. coli* at pressures of 3000–9000 psi. Malic and formic dehydrogenase activity showed a similar response (Morita, 1957). However, Haight and Morita (1962) found that the effect of pressure on aspartase activity in the whole cell was somewhat different from that exhibited in a cell-free system, inhibition by pressure occurring at or below 53°C in the former and 45°C in the latter. These results point out very clearly that a comparison of pressure effects on a whole cell system with those on a cell-free extract must be carefully approached (see also reviews of ZoBell, Chapter 4 and Morita and Becker, Chapter 3).

Finally, the effect of pressure on the biosynthesis of bacteriophage has been investigated by several workers. Foster and Johnson (1951) found that the ability of infected cells to support phage growth was inhibited. Rutberg (1964) reported that by using *E. coli* K-12, lysogenic for lambda phage, phage formation could be induced by a short pulse of relatively low pressure. The number of induced cells was shown to increase exponentially with increasing pressure. Hedén (1964) has reviewed most of this work in a highly speculative but extremely stimulating report. Unfortunately, to date, no one has applied biochemical techniques to measure pressure effects on the virus–host system. Considering current concepts and knowledge of virus infection, synthesis and aggregation, and the availability of sophisticated analytical methods, it would seem that this type of experimental system could serve as a model for the delineation of pressure effects on biosynthesis and aggregation.

II. Equipment and Methods

The pressure equipment used for these experiments is relatively uncomplicated. Pressure is generated by means of a hand-operated hydraulic pump (Enerpac Test Systems, Butler, Wisconsin). The pump is made of stainless steel, and the fluid phase of the system is water. The pressure is conveyed to the experimental chamber through a flexible hose. Levels of 10,000–12,000 psi may be attained in several seconds with a few strokes of the handle. Release is almost instantaneous upon opening of a relief valve. Earlier experiments (Landau and Peabody, 1963) had indicated that effects on biochemical processes could be instantaneously reversed upon return to atmospheric pressure, so chambers were designed either for fast disassembly or to allow for rapid chilling or addition of acid precipitants at any time during pressure treatment. Figure 1 shows the technique particularly suited, and now being used, for cell-free extract experiments, and Figs. 2 and 3 the mixing-chamber system used in the experiments reported here. All metal parts are composed of stainless steel. The smaller internal chamber fits within the larger outer chamber. A simple shaking motion succeeds in breaking the cover slip and mixing the contents of the two-compartment inner chamber. In this manner either labeled precursors or inducing agents can be added after pressure application or biochemical reactions may be stopped by addition of trichloroacetic acid (TCA) while under pressure. Although not shown in the diagram, a thermistor may be added to the chamber and internal temperatures may be monitored. Temperature is easily controlled by a water or air incubator for the chamber, and the entire chamber may be chilled from 37°C to 1°C in 30 seconds by immersion in a Dry Ice–acetone mixture. Disassembly of the apparatus takes from 20 to 35 seconds.

A chamber for the direct spectrophotometric analysis of reactants while under pressure treatment has been designed by Morita (1957), and its use is anticipated in future experiments.

The studies to be reported here were designed with certain specific limitations set to minimize indirect metabolic effects. (1) Pressures were limited to an upper level of 10,000 psi. All investigators have reported lethal effects at higher pressures. (2) The period of pressure application was a maximum of 20–35 minutes, with most experiments at a 10-minute exposure level. (3) Rapidly reversible effects were of primary interest. Therefore, the approach was to apply moderate pressures over short periods of time at carefully controlled temperatures and to note any reversible modification of the normal synthetic capabilities of the cell during this period.

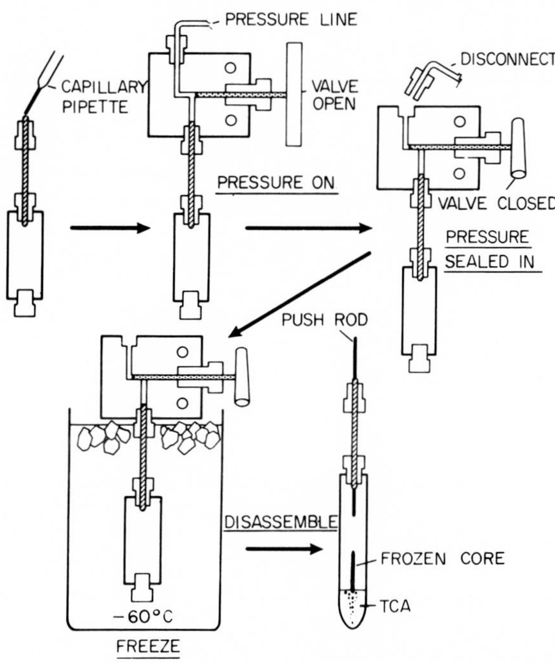

FIG. 1. The method used for quick freezing of pressurized samples. The tubelike chamber represented here holds about 0.4 cm^3 of sample. Redrawn from Landau and Peabody (1963).

III. The Incorporation of ^{14}C-Amino Acids into Protein in *Escherichia coli* K-12

The cells were grown to a density of 2.6×10^8/ml, and aliquots were taken for experimental and control samples. Each pressure sample had

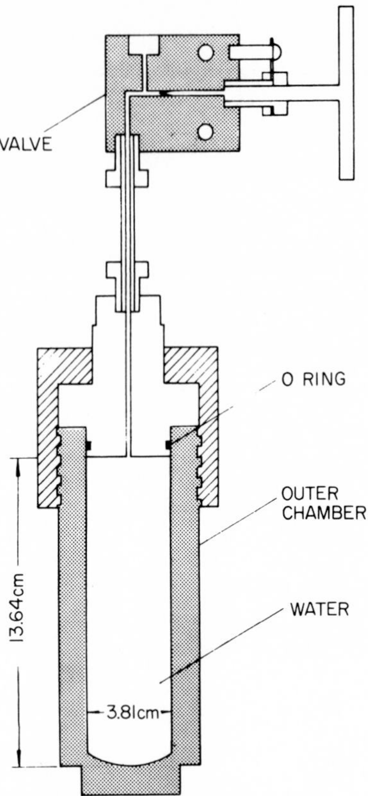

FIG. 2. The main pressure chamber utilized for addition of materials while under pressure. Disassembly may be accomplished in 20–30 seconds.

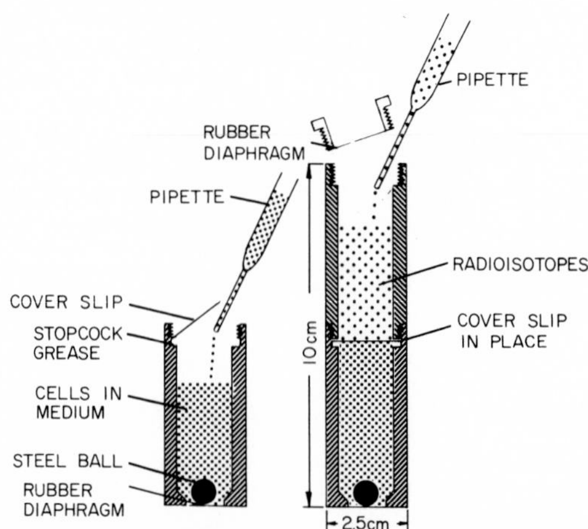

FIG. 3. The inner two-compartment chamber, which occupies a major portion or the water space shown in Fig. 2. Redrawn from Landau and Thibodeau (1962).

its individual atmospheric control. The initial procedure was to apply pressure to a suspension of cells, immediately add either ^{14}C-leucine or ^{14}C-glycine, maintain the pressure for a short specific period of time, and then measure the radioactivity incorporated into the TCA-precipitable protein fraction (Landau, 1966). Dilution in nutrient broth yielded a final radioactivity of 0.4 μCi ml. In nutrient broth, a pressure

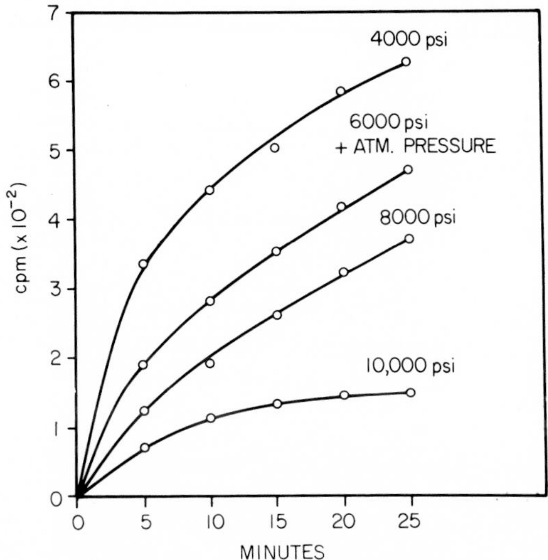

FIG. 4. Pressure effects on ^{14}C-amino acid incorporation into protein of *E. coli* at 37°C. Redrawn from Landau (1966). Copyright 1966 by the American Association for the Advancement of Science.

of 10,000 psi at 37°C for 25 minutes resulted in a negligible decrease of viability (approximately 6%). Since this was the maximum pressure level and time period employed, lethal effects could be ruled out as a major factor.

The results, shown in Fig. 4, were identical whether leucine or glycine was used. At 6000 psi and 37°C, incorporation is the same as that of controls at atmospheric pressure; at higher pressures there is an inhibition and at lower pressures a marked stimulation of labeled amino acid incorporation. At 10,000 psi, after a small initial uptake, almost complete inhibition occurs.

The data at 4000 and 10,000 psi are particularly intriguing. In several experiments, pressure was released after 10 minutes at 4000 psi and samples taken at successive 5-minute intervals. As shown in Fig. 5, the resultant incorporation curve is identical to the one shown where 4000 psi had been maintained over the full period of the experiment. This would seem to indicate that following an initial stimulation, this pressure has no further effect on the rate of amino acid incorporation during the time period indicated. The almost total inhibition at 10,000 psi posed some important questions. Previous workers (ZoBell and Cobet, 1962) had shown that pressure of this magnitude effectively prevents division of *E. coli* and that release of pressure is followed by an extended lag period prior to the resumption of the division process. Would release show a similar lag in incorporation? To test this, incorporation was completely

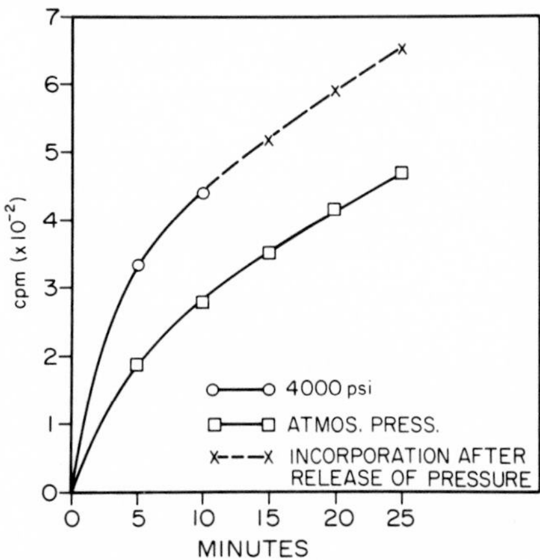

FIG. 5. The incorporation of ^{14}C-leucine into protein of *E. coli* at 37°C following application and release of 4000 psi.

inhibited by application of 10,000 psi, pressure was then released and subsequent determinations of incorporation were made at atmospheric pressure. The results are shown in Fig. 6. Release of pressure resulted in immediate resumption of a normal rate of protein synthesis, as measured by incorporation, indicating that inhibition is a direct effect of pressure and that no irreversible or even slowly reversible alterations in the protein synthesizing mechanism are produced. The lag in cell division subsequent to pressure release may then be ascribed to an interference in cell wall synthesis. It is possible that there is a permanent alteration of some component of this synthetic pathway and that new cell wall formation requires a restoration of this component.

The effect of temperature on incorporation over a 10-minute period is shown in Fig. 7. Depending on the temperature, 4000 psi can either stimulate, have no effect on, or inhibit protein synthesis. As an initial approach, a simplified theoretical interpretation has been proposed (Landau, 1966). A primary rate controlling reaction is postulated to involve an activated enzyme. A reversible thermal inactivation of this enzyme

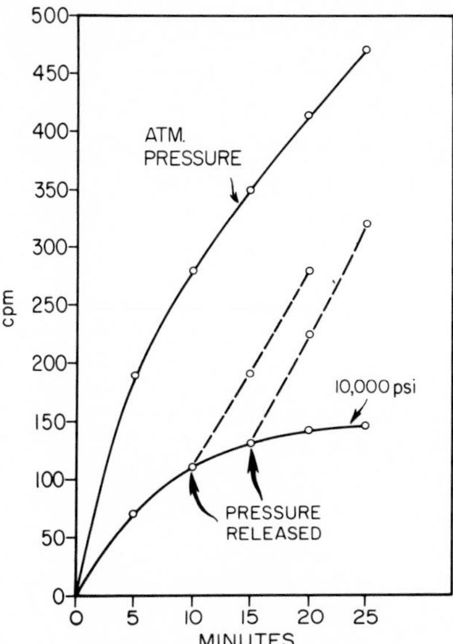

FIG. 6. The effect of release of inhibiting pressure (10,000 psi, 37°C) on ^{14}C-glycine incorporation in *E. coli*. The return to normal incorporation rate is almost immediate. Redrawn from Landau (1966). Copyright 1966 by the American Association for the Advancement of Science.

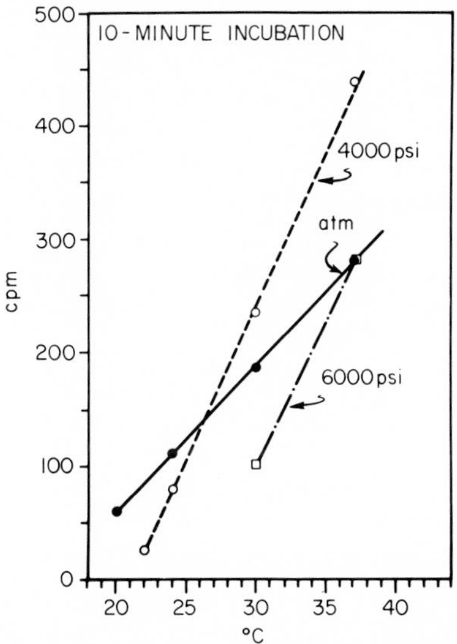

FIG. 7. ^{14}C-Leucine incorporation in *E. coli* after 10 minutes incubation as affected by pressure and temperature variation. Redrawn from Landau (1966). Copyright 1966 by the American Association for the Advancement of Science.

would begin at 25°–27°C. Both the primary reaction and the enzyme inactivation would be inhibited by pressure application, with the latter displaying a greater sensitivity. Thus the application of 4000 psi at 37°C would have little repressive effect on the primary reaction and result in a stimulation by affording a greater amount of activated enzyme. As minimum evidence for support of this theory, the secondary thermal inactivation step should be eliminated by a decrease in temperature, and

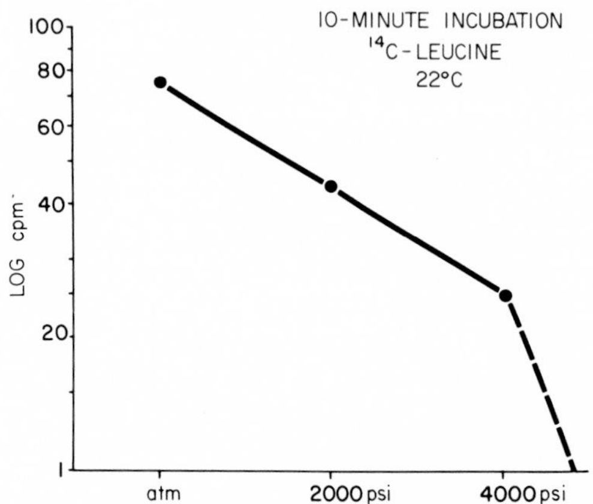

FIG. 8. The effect of pressure on ^{14}C-leucine incorporation in *E. coli* at 22°C. The slope of the line indicates the possibility of a rate limiting reaction with a $\triangle V^*$ of 100 cm^3/mole.

the inhibition of the postulated primary reaction by pressure should follow first order reaction kinetics. As is shown in Fig. 8, at 22°C the log of the rate of incorporation when plotted against the pressure level yields a straight line decrease from atmospheric pressure to 4000 psi, followed by a complete inhibition at pressures above 4000 psi.

IV. Nucleic Acid Synthesis

Incorporation of ^{14}C-adenine into nucleic acid as a result of pressure application is shown in Fig. 9. The use of ^{14}C-uridine yielded almost identical results, with slightly enhanced rates at the higher pressures. The results superficially parallel those on amino acid incorporation into protein, but some intriguing differences appear on close examination. After 10 minutes at 4000 psi the rate of incorporation is not identical to that of the control, and at 10,000 psi, although protein synthesis is completely

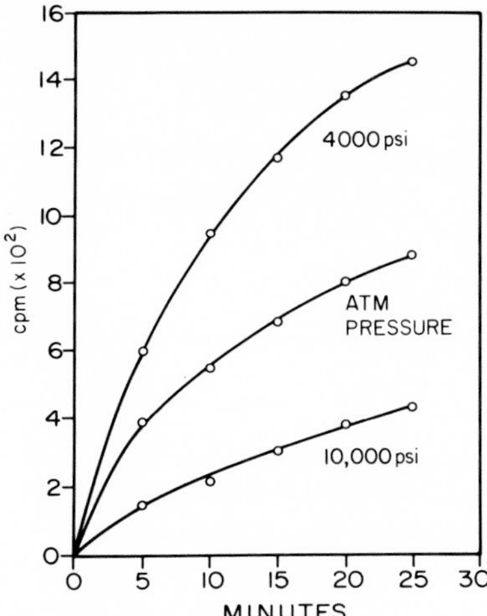

FIG. 9. ^{14}C-Adenine incorporation into nucleic acid of E. coli at various pressures and 37°C. The effect is similar when ^{14}C-uridine is used. Redrawn from Landau (1966). Copyright 1966 by the American Association for the Advancement of Science.

inhibited, nucleic acid synthesis continues at an easily measurable rate (Fig. 10). At 10,000 psi ^{14}C-thymidine incorporation is completely inhibited. There seems little doubt, therefore, that pressure has a differential effect on the synthesis of deoxyribonucleic acid (DNA), ribonucleic acid (RNA), and protein. RNA synthesis is maintained at pressures where protein and DNA synthesis are completely inhibited. The same results, under somewhat different conditions, have been reported by Pollard and Weller (1966).

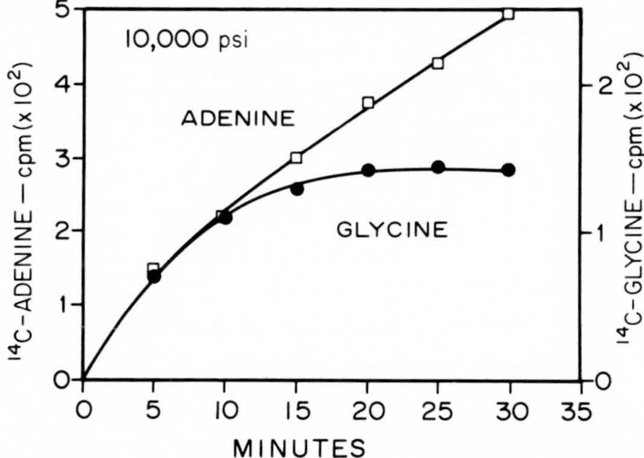

FIG. 10. A comparison of ^{14}C-adenine and ^{14}C-glycine incorporation into the nucleic acid and protein of *E. coli* at 10,000 psi.

V. The Question of Permeability

Is it possible that the entire pressure effect may be attributed to a reduction in the availability of amino acids (or other precursors) for the synthetic processes of the cell? Could the entry of such compounds be slowed up or stopped, and is the inhibition of protein synthesis at 10,000 psi really a measurement of effect on permeability? In order to test this possibility, 10,000 psi was applied to a suspension of cells at 37°C, ^{14}C-leucine was added, and pressure was maintained for 10 minutes. At this time the entire chamber was cooled to 1°C (within 45 seconds) and pressure released. The cells were rapidly washed several times at 1°C and then treated with cold 5% TCA. A measurement of TCA-soluble radioactive material was then made. A comparison with the soluble fraction of atmospheric controls showed about 10% less radioactivity in the pressurized cells. Comparable results were found with either glycine, alanine, or a mixture of amino acids. Although this would not rule out the permeability factor, it would seem highly improbable that it is of major consequence (see also Zimmerman, Chapter 10).

VI. The Induced Enzyme System

Although the results reported thus far indicate the possibility of a pressure effect on a rate limiting reaction, there is no indication of which reaction it may be. The next step was to utilize an *in vivo* system in which the two major processes of protein synthesis, i.e., transcription and translation, could be reasonably delineated (Landau, 1967).

The induced synthesis of β-galactosidase in *E. coli* provides such a system. It is ideal in that the operationally distinct transcription and translation phases may be experimentally separated by a variety of techniques. Hurwitz and Rosano (1965) have reported a method for the measurement of rate of transcription of β-galactosidase messenger RNA (mRNA), which involves the use of borate or proflavine, and is particularly suitable for pressure studies. *E. coli* ML-3 was used. The cells are genetically designated i^+, z^+, y^- (β-galactosidase inducible, permease negative). Isopropyl thiogalactopyranoside (IPTG) was used as the inducer, as it has the ability to enter the cell in the absence of permease.

The initial series of experiments consisted of addition of IPTG to the cells for a period of 20 minutes at atmospheric pressure at 37°C, followed by the application of pressure for 10 minutes, release of pressure and subsequent sampling. β-Galactosidase was assayed according to the method of Kepes (1963). The results are shown in Fig. 11. At 4000 psi there is no effect on enzyme synthesis, the rate being identical with that of cells kept at atmospheric pressure. Above 4000 psi the rate of synthesis decreases with increasing pressure until at 10,000 psi synthesis is totally inhibited. Upon release of pressure there is an immediate return to a normal rate of synthesis. At the point of pressure application, maximum levels of β-galactosidase mRNA are available and enzyme synthesis would normally continue for a limited time in the absence of further mRNA

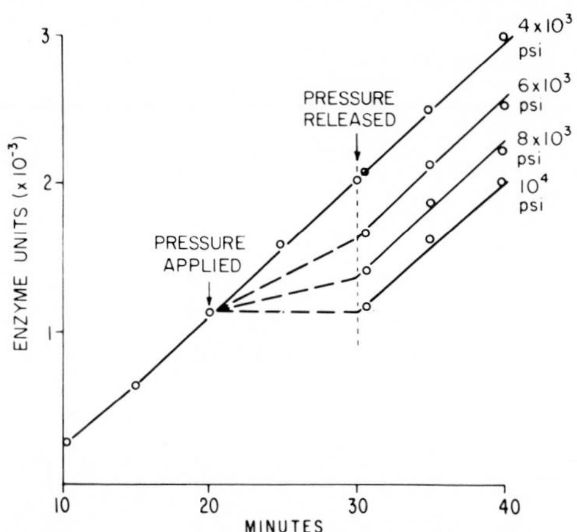

FIG. 11. The synthesis of β-galactosidase at various pressures and 37°C. Normal rate of synthesis is resumed immediately upon release of pressure. Redrawn from Landau (1967).

synthesis (Kepes, 1963). Considering the abruptness of the change in rate upon application of pressure, it would seem that the inhibition can be considered to be at the level of translation. Once again, the inhibition of protein synthesis, as indicated by the data between 4000 and 8000 psi, could possibly be considered as conforming to first-order reaction kinetics (Fig. 12). A reduction in temperature from 37°C to 22°C yielded the same slope as Fig. 12, but reduced the threshold for inhibition to levels over 1000 psi. In effect, the entire curve was shifted by an equivalent of about 3000 psi.

A second series of experiments involved pressure effects on the synthesis and stability of β-galactosidase mRNA. It has been shown that RNA synthesis continues at 10,000 psi, even though protein synthesis is effectively stopped. It seemed possible, however, that mRNA was not a component of the total RNA synthesis measured in the incorporation studies. The β-galactosidase system has been used to test this possibility. When cells are induced by IPTG and allowed a short period of time for mRNA production, then subjected to 10,000 psi for 5 minutes, all enzyme synthesis will stop for this period of time. Since proflavine and borate effectively block mRNA synthesis with little or no effect on subsequent enzyme production by presynthesized mRNA (Hurwitz and Rosano, 1965), either of these compounds can be added immediately at release of pressure, and the subsequent production of enzyme will be a measure of the amount

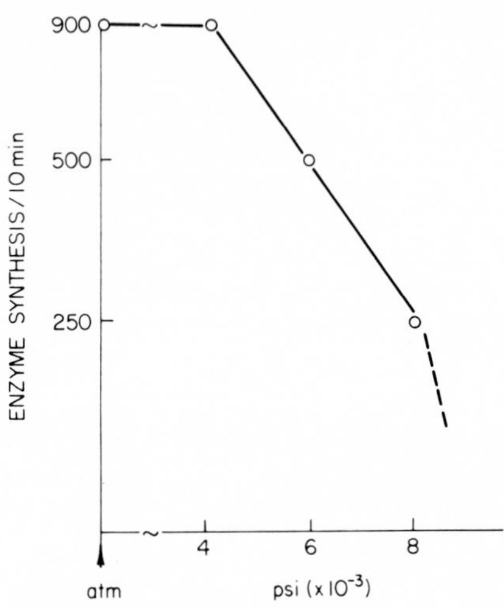

FIG. 12. The rate of enzyme synthesis as a function of pressure at 37°C. Redrawn from Landau (1967).

of β-galactosidase mRNA present at the time of pressure release. Cells were induced for 4 minutes and then divided into three groups. The first group was the control, maintained at atmospheric pressure. Borate was added to this control group and subsequent enzyme determinations were made. The amount of enzyme activity is an indication of the amount of β-galactosidase mRNA synthesized during the 4 minutes of induction. In the second group 10,000 psi was applied after 4 minutes of induction and borate added at this time. After 5 minutes the pressure was released and enzyme levels were measured. This determined the effect of pressure on presynthesized mRNA. The third group was pressurized at 10,000 psi after 4 minutes of induction, pressure was maintained for 5 minutes, borate was then added, pressure was released and enzyme content determined. In this case the data reflect any new synthesis of mRNA under pressure. The results are shown in Fig. 13. They were identical when the experiments were repeated with proflavine in place of borate.

Since β-galactosidase mRNA has a half-life period of about 2.5 minutes, normal decay without concomitant synthesis over the 5-minute pressure period should be reflected in a 75% decrease in maximum enzyme production. The data show about a 90% decrease, indicating the possibility of a somewhat enhanced breakdown. On the other hand, when synthesis of mRNA is not chemically blocked, the 5-minute pressure period results in only a 30% decrease in enzyme production. It is likely, therefore, that both decay and synthesis occur simultaneously at 10,000 psi.

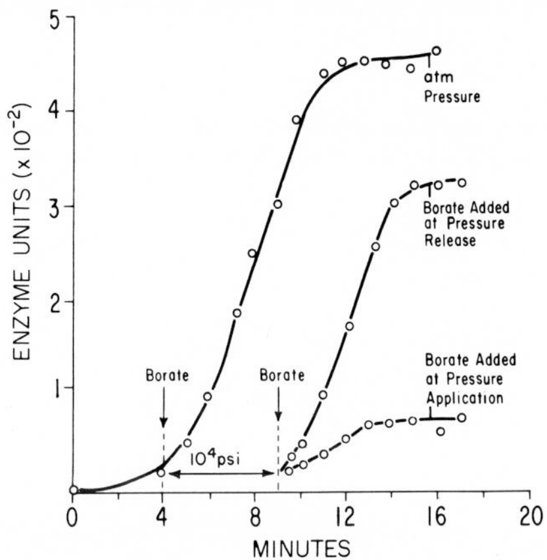

FIG. 13. The effect of pressure on β-galactosidase mRNA synthesis. Borate is utilized as an inhibitor of mRNA synthesis. Identical results are obtained when proflavin is utilized as an inhibitor in place of borate. Redrawn from Landau (1967).

The final series of experiments on this system concerned the effect of pressure on induction of the *lac* operon. In terms of the operon hypothesis (Jacob and Monod, 1961) inducer enters the cell and forms a complex with the repressor molecule, thus activating the operator locus and permitting specific mRNA synthesis. In the experiments reported above, pressure was not applied until after inducer had been added and transcription was already established. In the following experiments, pressure was first applied and inducer was added immediately afterward, thus permitting a possible delineation of effects on the initial events of inducer entry and complex formation. Pressure was maintained for 10 minutes, released, and the cell suspension was analyzed for β-galactosidase activity. A decrease in the synthesis of the enzyme during the pressure period was found as pressure was increased up to 4000 psi, above which total inhibition occurred (Fig. 14). The data yield a straight line, indicating an effect on a rate-limiting reaction with a volume change of activation (ΔV^*) of about 55 cm^3/mole.

Since pressure interfered with the synthesis of β-galactosidase under these specific experimental conditions, at levels too low to inhibit either transcription or translation, the effect can be considered to be either on the entry of IPTG or on inducer–repressor complex formation.

To investigate the former possibility, cells were placed at 10,000 psi and IPTG added, the pressure was maintained for 5 minutes, released,

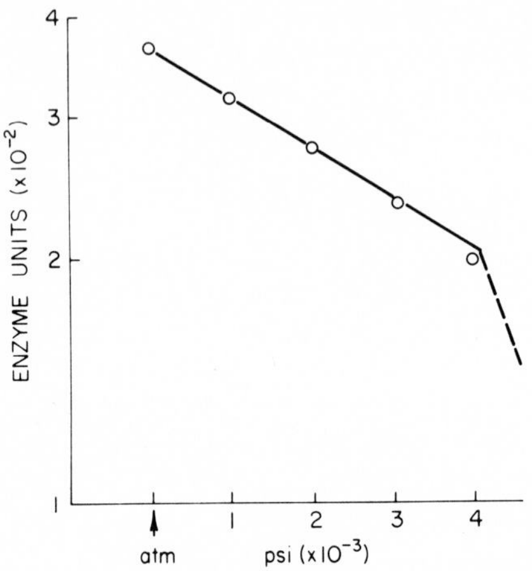

FIG. 14. The effect of pressure on the induction of β-galactosidase synthesis. The data indicate that inducer–repressor complex formation is accompanied by a $\triangle V^{*}$ of 55 cm^3/mole. Redrawn from Landau (1967).

and the cell suspension washed free of extracellular IPTG on an HA Millipore filter with two 10-ml portions of cold saline. Dismantling and emptying the pressure chamber required 35 seconds. The cells were than resuspended in warm (37°C) nutrient broth without IPTG and samples taken every 2 minutes for β-galactosidase assays. The results are shown in Fig. 15A. As a control, cells were exposed to 10,000 psi for 5 minutes and IPTG added at release for 35 seconds (Fig. 15B). The results clearly show that IPTG enters the cell even at this maximum pressure level (10,000 psi).

This led to a consideration of the second possibility, namely, that IPTG continues to be available for inducer-repressor complex formation, but that such binding is inhibited under pressure. If the experiment is repeated and proflavine added at release of pressure, there is no subsequent β-galactosidase synthesis (Fig. 15E), indicating that even though IPTG is internally available, it has not bound to the repressor and transcription has not occurred. However, if the IPTG was not bound to the repressor molecule within the cell during pressure, it should have been readily washed out on the filter, unless the 35-second period at atmospheric pressure and 37°C was sufficient for complex formation. On the assumption that this binding may be prevented by low temperature, the experiments were repeated with rapid chilling to 1°C just prior to the release of

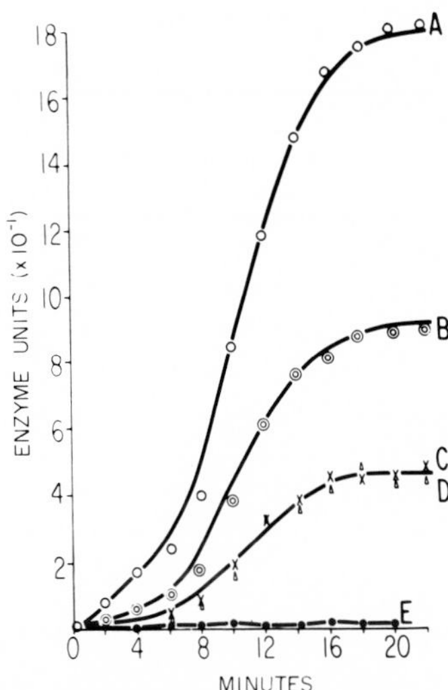

FIG. 15. The effect of 10,000 psi on permeability to IPTG and on inducer–repressor complex formation. (A) Cells were subjected to 5 minutes at 10,000 psi in the presence of inducer, followed by 35 seconds at 1 atm at 37°C, then washed free of extracellular inducer. (B) Cells were subjected to 5 minutes at 10,000 psi without inducer followed by 35 seconds at 1 atm with inducer at 37°C and washed. (C) and (D) represent the same experiments as (A) and (B) with the 35-second period at 1°C. (E) indicates results when proflavine is added at pressure release. The difference between (A) and (B) represents the entry of IPTG during the pressure period. Landau (1967).

pressure. Under these conditions subsequent β-galactosidase production on resuspension at 37°C was markedly reduced and was the same regardless of time of exposure to IPTG (Fig. 15C and D). This would indicate that the binding of the inducer to the repressor is inhibited at low temperature as well as high pressure and that intracellular IPTG can be washed out to a large extent if the complex has not formed.

Since IPTG is a comparatively small, apolar molecule, it is implicit that the volume increase of activation of 55 cm³/mole for inducer–repressor complex formation involves an induced configurational change of the repressor protein. The *lac* repressor has been estimated to have a molecular weight of about 150,000–200,000 (Gilbert and Müller-Hill, 1966), which would seem to make the magnitude of the volume change feasible. It is also doubtful that the inducer competes with the operon for the same sites on the repressor, making a configurational change in the repressor a most plausible means for inactivation of repressor–operon binding.

In the β-galactosidase system the pressure data seem amenable to the analysis presented here, but in another well-investigated system, that of lambda phage activation, a similar analysis yields less clear results. Rutberg (1964) has shown that a short pressure pulse will activate or derepress lambda phage. The data indicate a ΔV^* of about -300 cm^3/mole associated with derepression. However, although the repressor itself has been well investigated, little is known about the "induction" or "activation" mechanism. The calculated volume change is very large and the components involved may be, in this instance, the operon and repressor. Activation is stimulated at pressures low enough to eliminate the possibility of simple inhibition of repressor synthesis. A meaningful interpretation of the pressure data must await a more complete delineation of the events involved in lysogenic phage activation.

VII. Synthesis of Protein and RNA in HeLa Cells

All the effects of pressure thus far reported have been on microbial systems. The question was posed as to whether this effect on protein synthesis was unique in *E. coli* or whether cells of a higher form would respond similarly. A culture of S-3 HeLa cells was grown in monolayer culture in roller bottles for 1 week, stripped from the glass, and suspended in serumless medium. Radioactive amino acids were added to the cell suspension for 15 minutes and then pressure was applied. On release of pressure the cells were fixed in 5% cold TCA. Measurements of total sample protein were made and incorporated label was indicated as cpm per mg protein. The experimental method was basically that used for induced enzyme synthesis in *E. coli*. The results are shown in Fig. 16 and offer a striking parallel to those on bacterial enzyme synthesis (Fig. 11). At 37°C there seems to be no effect of pressure until above 4000 psi and an exponential decrease with increasing pressure to 8000 psi, above which complete inhibition of protein synthesis occurs. When the temperature is lowered to 22°C, synthesis becomes increasingly sensitive to pressure at each increment (Fig. 17). At this temperature, the inhibitory effect is first indicated at pressures above 2000 psi, and complete inhibition occurs at pressures above 5000 psi.

The identical results of pressure application on HeLa cells and *E. coli* strongly suggest that a common mechanism in protein synthesis is affected. Further, the minimum pressure level for the appearance of the inhibitory effect is a function of the temperature. Once again, the pos-

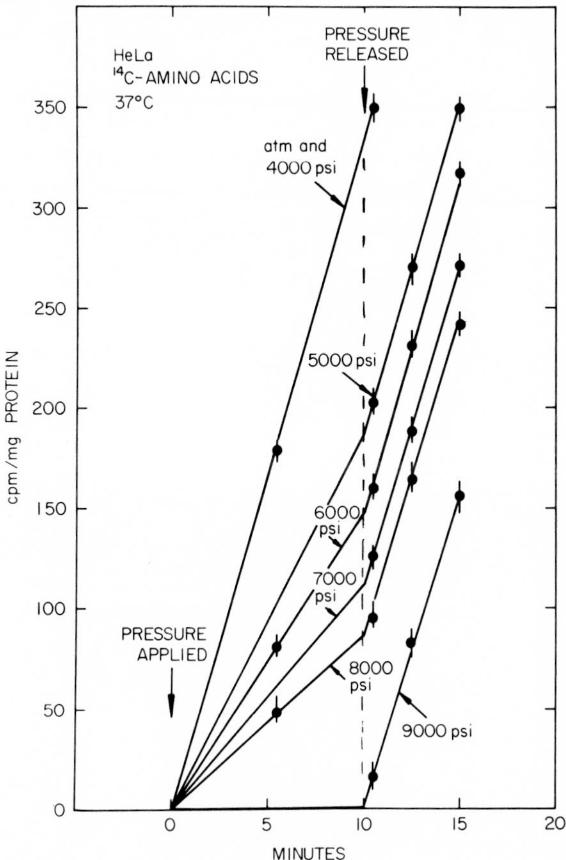

FIG. 16. The effect of pressure on the incorporation of ^{14}C-amino acids into protein of HeLa cells at 37°C. Return to normal rate of incorporation is immediate upon release of pressure (Landau, 1970).

sibility of an indirect effect on synthesis mediated through the inhibition of the permeability of amino acids was at least partially eliminated by experiments similar to those done on *E. coli*. The intracellular acid soluble label found after 10 minutes of exposure to ^{14}C-amino acids at 10,000 psi was at most 20% less than that of atmospheric controls and in several experiments was within 5%. There would seem to be sufficient overall permeability to amino acids for protein synthesis to continue, unless the slightly lower label found in the free amino acid content of the pressure treated cell indicates the lack of penetration of certain specific amino acids.

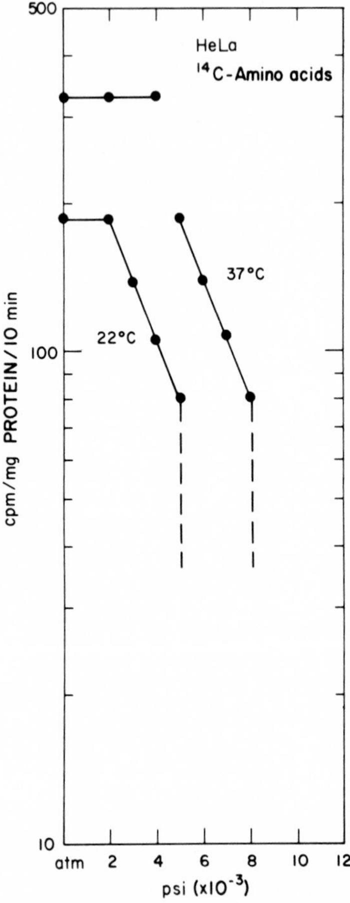

FIG. 17. The effect of variations in pressure and temperature on the rate of protein synthesis in HeLa cells (Landau, 1970).

The incorporation of uridine into the nucleic acid fraction of the cell was used to indicate the effect of pressure on net RNA synthesis. The experimental design was the same as that used for amino acid incorporation (Fig. 18). Interestingly, at 37°C, all pressures greater than atmospheric pressure inhibit RNA synthesis, but total inhibition is not achieved at the maximum pressure applied, namely, 10,000 psi. The response of the rate of incorporation to increasing pressure is shown in Fig. 19. The data reveal a bimodal curve, which may be indicative of two distinct reactions. Whether the reaction below 3000 psi may be eliminated by decreasing the temperature remains to be seen. The calculated ΔV^* of the major re-

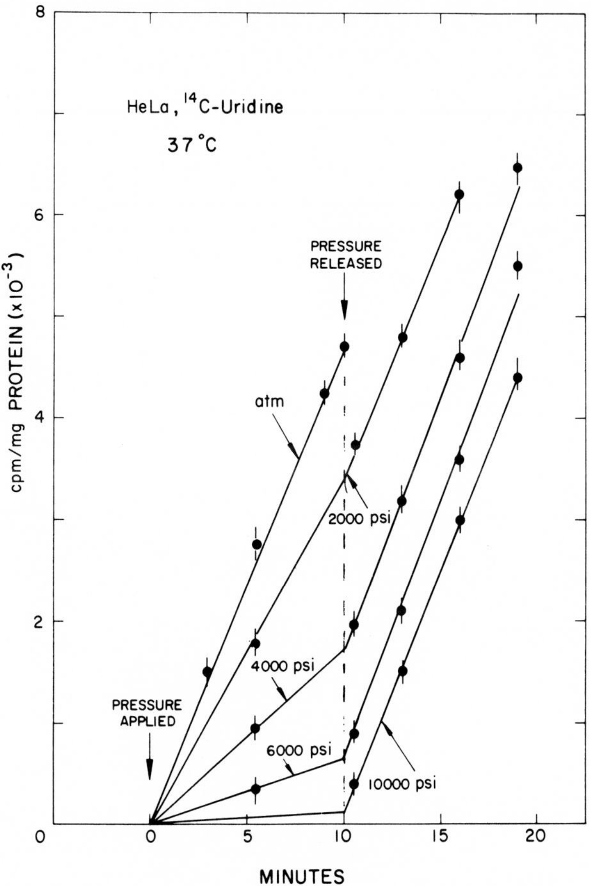

FIG. 18. The incorporation of ^{14}C-uridine into RNA as a result of pressure application at 37°C. A normal rate of incorporation is resumed immediately on release of pressure (Landau, 1970).

action between 3000 psi and 10,000 psi at 37°C, is 175 cm^3/mole. Whatever this may mean, it is significantly different from any result obtained for amino acid incorporation into protein and suggests that pressure effects on nucleoside incorporation into RNA can be adequately distinguished from those on the translational phase of synthesis.

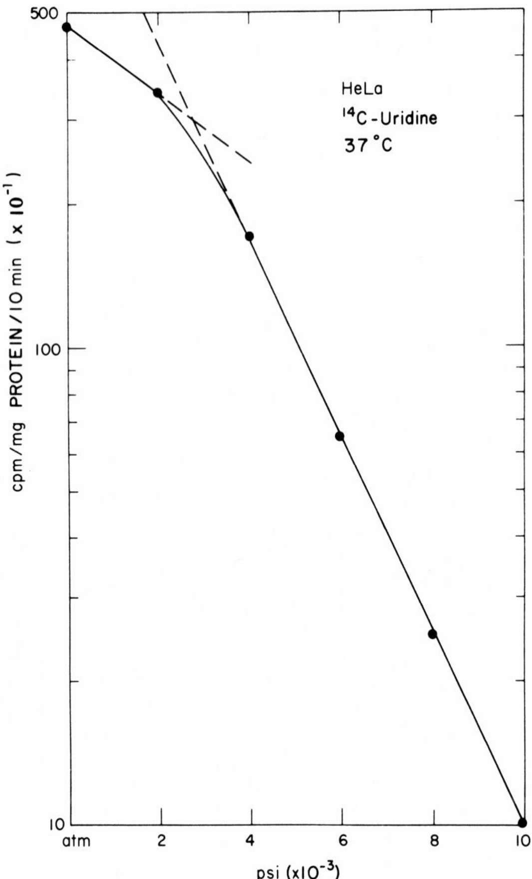

FIG. 19. The effect of pressure on the rate of ^{14}C-uridine incorporation in HeLa cells at 37°C (Landau, 1970).

VIII. Some Questions and Conclusions

There are several common factors that are strikingly apparent in an analysis of the data presented here. Those reactions which may reasonably be considered as part of the translational process of protein synthesis show an identical response to the application of pressure in both bacteria and HeLa cells. When primary reaction rate theory is applied to the data of Figs. 8, 12, and 17, the ΔV^* may be calculated as about 100 cm^3/mole in each instance. What does this figure mean? Can it be used as an indicative value in terms of considering the path of future investigations?

Lamanna and Mallette (1959) have pointed out the lack of success in application of absolute reaction theory to complex biological systems, but with appropriate experimental design such application may offer a pattern amenable to interpretation, particularly if a single or at most a few reactions in a pathway can be isolated from the complex system.

If the ΔV^* associated with translation is really involved in the formation of an active complex in a single rate limiting reaction, can the reaction be delineated? The calculated volume change is relatively large and the molecules involved would have to be of reasonably high molecular weight. There would seem to be at least two areas for primary consideration. The first would be those reactions involving the synthetase molecules concerned with protein synthesis. Such steps as amino acid activation and the formation of aminoacyl transfer RNA can be readily investigated as a function of pressure. The second area would be those reactions involving ribosomal subunit aggregation and polysome formation, the effect of pressure on these reactions is also amenable to test. Should the effect of pressure application on any one delineated reaction be quantitatively similar to the effect on the whole cell, the initial theoretical approach would be justified. Should the quantitative aspects differ to a large degree, at least some important facts concerning function changes in molecular configuration may be revealed.

Of particular interest is a comparison of these data on protein synthesis with those of Marsland (1942) concerning pressure effects on cytoplasmic sol–gel phenomena in a wide range of organisms. A volume change of about 100 cm^3/mole is also found when the log of the change in relative viscosity is plotted as a function of pressure (Johnson *et al.*, 1954). Also, temperature–pressure experiments on cytoplasmic gels have shown that in almost every instance a 5°C decrease in temperature, within certain limitations, yields a decrease in gel strength equivalent to that brought about by a 1000-psi increase in pressure. The data of pressure–temperature manipulations on protein synthesis (Figs. 8 and 17) show an identical equivalence.

This comparison of data leads to some interesting speculation.

1. Do the pressure data reflect the effect on some overall or specific biochemical process which would in turn produce similar quantitative results whether measuring protein synthesis or relative viscosity? One possibility may lie in the effect of pressure on pH. Kauzmann (1954) has pointed out that a phosphate buffer solution at about pH 7 at atmospheric pressure can be expected to become about 0.4 unit more acid at 10,000 psi. Although organic buffers in general would show a smaller shift in pH, this may be an important factor.

2. Is protein synthesis *immediately* necessary or responsible for gel network formation? This can be readily approached by means of fast acting, specific inhibitors of protein synthesis (e.g., puromycin). Inhibition of synthesis should, if the question is to be answered affirmatively, cause an immediate decrease in gel viscosity. Studies on the viscosity of crude rabbit myosin extracts (Marsland and Brown, 1942) conform with the pressure data presented here and do not indicate the necessity for a functional protein synthesizing system in order to maintain relative gel strength.

3. Is it possible that the maintenance of a minimum gel state is one of the prerequisites for the efficient synthesis of protein in a living cell? Could it be that the major pressure effect is on the structure of a cytoplasmic matrix and that the data ascribed to the process of translation merely reflect a loss in matrix structure? Such an hypothesis seems most attractive to this author. It would mean that while a level of synthesis could be performed under optimal conditions in cell free systems, within the cell a structural component is necessary for the efficient fulfillment of the functional demands of the cell. It may even be possible that such molecules as the various synthetases could be structural components of the extremely labile gel matrix. The nature of the gel network may also allow for the localized concentration of molecular components of the protein synthesizing machinery. Quantitative pressure data on cell-free synthesizing systems should indicate whether the whole cell reflects a direct relationship between relative gel strength and protein synthesis.

On the basis of the material presented here, it can readily be seen that only a beginning has been achieved in regard to effect of pressure on macromolecular biosynthesis and that any logical interpretation of data requires some initial broad assumptions. It should be apparent, however, that these assumptions are amenable to test. Certainly the theory and method are available for a more definitive analysis of pressure effects on the interaction of biological macromolecules and thereby on the physical nature of these interactions.

ACKNOWLEDGMENT

The author wishes to thank Libby Fadal, Sophia Horbat, Ronald Scheinzeit, Linda Karig, and John Schwarz for their able assistance in various phases of the experimental work. Portions of the work presented here were performed at the Albany Veterans Administration Hospital and supported in part by USPHS Grants No. CA-02664 and GM-14371 to Albany Medical College, Albany, New York. The most recent work is supported in part by USPHS Grant No. GM-16205 to Rensselaer Polytechnic Institute.

REFERENCES

Berger, L. R. (1959). The effect of hydrostatic pressure on cell wall formation. *Bacteriol. Proc.* **12**, 129.

Brown, D. E. (1934). The pressure coefficient of "viscosity" in the eggs of *Arbacia punctulata. J. Cellular Comp. Physiol.* **5**, 335–346.

Brown, D. E. (1936). The sequence of events in the isometric twitch at high pressure. *Cold Spring Harbor Symp. Quant. Miol.* **4**, 242–251.

Foster, R. A. C., and Johnson, F. H. (1951). Influence of urethane and of hydrostatic pressure on the growth of bacteriophages T2, T5, T6 and T7. *J. Gen. Physiol.* **34**, 529–550.

Gilbert, W., and Müller-Hill, B. (1966). Isolation of the lac repressor. *Proc. Natl. Acad. Sci. U.S.* **56**, 1891–1898.

Haight, R. D., and Morita, R. Y. (1962). Interaction between the parameters of hydrostatic pressure and temperature on aspartase of *E. coli. J. Bacteriol.* **83**, 112–120.

Hedén, C. G. (1964). Effects of hydrostatic pressure on microbial systems. *Bacteriol. Rev.* **28**, 14–29.

Hurwitz, C., and Rosano, C. L. (1965). Measurement of rates of transcription and translation by means of proflavine or borate. *Biochem. Biophys. Acta* **108**, 697–700.

Jacob, F., and Monod, J. (1961). Genetic regulatory mechanisms in the synthesis of proteins. *J. Mol. Biol.* **3**, 318–356.

Johnson, F. H., and Lewin, I. (1946). The influence of pressure, temperature, and quinine on the rates of growth and disinfection of *E. coli* in the logarithmic growth phase. *J. Cellular Comp. Physiol.* **28**, 77–97.

Johnson, F. H., Eyring, H., and Polissar, M. J. (1954). "The Knietic Basis of Molecular Biology." Wiley, New York.

Kauzmann, W. J. (1954). Cited in "The Kinetic Basis of Molecular Biology" (F. H. Johnson, H. Eyring, and M. J. Polissar), p. 302, Wiley, New York.

Kepes, A. (1963). Kinetics of induced enzyme synthesis. *Biochim. Biophys. Acta* **76**, 293–309.

Lamanna, C., and Mallette, M. F. (1959). "Basic Bacteriology," 2nd ed. Williams & Wilkins, Baltimore, Maryland.

Landau, J. V. (1966). Protein and nucleic acid synthesis in *Escherichia coli*: Pressure and temperature effects. *Science* **153**, 1273–1274.

Landau, J. V. (1967). Induction, transcription and translation in *Escherichia coli:* a hydrostatic pressure study. *Biochim. Biophys. Acta* **149**, 506–512.

Landau, J. V. (1970). Protein and nucleic acid synthesis in HeLa cells: a hydrostatic pressure study. In preparation.

Landau, J. V., and Peabody, R. A. (1963). Endogenous adenosine triphosphate levels in human amnion cells during application of high hydrostatic pressure. *Exptl. Cell Res.* **29**, 54–60.

Landau, J. V., and Thibodeau, L. (1962). The micromorphology of *Amoeba proteus* during pressure-induced changes in the sol-gel cycle. *Exptl. Cell Res.* **27**, 591–594.

Marsland, D. A. (1942). Protoplasmic streaming in relation to gel structure in the cytoplasm. *In* "The Structure of Protoplasm" (W. Seifriz, ed.), pp. 127–161. Iowa State Coll. Press, Ames, Iowa.

Marsland, D. A. (1956). Protoplasmic contractility in relation to gel structure: Temperature-pressure experiments on cytokinesis and amoeboid movement. *Intern. Rev. Cytol.* **5**, 199–227.

Marsland, D. A., and Brown, D. E. (1942). The effects of pressure on sol-gel equilibria with special reference to myosin and other protoplasmic gels. *J. Cellular Comp. Physiol.* **20**, 295–305.

Morita, R. Y. (1957). Effect of hydrostatic pressure on succinic, formic and malic dehydrogenaces in *Escherichia coli. J. Bacteriol.* **74**, 251–255.

Morita, R. Y., and ZoBell, C. E. (1956). Effect of hydrostatic pressure on the succinic dehydrogenase system in *E. coli J. Bacteriol.* **71**, 668–672.

Pollard, E. C., and Weller, P. K. (1966). Pressure on the synthetic processes in bacteria. *Biochim. Biophys. Acta* **112**, 573–580.

Rutberg, L. (1964). On the effects of high hydrostatic pressure on bacteria and bacteriophage. I. Action on the reproduction of bacteria and their ability to support growth of bacteriophage T2. *Acta Pathol. Microbiol. Scand.* **61**, 81–90.

ZoBell, C. E., and Cobet, A. B. (1962). Growth, reproduction, and death rates of *Escherichia coli* at increased hydrostatic pressures. *J. Bacteriol.* **84**, 1228–1236.

ZoBell, C. E., and Oppenheimer, C. H. (1950). Some effects of hydrostatic pressure on the multiplication and morphology of marine bacteria. *J. Bacteriol.* **60**, 771–781.

CHAPTER 3

HYDROSTATIC PRESSURE EFFECTS ON SELECTED BIOLOGICAL SYSTEMS

Richard Y. Morita and Robert R. Becker

I. Introduction[1]

All organisms living in an aquatic environment are subjected to varying degrees of hydrostatic pressure. Because of this situation the effects of hydrostatic pressure on biological systems becomes extremely important in the marine environment. Since most biologists, not to mention biochemists and biophysicists, do not consider hydrostatic pressure as an experimental variable, the progress in this area of study has been slow. In addition, the tools needed for this type of research require greater sophistication. Only when sufficient input has been made in this area of research can we expect significant progress. As a result of the amount of effort in the scientific community in this field, we find that there are great gaps in our knowledge.

There are two main approaches to the study of hydrostatic pressure

[1] This chapter deals mainly with research in our laboratories.

on biological systems: the molecular and the cellular levels. The molecular approach has certain drawbacks (especially when working with purified preparations), since it does not take into consideration the other functions or components of the cell. As a result, permeability factors, metabolic control mechanisms, protein–protein interactions, protein–nucleic acid interactions, protein–carbohydrate interactions, etc., may be missing. Temperature has been widely used as a parameter in research, and it is only a matter of time when pressure will be used as an additional factor which should provide valuable information. It is well known that various species of bacteria are affected differently by the application of hydrostatic pressure. However, much valuable data can be obtained using the molecular approach. The use of whole cells also has many drawbacks because of permeability factors, effect of added substrates or cofactors (especially with enzymes since the cofactor or substrate may protect the enzyme by combining with it to form a more stable state), as well as not being able to determine the effect of pressure on the various cellular constituents. There is even a difference in aspartase activity (in relation to temperature and pressure) when washed cells and cell-free extracts are used as the source of enzyme (Haight and Morita, 1962).

Although we recognize the above difficulties, our approach to the overall problem is from both levels. At the molecular level, it appeared that a study of systems involving hydrophobic interactions would provide information relevant to at least some of the noncovalent interactions that provide stabilization to native proteins. Since much is known about the structure of bovine pancreatic ribonuclease, the effects of hydrostatic pressure on the interactions of this enzyme and some of its derivatives have been carried out.

II. Complicating Factors in the Study of Hydrostatic Pressure as a Variable in Biological Systems

Generally speaking, when studies are undertaken employing pressure as a variable, the results are analyzed mainly in relation to the amount of pressure applied to the system. Although the various investigators are cognizant of the complicating factors involved when using pressure in their research, the data are generally interpreted in relation to the amount of pressure applied to the system. It would be impossible to control all the changes brought about by the application of pressure to a cellular system. Changes in pH due to pressure take place in buffered solutions as well as in sea water (Johnson *et al.*, 1954; Pytkowicz and Conners,

1964; Buch and Gripenberg, 1932; Distèche, 1959). The ionization of water becomes greater under pressure (Owen and Brinkley, 1941; Hamann, 1963), and viscosity also changes (Horne and Johnson, 1966). In addition, water loses certain structural qualities when pressure is applied, but some of this is regained when sodium chloride is added (Horne and Johnson, 1966, 1967). If sodium chloride aids the structural regions in water, then it can readily be seen that one cannot neglect salt as a complicating factor when pressure studies are undertaken. Chemical reaction rates, ionization of various substances (Hamann, 1964; Owen and Brinkley, 1941; Distèche and Distèche, 1965; Ewald and Hamann, 1956; Hamann and Strauss, 1965; Pytkowicz and Fowler, 1967; Pytkowicz *et al.*, 1967), partial specific volume of various compounds, as well as partial equivalent volume of salts (Anderson, 1963; Duedall and Weyl, 1967) are also affected by pressure. Molecular volume changes discussed at great length by Johnson *et al.* (1954) and in this book (cf. Johnson and Eyring, Chapter 1; Landau, Chapter 2; and ZoBell, Chapter 4). Linderstrøm-Lang and Jacobsen (1941) have calculated volume changes accompanying the formation of ion pairs, amino acid dipoles, and peptide dipoles, but those accompanying precise conformational changes are not known. Since ionization of water as well as water structure can change due to pressure, how water molecules group themselves around a particular molecule (especially proteins) becomes an important problem. All of these factors arise when one deals with the application of pressure on biological systems.

Many molecular volume changes have been calculated for molecules, but the effects of the perturbing force on any macromolecule may be unique to the molecule under study. Morita and Haight (1962) and Morita and Mathemeier (1964) demonstrated that increase in molecular volume could be counteracted by hydrostatic pressure so that malic dehydrogenase and inorganic pyrophosphatase could function above 100°C. The data should not be interpreted to mean that the conformational changes due to temperature increases are acted on by hydrostatic pressure in exactly the opposite way. The difference in the conformational changes in ovalbumin, when measured in terms of optical rotation, is $-47.9°$ when pressured-denatured at 9000 kg/cm^2, for 2 hours at 30°C, $-100°$ when urea-denatured, and $-60°$ when heat-denatured at 100°C for 30 minutes, whereas the native state has an optical rotation of $-27.6°$ (Suzuki and Suzuki, 1962). According to Suzuki and Suzuki (1962), pressure denaturation of ovalbumin was due to the rupture of H bonds responsible for secondary structure.

In addition to the above complicating factors is the problem of instrumentation. Owing to the lack of investigators interested in this subject matter, most of the apparatus employed lacks sophistication (Morita, 1967, 1969). One of the major difficulties is probably due to insufficient funds, coupled with the lack of engineering expertise to design specific equipment for hydrostatic pressure research. Some of the present methodology employed for studying hydrostatic pressure in relation to biology is presented by Morita (1969).

III. Present Studies

A. Baroduric Bacteria

During Expedition Dodo, sediment samples were obtained from the Challenger Deep and the Philippine Trench by Jean ZoBell, C. E. ZoBell, and R. Y. Morita. From these sediment samples taken from depths greater than 10,000 m, quite a few bacteria were isolated. These bacteria have been classified by Quigley and Colwell (1968). When these isolates were placed in nutrient medium under the isobaric and isothermic conditions from whence the sediment was taken, the cells expired. This same situation has been noted before with sediment samples taken during the Galathea Deep Sea Expedition.

The bacteria, since they originated from source material obtained from great depths, could be termed "baroduric." It is possible that these isolates may be contaminants. However, the experimental techniques employed would certainly rule out that all of the isolates were contaminants. The question that must be answered is the following: What physiological state must these bacterial cells be in for them to withstand the environmental conditions at the ocean depth? The following may be partial answers to this question based on data obtained with these deep sea isolates.

1. Bacteria in the log growth phase expire more quickly when placed under pressure than do cells in the stationary phase of growth.
2. Bacteria obtained from a slow growing culture (nonoptimal conditions for growth) are more resistant to 1100 atm and 5°C than are cells grown under optimal conditions.
3. Bacteria obtained from growth medium in which a limited amount of surface area is exposed are more resistant to 1100 atm and 5°C than are cells grown under good aeration. This limited surface area is related to the oxygen tension of the medium.
4. By subjecting these isolates to a medium used for halophilic bacteria, it is again possible to prolong their duration of survival to pressures

and low temperature. When these isolates are subjected to halophilic medium at 1 atm, they expire quite rapidly.

By observing the above, it is possible, in the laboratory, to prolong the time required for a culture to expire when placed at 1100 atm and 5°C.

However, the above still does not answer the question as to what physiological state the deep isolates were in so that they survived in the environmental conditions found in the hadal portions of the sea. Many more data are needed to verify the concept of a physiological state for some bacteria so that they can survive the high pressure and low temperature conditions found in the deeps. Whether or not it is possible to duplicate nature in this investigation remains to be seen.

B. Hydrostatic Pressure Studies on RNase

Studies on the effects of hydrostatic pressure on macromolecules from living systems have been done mainly on enzymes, since there is an easy way to measure the pressure effects (i.e., end product formed, decrease of substrate, oxidation or reduction of cofactor, etc.).

Most of the research done in our laboratory on the metabolic functions of microbes has been summarized in review articles by Morita (1967, 1970), as well as mentioned by ZoBell and others in this book. The more molecular aspects of our research have not been the subject of any review articles and as a result some of the background literature obtained in the past are necessary to lay the bases for much of our present studies.

In 1965 we reported the effect of hydrostatic pressure on the aggregation reaction of poly-L-valyl-ribonuclease (PVRNase) (Kettman *et al.*, 1965). PVRNase exhibits a turbidity at temperatures as low as 30°C, whereas RNase aggregates at higher temperatures. This low temperature aggregation of PVRNase is attributed to increased apolar interactions, which is due to the polyvalyl chains attached to the enzyme (Becker and Sawada, 1963). The effect of hydrostatic pressure on PVRNase in relation to thermal aggregation (39°C) is shown in Fig. 1. Although the turbidity rate is decreased at 150 and 300 atm, the release of pressure immediately allows the system to aggregate at the 1-atm rate. The diminution of the aggregation rate by pressure indicates that the transition state of the rate-limiting step exhibits a positive volume of activation. Calculation of the ΔV^* for the system shown in Fig. 1 is 203 ml/mole. It might also be mentioned that the ΔV^* for PVRNase is dependent upon the ionic strength of the system employed (higher ΔV^* at lower ionic strengths). The effects of solvent medium (pH, various buffers, concentration of PVRNase, time, nonelectrolytes, ionic strength, polyanions, etc.) on PVRNase is discussed by Nishikawa *et al.* (1968).

Gill and Glogovsky (1965) studied the thermal transition of bovine pancreatic ribonuclease (RNase) under high pressure by optical rotation methods. A molar volume change of −30 ml/mole was noted for this intramolecular conformation change. Pressure-induced spectral shifts of hemoglobin and its derivatives have been noted by Fabry and Hunt (1968), which suggest the effect of pressure may not be on the solvent

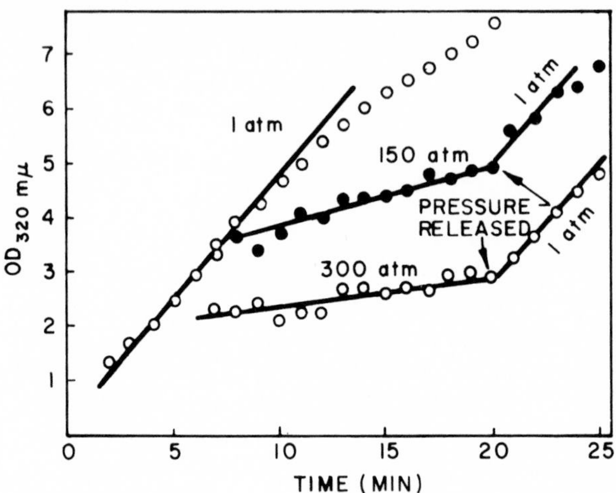

FIG. 1. Time course of PVRNase aggregation. Conditions as indicated in text with solutions of 0.6 *M* in sodium chloride. (After Kettman *et al.*, 1965.)

but on the protein structure. Murayama (1966) studied the polymerization of sickle-cell hemoglobin. The aggregation appears to involve hydrophobic interactions of the N-terminal valine-rich portion of the β-chains. This aggregation was prevented by pressure. The aggregation of sickle cell hemoglobin at 37°C was found to have a ΔV^* of 400 ml/mole from 1 to 50 atm; but between 50 and 150 atm, the ΔV^* is −500 ml/mole, changing its algebraic sign (Murayama and Hasegawa, 1969). It appears that a phase change takes place at 50 atm. Josephs and Harrington (1968) have demonstrated a positive volume change of 380 ml/mole on the aggregation of myosin monomer units by studying the effects of hydrostatic pressure on the equilibrium in an ultracentrifuge.

It should be realized that hydrostatic pressure can become an important factor in ultracentrifugation studies of macromolecular systems, especially when hydrostatic pressure can affect the aggregation of macromolecules (Josephs and Harrington, 1966, 1967; TenEyck and Kauzmann, 1967; Kegeles *et al.*, 1967).

1. *Pressure Effects on RNase-A*

An absorbance increase in the spectra below 280 nm on the thermal unfolding of RNase-A due to hydrostatic pressure was shown by Nishikawa (unpublished data). According to Timasheff (1966), this increased absorbance indicates a spectral change rather than an increase in turbidity. The circular dichroic (CD) spectrum of native RNase-A exhibits a small positive band at 240 nm and a larger negative band at 275 nm (Isemura *et al.*, 1968; Beychok, 1966). As a result of the unpublished observations by Nishikawa, further CD studies were made on RNase-A under hydrostatic pressure.

TABLE I

CHANGES IN OBSERVED OPTICAL PARAMETERS OF RNASE

Condition	$[\theta]_{240}$	$[\theta]_{216}$	$[\theta]_{275}$	$\Delta\varepsilon_{287}$
pH 7 + 1000 atm	+73	−9890	−228	0
pH 2	+35	−8630	−217	−246
pH 2 + 1000 atm	+ 8	−8570	−202	−594
pH 2 + 1000 atm[a]	+74	−8570	—	—

[a] Material neutralized, then observed again.

When native RNase-A at pH 2 and 30°C was pressurized to 1000 atm, two distinct changes in the ultraviolet spectra were observed. Under pressure the denaturation blue shift of the 278-nm band was enhanced but was reversed when depressurized. A nonreversible general hyperchromicity was observed below 280 nm, the largest CD change taking place at the 240-nm band. When the pH of the system was 7.2 no changes in the CD spectra could be observed between the 1000- and 1-atm conditions.

Observed changes in ellipticities can be noted for the RNase at pH 2 and the RNase at pH 2 and 1000 atm (Table I). The reversibility of the 240-nm band of RNase at pH 2 and 1000 atm, when neutralized, suggests that pressure induces a metastable conformational state in RNase expanded by low pH, but probably does not involve changes in most of the peptides situated in the α-helix or β-structure regions. The latter is suggested by the ellipticity observed at 216 nm. Further unpublished data by Nishikawa indicate that 1000 atm on RNase at pH 2 enhances the exposure of abnormal tyrosines in a manner similar to thermally induced unfolding.

2. *Pressure Effects on RNase-S*

Kauzmann (1959), in reference to the earlier work of Masterton (1954) on the transfer of small hydrocarbons into an aqueous environment, states that the formation of hydrophobic bonds should be accompanied by an expansion of 20 ml/mole of aliphatic side chains, and a lesser value for phenylalanine residues. Nemethy and Scheraga (1962a,b,c), in a series of papers concerning the thermodynamics of hydrophobic bonding, indicate that the formation of "hydrophobic bonds" can be regarded as a partial reversal of the solvation process with a resulting calculated increase in volume of 3–8 ml/mole depending upon the nature of the side chain group and the extent of interaction. Since certain limits have been established for the magnitude and direction of the volume change resulting from "hydrophobic bonding," it was decided to seek a protein system that might allow experimental determination of these parameters.

The recently published x-ray structure of ribonuclease-S at 3.5-Å resolution by Wyckoff *et al.* (1967) indicates that the S-peptide portion is largely held in place by hydrophobic interactions of Ala 4, Phe 8, His 12, and Mete 13. For this reason ribonuclease-S and its component parts, S-protein and S-peptide, have been selected as a model system in an attempt to obtain some gross indication as to the magnitude and direction of the volume change accompanying "hydrophobic bonding." From the knowledge of the interactions involved as indicated by the x-ray data, one can make a gross estimation of the magnitude of volume change that should accompany the formation of ribonuclease-S.[2] If one assumes a volume increase of 8 ml/mole of aliphatic side chain interaction and 3 ml/mole of aromatic side chain interaction as suggested by Nemethy and Scheraga (1962b), and estimating a volume increase of 10–20 ml/mole of side chain electrostatic interaction, a total volume increase of approximately 30–40 ml/mole S-protein–S-peptide interaction should accompany the formation of ribonuclease-S′. Correspondingly, if one assumes a volume increase of 20 ml/mole aliphatic side chain interaction as suggested by Kauzmann (1959), a lesser value of 10 ml/mole aromatic side chain interaction, and again estimating 10–20 ml/mole increase for the electrostatic interaction, a total volume increase of approximately 70–80 ml/mole S-protein–S-peptde interaction should accompany the formation of ribonuclease-S′.

Both direct and indirect volume change measurements have been made in this laboratory in the form of dilatometric measurements and indirect determinations of volume change through the study of pressure effects on

[2] Ribonuclease-S′ is the commonly used notation indicating the active enzyme formed from the interaction of S-protein and S-peptide. Ribonuclease-S is the commonly used notation indicating the limited subtilisin-digested bovine pancreatic ribonuclease-A. Ribonuclease-S is fully active and differs from ribonuclease-A only in the respect that the peptide linkage between residues 20 and 21 or 21 and 22 has been cleaved (Doscher and Hirs, 1967).

the equilibrium constant of the reversible thermal denaturation of ribonuclease-S and S-protein. Studies on the thermal transition of ribonuclease-S and S-protein over a range of acid pH were conducted because of the importance to the pressure studies of knowing the state of denaturation of ribonuclease-S and S-protein with respect to temperature and pH. Utilizing these data an equilibrium expression for the reversible thermal denaturation was set up in a manner used by previous workers (Hermans and Scheraga, 1961; Holcomb and Van Holde, 1962; Brandts, 1964; Gill and Glogovsky, 1965). From this equilibrium expression the enthalpy of denaturation was determined as a function of pH for ribonuclease-S, and the effect of pressure on the equilibrium constant for the reversible denaturation was followed for ribonuclease-S and S-protein at a specified temperature and pH. The change in the equilibrium constant with pressure was related to a change in the partial molar volume of the system through the use of an expression first derived by Planck (1887). The difference in the effect of pressure on the denaturation state of ribonuclease-S and S-protein under similar conditions is assumed to reflect the dissociation of the S-peptide portion of ribonuclease-S. Furthermore, the difference in the change in partial molar volume of the two systems is interpreted to give a direct indication of the magnitude and direction of volume change occurring when the S-peptide dissociates from ribonuclease-S.

From the change in molar absorbance at 287 nm and 286 nm, respectively, equilibrium constants as a function of pressure were computed for ribonuclease-S and S-protein. A plot of log K vs pressure at 25° was used to calculate a volume change of -80 ± 2 ml/mole which was obtained in the pressure range of 300–900 atm for ribonuclease-S, while S-protein undergoes an apparent volume change of -45 ± 2 ml/mole under similar conditions.

The volume change accompanying reactions as well as phenomena associated with conformational changes in proteins can be followed by dilatometry. Linderstrøm-Lang and Jacobsen (1941) used the technique of dilatometry to monitor the hydrolysis of β-lactoglobulin by trypsin or chymotrypsin. Their studies indicated the volume change associated with the hydrolysis of the peptide, which was previously found to be accompanied by a contraction of 20 ml/mole of peptide linkage hydrolized, could not account for the total contraction observed in β-lactoglobulin. This additional contraction ranged from 100 ml/mole to 700 ml/mole protein at 0°. Their results indicated that the initial hydrolysis of β-lactoglobulin must involve a considerable rearrangement in the protein structure. More important, they demonstrated that dilatometric studies can contribute information concerning the magnitude and volume changes accompanying protein conformational changes. Other dilatometric studies have been made to determine such phenomena as the volume change resulting from ionization reactions in proteins (Rasper and Kauzmann, 1962), the volume change accompanying the helix coil transition (Noguchi and Yang, 1962), and the volume change associated with the thermal determined by hydrostatic pressure experiments. The values obtained for metric studies related to the determination of the volume change due to hydrophobic bonding have been conducted by Lauffer (1964) and by Gerber and Noguchi (1967). These studies were performed with aggregating protein systems in which the exact nature and number of binding sites involved in the aggregation were not known. Both studies were entrophy driven, endothermic processes, resulting in an increase in volume with aggregation. We have attempted to correlate dilatometric data and the value determined from the indirect volume change measurements with the estimated magnitudes and direction of the volume change as denaturation of ribonuclease-A (Holcomb and Van Holde, 1962). Dilatothe $\Delta \overline{V}^{\infty}_{av}$ of ribonuclease-S and S-protein calculated from the effect of pressure on the denaturation equilibrium expression were —80 ± 2 ml/mole and −45 ± 2 ml/mole, respectively, under similar conditions of pH, temperature, pressure, and protein concentration. This is taken to indicate that an approximate volume decrease of −35 ml/mole occurs upon dissociation of the S-peptide, and the concomitant reversible denaturation of the S-protein portion involves a volume change of approximately −45 ml/mole.

The results of the dilatometer studies tend to agree in a gross manner with the results of the pressure studies. The determination of the volume change due to the interaction of S-peptide and S-protein was found to be +32 ± 6 ml/mole S-protein and would appear on the surface to agree quite well with the value of −35 ± 2 ml/mole ribonuclease-S obtained from the pressure studies for the assumed dissociation of S-peptide and S-protein.

From these studies it is concluded that a volume increase of 17 ml/mole to 35 ml/mole accompanies the interaction of S-peptide with S-protein. Furthermore, it is suggested that a volume change of −27 ml/mole protein to −45 ml/mole protein must accompany the reversible denaturation of S-protein. From the data obtained it is difficult to speculate on the individual contributions to the volume change from the various hydrophobic groups. In a gross manner, it is felt that a value of +8 ml/mole as suggested by Nemethy and Scheraga (1962c) for a volume change in a hydrophobic environment is in the proper range. For any case, the general agreement of both hydrostatic pressure experiments and the direct dilatometric measurements in this system provides a basis for studies on more complex systems where hydrophobic interactions are implicated.

ACKNOWLEDGMENT

This paper was supported in part by the Office of Naval Reesarch contract NR 108-708 and the National Aeronautics and Space Administration contract NRG-38-002-017. Special thanks are due to our colleagues and associates (Dr. A. H. Nishikawa, Dr. T. O. Tiffany, and Miss Janet M. Hall) for use of data that has not been published.

REFERENCES

Anderson, G. R. (1963). A study of the pressure dependence of the partial specific volume of macromolecules in solution by compression measurements in the range of 1–8000 atm. *Arkiv Kemi* **20**, 513–571.

Becker, R. R., and Sawada, F. (1963). Enzymatic properties of polypeptidyl-ribonuclease. *Federation Proc.* **22**, 419.

Beychok, S. (1966). Circular dichroism of biological macromolecules. *Science* **154**, 1288–1299.

Brandts, F. (1964). The thermodynamics of protein denaturation. I. The denaturation of chymotrypsinogen. *J. Am. Chem. Soc.* **85**, 4291–4301.

Buch, K., and Gripenberg, S. (1932). Uber den Einfluss des Wasserdruckes auf pH das Kohlensaure gleichgewicht in grosseren Meerestiefen. *J. Conseil, Conseil Perm. Intern. Exploration Mer* **7**, 233–245.

Distèche, A. (1959). pH measurements with a glass electrode withstanding 1500 kg/cm^2 hydrostatic pressure. *Rev. Sci. Instr.* **30**, 474–478.

Distèche, A., and Distèche, S. (1965). The effect of pressure on pH and dissociation constants from measurements with buffered and unbuffered electrode cells. *J. Electrochem. Soc.* **112**, 350–354.

Doscher, M., and Hirs, C. H. W. (1967). The heterogeneity of bovine pancreatic ridonuclease-S. *Biochemistry* **6**, 304–311.

Duedall, I. W., and Weyl, P. K. (1967). The partial equivalent volumes of salts in seawater. *Limnol. Oceanog.* **12**, 52–59.

Ewald, A. H., and Hamann, S. D. (1956). The effect of pressure on complex ion equilibria. *Australian J. Chem.* **9**, 54–60.

Fabry, T. L., and Hunt, J. W., Jr. (1968). Pressure-induced spectral shifts in hemoproteins. *Arch. Biochem. Biophys.* **123**, 428–429.

Gerber, B. R., and Noguchi, H. (1967). Volume change associated with the G–F transformation of flagellan. *J. Mol. Biol.* **26**, 197–210.

Gill, S. J., and Glogovsky, R. L. (1965). Influence of pressure on the reversible unfolding of ribonuclease and poly-γ-benzyl-L-glutamate. *J. Phys. Chem.* **69**, 1515–1519.

Haight, R. D., and Morita, R. Y. (1962). Interaction between the parameters of hydrostatic pressure and temperature on aspartase of *Escherichia coli*. *J. Bacteriol.* **83**, 112–120.

Hamann, S. D. (1963). The ionization of water at high pressures. *J. Phys. Chem.* **67**, 2233–2235.

Hamann, S. D. (1964). High pressure chemistry. *Ann. Rev. Phys. Chem.* **15**, 349–370.

Hamann, S. D., and Strauss, W. (1965). The chemical effect of pressure. Past 3—

Ionization constants at pressures up to 1200 atm. *Trans. Faraday Soc.* **51**, 1684–1690.

Hermans, J., Jr., and Scheraga, H. A. (1961). Structural studies of ribonucleases. V. Reversible change of configuration. *J. Am. Chem. Soc.* **83**, 3283–3292.

Holcomb, D. N., and Van Holde, K. E. (1962). Ultracentrifugal and viscometric studies of the reversible thermal denaturation of ribonuclease. *J. Phys. Chem.* **66**, 1999–2006.

Horne, R. A., and Johnson, D. S. (1966). The viscosity of water under pressure. *J. Phys. Chem.* **70**, 2182–2190.

Horne, R. A., and Johnson, D. S. (1967). The effect of electrolyte addition on the viscosity of water under pressure. *J. Phys. Chem.* **71**, 1147–1149.

Isemura, T., Yutani, K., Yutani, A., and Imanishi, A. (1968). The reoxidation of reduced ribonuclease to the native conformation as determined by circular dichroism. *J. Biochem (Tokyo)* **64**, 411–413.

Johnson, F. H., Eyring, H., and Polissar, M. J. (1954). "The Kinetic Basis of Molecular Biology." Wiley, New York.

Josephs, R., and Harrington, W. F. (1966). Studies on the formation and physical chemistry properties of synthetic myosin filaments. *Biochemistry* **5**, 3474–3487.

Josephs, R., and Harrington, W. F. (1967). An unusual pressure dependence for a reversibly associating protein system; sedimentation studies on myosin. *Proc. Natl. Acad. Sci. U.S.* **58**, 1587–1594.

Josephs, R., and Harrington, W. F. (1968). On the stability of myosin filaments. *Biochemistry* **7**, 2834–2847.

Kauzmann, W. (1959). Factors in interpretation of protein denaturation. *Advan. Protein Chem.* **14**, 1–63.

Kegeles, G., Rhodes, L., and Bethune, J. L. (1967). Sedimentation behavior of chemically reacting systems. *Proc. Natl. Acad. Sci. U.S.* **45**, 45–51.

Kettman, M. S., Nishikawa, A. H., Morita, R. Y., and Becker, R. R. (1965). Effect of hydrostatic pressure on the aggregation reaction of poly-L-valyl-ribonuclease. *Biochem. Biophys. Res. Commun.* **22**, 262–267.

Lauffer, M. A. (1964). Protein–protein interaction: endothermic polymerization and biological processes. *In* "Symposium on Foods: Proteins and their Reactions" (H. W. Schultz and A. F. Anglemeir, eds.), pp. 87–116. Avi Publ. Co., Inc., Westport, Connecticut.

Linderstrøm-Lang, K., and Jacobsen, C. F. (1941). The contraction accompanying enzymatic breakdown of proteins. *Compt. Rend. Trav. Lab. Carlsberg, Ser. Chim.* **24**, 1–46.

Masterton, W. L. (1954). Partial molar volumes of hydrocarbons in water solution. *J. Chem. Phys.* **22**, 1830–1833.

Morita, R. Y. (1967). Effects of hydrostatic pressure on marine microorganisms. *Oceanog. Marine Biol. Ann. Rev.* **5**, 187–203.

Morita, R. Y. (1969). Application of hydrostatic pressure to microbial cultures. *Methods Microbiol.* **2** (in press).

Morita, R. Y. (1970). Pressure—Bacteria, fungi and blue green algae. *In* "Marine Biology—Environmental Factors" (O. Kinne, ed.). Wiley (Interscience), New York (in press).

Morita, R. Y., and Haight, R. D. (1962). Malic dehydrogenase activity at 101°C under hydrostatic pressure. *J. Bacteriol.* **83**, 1341–1346.

Morita, R. Y., and Mathemeier, P. F. (1964). Temperature–hydrostatic pressure studies on partially purified inorganic pyrophosphatase activity. *J. Bacteriol.* **88**, 1667–1671.

Murayama, M. (1966). Molecular mechanism of red cell "sickling." *Science* **153**, 145–149.

Murayama, M., and Hasegawa, F. (1969). Effect of hydrostatic pressure on the aggregation reaction of human sickle cell hemoglobin at 37°. *Federation Proc.* **28**, 536.

Nemethy, G., and Scheraga, H. A. (1962a). Structure of water and hydrophobic bonding in proteins. I. A model for the thermodynamic properties of liquid water. *J. Chem. Phys.* **36**, 3382–3401.

Nemethy, G., and Scheraga, H. A. (1962b). Structure of water and hydrophobic bonding in proteins. II. Model for the thermodynamic properties. *J. Chem. Phys.* **36**, 3402–3417.

Nemethy, G., and Scheraga, H. A. (1962c). The structure of water and hydrophobic bonding in proteins. III. The thermodynamic properties of hydrophobic bonds in proteins. *J. Phys. Chem.* **55**, 1773–1789.

Nishikawa, A. H., Morita, R. Y., and Becker, R. R. (1968). Effects of the solvent medium on polyvalyl-RNase aggregation. *Biochemistry* **7**, 1506–1513.

Noguchi, H., and Yang, J. T. (1962). Dilatometric and refractometric studies of the helix-coil transition of poly-L-glutamic acid in aqueous solution. *Biopolymers* **1**, 359–370.

Owen, B. B., and Brinkley, S. R. (1941). Calculation of the effect of pressure upon ionic equilibria in pure and salt solutions. *Chem. Rev.* **29**, 461–473.

Planck, M. A. (1887). Ueber das princip der vermehrung der entropie. *Ann. Phys. Chem.* **32**, 462–503.

Pytkowicz, R. M., and Conners, D. N. (1964). High pressure solubility of calcium carbonate in sea water. *Science* **144**, 840–841.

Pytkowicz, R. M., and Fowler, G. A. (1967). Solubility of foraminifera in seawater at high pressures. *Geochem. J. (Nagoya)* **1**, 169–182.

Pytkowicz, R. M., Distèche, A., and Distèche, S. (1967). Calcium carbonate solubility in seawater at *in situ* pressures. *Earth Planetary Sci. Letters* **2**, 430–432.

Quigley, M. M., and Colwell, R. R. (1968). Properties of bacteria isolated from deep-sea sediments. *J. Bacteriol.* **95**, 211–220.

Rasper, J., and Kauzmann, W. (1962). Volume changes in protein reactions. I. Ionization reactions of proteins. *J. Am. Chem. Soc.* **84**, 1771–1777.

Suzuki, C., and Suzuki, K. (1962). Protein denaturation by high pressure. Changes of optical rotation and susceptibility to enzymic proteolysis with ovalbumin denatured by pressure. *J. Biochem. (Tokyo)* **52**, 67–71.

TenEyck, L. F., and Kauzmann, W. (1967). Pressure and hydration effects on chemically reacting systems in the ultracentrifuge. *Proc. Natl. Acad. Sci. U.S.* **58**, 888–894.

Timasheff, S. N. (1966). Turbidity as a criterion of coagulation. *J. Colloid Interface Sci.* **21**, 489–497.

Wykoff, H. W., Hardman, K. D., Allewell, N. M., and Richards, F. M. (1967). The structure of ribonuclease-S at 3.5 A resolution. *J. Biol. Chem.* **342**, 3984–3988.

CHAPTER 4

PRESSURE EFFECTS ON MORPHOLOGY AND LIFE PROCESSES OF BACTERIA

Claude E. ZoBell

I. Introduction

In view of the vast extent of the biosphere having a hydrostatic pressure higher than atmospheric, it seems paradoxical that so little is known about the effects of increased pressure on the structure, survival, growth, and biochemical activities of organisms. Besides soil and sedimentary deposits in which living bacteria occur to depths measured in hundreds of meters (ZoBell, 1958), more than two-thirds of the earth is covered with water inhabited by animals (Wolff, 1960) and bacteria (ZoBell, 1968) to depths exceeding 10,000 m. At the greatest known depth in the sea, 11,034 m in the Marianas Trench (Fisher and Hess, 1963), the hydrostatic pressure is about 1160 atm. The pressure–depth gradient in the sea ranges from 0.099 to 0.105, depending upon salinity, temperature, latitude, compression, and other factors which affect the density of sea water. Nearly 90% of the sea floor is covered with more than 1000 m of water, meaning that nearly 90% of the marine environment, or about 56% of the earth's biosphere by area, has pressures ranging from 100 atm to more than 1100 atm.

Biologists use various units to designate pressure: 1.0 atm = 1.01325 $\times 10^5$ N/m^2 = 1.033227 kg/cm^2 = 1.01325 bars = 1,013,250 dynes/cm^2 = 760 mm Hg at 0°C = 14.696 psi = 760 Torr.

A study of the effects of pressure on organisms is known as *barobiology*. Pressures higher than standard atmospheric are designated *hyperbaric*, a term commonly applied to pressures between 2 and 20 atm. Rarefied atmospheres such as occur at great elevations or in partially evacuated receptacles are described as being *hypobaric*. Pressures of less than 1.0 mm Hg are usually expressed in Torrs (1.0 Torr = 1.0 mm Hg at 0°C = 1/760 atm). Certain bacteria survive for prolonged periods of time at pressures as low as 10^{-9} Torr (Silverman and Davis, 1964), but the growth of most bacteria is retarded by pressures as low as 1 Torr.

The term *barophile* has been coined to describe organisms which grow well at pressures higher than 400 or 500 atm. Most bacteria are capable of some growth at pressures up to 200 or 300 atm. Those that grow poorly or not at all at pressures higher than 300 to 400 atm are termed *barophobic*, meaning that they dislike high pressure. A few *eurybaric* varieties are able to grow over a wide range of pressures, say at 1 atm and also at 500 atm or higher. *Baroduric* or *barotolerant* organisms survive for prolonged periods at high pressures, say 500 to >2000 atm, but may not be able to grow at such pressures.

Morita (1969a) has given a good account of the instruments and methods employed by various investigators for studying the effects of pressure on the growth, viability, morphology, and biochemical activities of microorganisms. Much useful information on instrumentation and methodology in barobiology has also been given by Aronson (1967), Bradley (1963), Comings (1956), Hamann (1957), Knight-Jones and

Morgan (1966), Marsland (1950), Morita (1967), Weale (1967), and ZoBell (1959).

II. Morphological Variants

In size, shape, arrangement of cells, cell wall structure, capsulation, flagellation, and internal structure, bacteria from the deep sea grown at *in situ* pressures appear to be very much like their terrestrial counterparts cultured at atmospheric pressure. However, when either terrestrial or marine bacteria are cultured at pressures 200 to 400 atm higher than their normal habitat, numerous morphological aberrations have been observed. The most obvious aberration is filament formation. Closer examination reveals, at increased pressures, the development of branching forms, large globular forms, loss of flagella, loss of pigmentation, decreased sporulation, and changes in fine structure.

The generalization concerning the morphology of deep sea bacteria is based on rather scanty observations made mostly under atmospheric conditions after decompression. Although deep sea samplers are available for hauling to the surface samples of water or bottom sediment at *in situ* pressures, practical procedures have not been perfected for the microscopic examination of bacteria in such samples at *in situ* pressures. The microscope pressure chamber, described by Marsland and Brown (1936) and Marsland (1950), provides for observing large cells, e.g., *Arbacia* eggs or protozoans, at deep sea pressures at magnification up to 600 diameters. Such apparatus has been extensively used by cytologists, but it has not proven to be practical for studying the morphology of bacteria.

In an effort to learn more about the morphology of bacteria at deep sea pressures, Boatman (1967) fixed bacteria during growth at pressures up to 680 atm with buffered 0.8% osmium tetroxide. He used apparatus similar to that described by Landau and Thibodeau (1962) for fixing *Amoeba proteus* under high pressure. This was achieved by compressing the microorganisms in the upper half of a two-chambered Lucite vessel. The lower half, containing the buffered fixative, was separated from the upper half by a circular glass cover slip sealed in place with a rubber O-ring. The upper compartment was closed with a rubber stopper, which functioned as a piston to compress the contents with hydraulic fluid in a steel pressure cylinder similar to apparatus described by ZoBell and Oppenheimer (1950) and ZoBell (1959). After the desired period of incubation at the desired pressure and temperature, the pressure cylinder containing the compressed culture vessel is inverted and shaken vigorously, causing a steel ball to break the glass cover slip and to mix the fixative with the cells. Aliquots of the bacterial cells thus fixed were examined by Boatman (1967) by phase-contrast microscopy for gross morphology. Ultramicroscopic sections of cells in other aliquots were prepared for examination by electron microscopy.

A. Filament Formation and Fine Structure

When incubated in nutrient medium at 270 to 400 atm, *Escherichia coli* was found by Boatman (1967) to be highly filamentous. Electron micrographs revealed no evidence of constrictive divisions in the filaments. Ribosomes and nuclear material appeared to be fairly evenly distributed at 270 atm. Cells grown at 400 atm exhibited marked pleomorphism with obvious disorganization of the cytoplasm. Ribosomes were reduced in number and numerous invaginations occurred in the cytoplasm.

An *Aerobacter* species grown at 400 atm was found by Boatman (1967) to form filaments, many being from four to sixteen times longer than cells which developed in nutrient medium at 1 atm. Similarly, a *Vibrio* species was found to form filaments, some being from five to eight times longer than normal cells which developed at 1 atm. Most of the cells which developed at 270 atm contained masses of ribosomes but little nuclear material. Cells of a *Corynebacterium* species which developed at 400 atm contained large quantities of nuclear material and decreased numbers of ribosomes.

Bacillus mycoides formed cells which were two or three times longer at 270 atm than at 1 atm. Mesosomes were scarcely discernible and the cell walls were thicker. Nearly all of the mesosomes in a *Bacillus* species, which Boatman (1967) designated 871_M, appeared to be "trapped" between the cell wall and cytoplasmic membrane completely apart from the nuclear material. Compression to 400 atm caused cell lysis, with some disorganization of cytoplasmic components.

Although most of the barophobic bacteria examined by Boatman (1967) formed filaments and highly pleomorphic cells at 270 to 400 atm, a barophobic species of *Pseudomonas* grown at 400 atm exhibited no detectable change in morphology. All of the cells were the same in size, shape, and structure as those which developed much more rapidly in nutrient medium at 1 atm.

The majority, but not all, of the barophobic and barophilic marine bacteria studied by Oppenheimer and ZoBell (1952) formed long filaments when cultivated in nutrient medium at pressures near the threshold of their tolerance. For example, *Serratia marinorubra,* whose growth was retarded at 400 atm and failed to grow enough at 600 atm to render nutrient medium perceptibly turbid, was found by ZoBell and Oppenheimer (1950) to form filaments up to 200 μ in length with no evidence of cross wall septa at 600 atm. When cultivated at normal atmospheric pressure, S. *marinorubra* rendered nutrient medium turbid in a day or two at 25°C with small rods, 0.6 to 1.5 μ in length, occurring

TABLE I

SOME MEASUREMENTS OF CELLS FORMED BY THREE STRAINS OF *Escherichia coli* AT DIFFERENT HYDROSTATIC PRESSURES IN 12 HOURS AT 30°C[a]

Strain	Pressure (atm)	Range of cell length (μ)	% of cells > 5 μ long	% of biomass in filaments
B	1	1–8	1.2	2.1
	200	1–10	1.9	8.0
	400	1–45	10.3	51.8
	475	1–160	23.4	89.8
S	1	1–6	0.3	0.5
	200	1–10	0.7	2.5
	400	1–35	8.4	29.4
	475	1–110	14.5	74.7
R_4	1	1–4	0	0
	200	1–4	0	0
	400	1–12	3.1	6.6
	475	1–35	8.2	16.4

[a] From ZoBell and Cobet (1964).

singly or in pairs. Similarly, at 300 to 600 atm *Bacillus borborokoites, Bacillus abysseus, Flavobacterium okeanokoites, Vibrio phytoplanktis,* and several other marine species formed filaments from 5 to 50 times longer than normal cells grown at 1 atm. On the other hand, increased pressure had no detectable effect on the cell length of *Vibrio haloplanktis, Achromobacter stationis,* and certain other marine species which grew at pressures no higher than 400 to 600 atm.

Not only do species differ in their tendency to form filaments at increased pressures, various strains of the same species exhibit marked differences in this respect as illustrated by data in Table I. When cultured near the threshold of their pressure tolerance (about 500 atm at 30°C), many cells formed by certain strains of *Escherichia coli* were from 10 to 100 times longer than the mean length of cells which developed normally at 1 atm, whereas other strains exhibited little tendency to form filaments under similar conditions of cultivation (ZoBell and Cobet, 1962, 1964). The filaments were normal in width, i.e., 0.5 to 1.0 μ. Neither phase-contrast nor electron microscopy of filaments fixed with osmium tetroxide revealed any constrictions, cell wall formation, or other evidence suggestive of fission.

As shown by data in Table II, substantially less biomass was produced in nutrient medium by *E. coli* at increased pressures. The mean cell length and weight per cell were increased by increased pressure, but the amount of protein produced per unit length of cell was about the same at all pressures. The amount of RNA produced per unit length of

TABLE II
CELL SIZE AND CHEMICAL COMPOSITION OF *Escherichia coli* (STRAIN B) CULTURED FOR 7 DAYS AT 30°C AT DIFFERENT HYDROSTATIC PRESSURES[a]

Pressures (atm):	1	200	400	425	450
Biomass, wet weight (mg/ml):	8.85	7.62	2.04	0.63	0.18
Mean cell length (μ):	1.97	2.28	4.56	4.77	5.39
Dry weight/cell (10^{-7} μg):	1.06	1.23	2.48	2.58	2.94
Protein/cell (10^{-7} μg):	0.79	0.94	1.85	1.88	2.14
Protein/μ of length (10^{-7} μg):	0.69	0.75	0.74	0.73	0.75
RNA/cell (10^{-8} μg):	0.51	0.68	1.39	1.54	1.76
RNA/μ of length (10^{-8} μg):	0.26	0.30	0.30	0.32	0.33
DNA/cell (10^{-8} μg):	0.34	0.44	0.49	0.41	0.34
DNA/μ of length (10^{-8} μg):	0.17	0.19	0.11	0.08	0.06

[a] From ZoBell and Cobet (1964).

cell was significantly more at increased pressures, but the amount of DNA was markedly less. This led ZoBell and Cobet (1964) to believe that the suppression of DNA production might be responsible for suppressed cell division and the formation of long forms at increased pressures. Berrah and Konetzka (1962) noted a relationship between filament formation by cells of *E. coli* (strain H), *Pseudomonas fluorescens*, and certain other gram-negative bacteria and the inhibition of DNA synthesis caused by treating cultures with 0.25 to 0.5% phenethyl alcohol. When the alcohol was removed from the nutrient medium, the filamentous forms started to synthesize DNA with concomitant cell division.

Using the uptake of ^{14}C-labeled thymine by pregrown (at 1 atm) cells to indicate synthetic activity, Pollard and Weller (1966) and Yayanos (1967) observed that increased pressure suppressed the synthesis of DNA by *E. coli* (strain 15_T). The incorporation of uracil in pregrown *E. coli* cells to form RNA was retarded to a lesser extent by a pressure of 450 atm (Pollard and Weller, 1966). The incorporation of ^{14}C-labeled glycine, leucine, and adenine into the respective protein and nucleic acid fractions of pregrown cells of *E. coli* (strain K-12) was found by Landau (1966) to be markedly inhibited by pressures of 544 and 680 atm but not at 400 atm at 37°C. But at 27°C a pressure of 400 atm inhibited the synthesis of nucleic acids. A pressure of 272 atm at 27°C appeared to be stimulatory. The latter result is at variance with the observations of Yayanos (1967) and with more recent observations of Landau (1967), who noted the suppression of protein synthesis by *E. coli* strain ML-3 at 37°C. Landau attributed the apparent discrepancy to temperature differences in the two experiments and not to differences in the pressure response of the two strains. The experiments, lasting only 30 to 240 minutes, were not designed to demonstrate the effects of compression on cell division or filament formation.

The synthesis of β-galactosidase by pregrown *E. coli* cells was found by Landau (1967) to be suppressed by moderate pressure. Each phase of the induced enzyme synthesis was distinct in its response to the level of pressure applied. Transcription seemed to be least affected, continuing at 670 atm at 37°C. Translation was totally inhibited at 670 atm and unaffected at 265 atm, while induction was totally inhibited above 265 atm (for further discussion, see Landau, Chapter 2).

Observations of Albright (1968) on the obligate psychrophile *Vibrio marinus* suggest that pressures of 400 to 500 atm at 15°C inhibit or stop RNA translation, which in turn seems to retard or stop DNA transcription and replication. At 1000 atm, the synthesis of protein, DNA, and RNA by cells of *V. marinus* was completely inhibited (Albright and Morita, 1968). Pressures of 400 and 500 atm substantially lowered the rates of protein and DNA synthesis, whereas RNA synthesis was unaffected at 400 atm. At 500 atm, RNA synthesis continued at the 1 atm rate for 30 minutes and then shifted to a lower rate. The maximum pressure at which cultures grew was 425 atm at 15°C. All the cells in cultures were killed within 24 hours at 475 atm.

Fertilized eggs of the sea urchin *Arbacia punctulata* were found to incorporate ^{3}H-thymidine into DNA during compression to 340 atm for 30 to 60 minutes, according to Zimmerman (1963), but DNA synthesis was interrupted when compression began at prophase or metaphase. The mitotic apparatus (spindle–aster chromosome complex) in *Arbacia* eggs was drastically disorganized by compression to 680 atm for 1 minute (Zimmerman and Marsland, 1964). All traces of fibrous structure were abolished and the chromosomes were clumped by such compression. There was an equivalent degree of disorganization in 3 to 4 minutes at 540 atm or in 5 to 6 minutes at 400 atm. A pressure of 500 atm for 1 to 5 minutes delayed the first cleavage of *Arbacia* eggs (Zimmerman and Silberman, 1965). Whereas at 1 atm (20°C) fertilized eggs incorporated ^{3}H-thymidine in 15 minutes after insemination, eggs in which pronuclear fusion had been inhibited at 340 atm did not incorporate the isotope (Zimmerman and Silberman, 1967). Although, even in the face of cessation of morphological mitotic activity, DNA synthesis was not blocked by 340 atm, pressures of 500 atm and higher did inhibit DNA synthesis (cf. Zimmerman, Chapter 10).

Rutberg (1964) attributed the inactivation of *E. coli* phages T2, T4D, and T5 by pressures of 1000 to 2000 atm in 5 minutes at 37°C to changes in the phage nucleic acids.

Katchmann *et al.* (1955) treated actively growing cultures of *E. coli* (strain B) with 0.3 μg/ml of 5-diazouracil to induce filament formation. Finding as much DNA per unit weight of the filaments as in normal cells led these workers to conclude that suppression of DNA synthesis is not the mechanism responsible for filament formation. Similarly, Zamenhof *et al.* (1956) found that greatly elongated cells of *E. coli* (strains I and II) resulting from treating cultures with halogenated uracils had practically normal content of DNA. Incidentally, strain B, which ZoBell and Cobet (1964) found to be most susceptible to filament formation and suppressed DNA synthesis when compressed, was not so affected by halogenated uracils, according to Zamenhof *et al.* (1956).

The fine structure and chemical composition of cells of a barotolerant *Pseudomonas* sp. (strain 8113) grown at 350 and 450 atm were found by Kriss *et al.* (1969) to differ markedly from normal cells grown at 1 atm. Long filaments were formed, the cell wall became separated from the cytoplasmic membrane, the cell wall was thickened in places as if frayed, and clear zones of spongy or reticular structure appeared in the cytoplasm with vast conglomerations of finely granular material. Where such conglomerations formed, the diameter of the cell was considerably larger. Intracytoplasmic membrane structures were more prominent in cells grown at 450 atm than at 1 atm. The RNA content decreased, but protein synthesis was not affected by pressures up to 450 atm. This agrees with the data of ZoBell and Cobet (1964) on protein synthesis by *E. coli* but is at variance with the results of Pollard and Weller (1966) and Landau (1966), who reported depressed protein synthesis at high pressure. Kriss *et al.* (1969a) noted a decrease in the synthesis of carbohydrates, phospholipids, and nucleic acids at 450 atm.

A eurybaric culture, described by Chumak (1959) as Kriss *No. 187*, grew slowly at 550 to 750 atm with the formation of filaments up to 300 μ long and only 0.6 μ wide. When cultured at 1 atm, the average cell size was 0.6 by 3.5 μ. Using the peloscope or capillary tube method of Perfil'ev and Gabe (1964), Mishustina (1968) found no abnormally long filaments in cultures of benthic bacteria from depths of 4000 to 6000 m, which were grown at 400 atm.

B. Nonfilamentous Pleomorphism

Although filament formation is the most spectacular morphological variation, the development of swollen bizarre shapes has been observed in numerous species of bacteria grown in nutrient media near the threshold of their pressure tolerance. According to Oppenheimer and ZoBell (1952), abnormally high pressures induced many cells of *Sarcina pelagia* and *Micrococcus aquivivus* to be highly pleomorphic and somewhat larger in size. *Vibrio phytoplanktis* developed as elongated, granular, pleomorphic rods. There was an overall general increase in the cell size of *Bacillus borborokoites*, some of which were multibulbous.

Most of the cells of *Vibrio* 3_M species, which Boatman (1967) incubated and fixed with osmium tetroxide at 270 atm, were bizarre in shape and had grossly convoluted cell walls. Many cells developed bulbous spheroplastlike ends. Cells of *Corynebacterium* species which developed at 270 atm were found by Boatman to be much more pleomorphic than those which developed at 1 atm. At 400 atm they were even more pleomorphic. The cell walls were markedly thicker and often convoluted. Many cells of *E. coli* showed marked increases in width. Pressure of 50 to 150 atm were observed by Berger (1959) to affect cell wall formation by *E. coli* and *Pseudomonas perfectomarinus*. Abnormally large cells formed at 200 to 400 atm often lysed, spilling out metabolites.

Gram-positive bacteria gradually became gram-negative at 3000 atm, a pressure which killed most bacteria examined by Larson *et al.* (1918).

In their review of the literature on the morphological effects of environmental stress, Duguid and Wilkinson (1961) do not mention hydrostatic pressure. They considered starvation, oxygen deficiency, accumulation of toxic metabolites, adverse pH, antibiotics, and hypertonicity of nutrient media as the main causes of morphological variants, including long filaments, branching forms, globular or bulbous forms, abnormal sizes, and bizarre shapes.

C. Spore Formation and Germination

Bacillus abysseus, B. borborokoites, B. epiphytus, B. mesentericus, B. mycoides, B. subtilis, Clostridium bifermentans, C. chauvei, C. histolyticum, C. septicum, C. sporogenes, and *C. welchii* are among the spore-forming bacteria which have been grown at pressures of 400 atm or higher

(ZoBell and Johnson, 1949; Oppenheimer and ZoBell, 1952). The spores of all these species germinated at 400 atm as indicated by the development of vegetative cells. Germination of spores did not appear to be delayed by pressure any more than could be accounted for by the slower rate of reproduction of these barophobic bacteria at increased pressure.

Bizarre pleomorphism was so common among cells which developed at 400 atm that it was extremely difficult to determine what proportion of the vegetative cells formed spores in nutrient media at 400 atm. Unquestionably, all species examined formed spores in nutrient media at 400 atm. Increased pressure seemed to suppress the sporulation of most species more than it suppressed reproduction under similar conditions of cultivation. The spores of *B. abysseus* which developed at 400 atm were definitely larger than the spores which developed at 1 atm and increased pressure seemed to favor the sporulation of this species.

When compressed in nutrient media to 270 to 400 atm at 12°C, spores of *Bacillus mycoides*, *Bacillus* 871_M, and *Bacillus* 874_M were found by Boatman (1967) to commence germinating within 3 days as compared with 6 hours at 1 atm. Spores of these species did not germinate at pressures of 540 or 680 atm, although all three species reproduced slowly and produced small numbers of spores at pressures up to 540 atm. The spores of *Bacillus* 871_M that developed at 540 atm were larger than those that developed at 1 atm.

Holding spores of *Bacillus pumilus* for 4 to 5 hours at 600 atm at 25°C did not affect their ability to germinate in nutrient medium at 1 atm, according to Clouston and Wills (1969), but at higher pressures the viability of spores decreased as pressure was increased to 1350 atm. The stainability of *B. pumilus* spores increased and their refractility decreased with increasing pressure in the range of 820 to 1085 atm. Prolonged compression above 600 atm resulted in the release of dipicolinic acid and calcium from spores. Up to 98% of their calcium content was lost in 4 hours at 820 atm. During 80 minutes at 1085 atm the dry weight of spores decreased 36%. The sensitivity of the spores to gamma radiation increased with increasing pressure in the range of 500 to 1350 atm.

Aerobic spore-forming rods of the genus *Bacillus* have been demonstrated in deep sea sediment samples (ZoBell and Morita, 1959). A few of these bacteria reproduced and formed spores very slowly at 1000 atm. If sustained efforts have been made to demonstrate *Clostridium* species from depths exceeding 4000 m, it has not come to my attention. Although relatively few bacteria found in sea water samples are spore formers, a disproportionately large percentage of the viable cells recovered from the deep sea floor are spore formers (ZoBell, 1968). The majority of the latter are barophobes of the genus *Bacillus* which grow best in laboratory media at pressures ranging from 1 to 300 atm.

D. Flagellation and Motility

It has not been determined whether flagella are produced by barophobic bacteria at high pressures. Neither phase-contrast nor electron microscopy has revealed any flagella on either *E. coli,* two species of *Vibrio,* or four *Pseudomonas* species at 400 atm, although identical procedures showed these bacteria to bear flagella when grown at 100 atm. Most motile bacteria are immobilized by prolonged compression at 200 to 400 atm. It has yet to be determined whether this is attributable to (a) pressure shock during compression or decompression, (b) mechanical manipulations which obscure the flagella, or (c) a failure of the bacteria to form flagella at increased pressures. Likewise, it is not known whether deep sea barophilic forms differ from better known barophobic bacteria with respect to flagellation.

The immobilizing effect of increased pressure on freshwater bacteria was first reported by Regnard (1891). He examined various bacteria found in milk, cheese, urine, and other natural products following prolonged compression at 650 to 700 atm.

The tentacles of the suctorian protozoan *Ephelota coronata* were observed by Kitching and Pease (1939) to undergo a sudden shortening and disintegration at critical pressures around 21°C. The critical pressure varied with individuals, but ranged from about 400 to 800 atm. According to Kitching (1957), flagella or ciliary movements of several different flagellates and ciliates ceased at pressures ranging from as low as about 300 to 500 atm for *Paramecium caudatum* individuals to a high of about 700 to 950 atm for *Polytoma uvella, Chlamydomonas pulsatilla, Spirostomum ambiguum,* and *Tetrahymena pyriformia. Astasia longa, Paramecium aurelia, Colpidium campylum, Colpoda cucullus,* and *Stentor polymorphus* were immobilized at pressures ranging from 550 to 750 atm. The flagella or cilia of all species were structurally damaged at slightly higher pressures.

Ciliary activity of the ciliates *Colpidium colpidium, Glaucoma chattoni,* and *Tetrahymena* spp. was observed by Fischer (1967) to be retarded at about 70 atm. Motility was stopped at about 340 atm. No lysis or other permanent injury resulted from compression of the ciliates. All of the ciliates grew at pressures up to 200 atm, but not as rapidly as at 1 atm. The normally oblong *Tetrahymena* spp. and *G. chattoni* became spheres within 15 to 30 minutes at 400 atm.

About 90% of the normally elongated flagellate *Euglena gracilis* examined by Byrne and Marsland (1965) became rounded in 15 minutes at about 990 atm at 25°C. Locomotion was completely abolished in all specimens compressed to 800 atm, but flagellar movement continued in most specimens following decompression (see also Zimmerman and Zimmerman, Chapter 8).

In their review of the literature, Knight-Jones and Morgan (1966) give numerous examples of barokinesis in marine animals, including planktonic as well as benthonic species. Rate of movement, vertical migrations, and orientation are among the many barokinetic responses. Certain animals respond to pressure changes as small as 2.5 mbar. Many shallow water forms are immobilized by deep sea pressures. Barokinesis is attributed to the effects of pressure on volume changes, mobility of ions, membrane permeability, sol–gel equilibria, compressibility of tissues, and the stability of enzymes.

E. Cells Ruptured by Explosive Decompression

The sudden release of high pressure may rupture bacteria, yeasts, and other microbial cells under certain conditions. Fraser (1951) found that compressing bacteria in solution to around 60 atm with argon, nitrogen, nitrous oxide, or carbon dioxide and then suddenly releasing the pressure resulted in as many as 90% of the cells being ruptured. J. W. Foster *et al.* (1962) have described apparatus for compressing cells in a solution saturated with nitrogen gas. When suddenly decompressed from 120 atm, from 10 to 60% of the cells of *Brucella abortus, Staphylococcus aureus, Serratia marcescens,* and other bacteria were ruptured. Increased pressure forces gas into solution and high concentrations are believed to diffuse into the microbial cells. When pressure is suddenly released, gas escaping from solution forms bubbles within cells, bursting cell walls and disrupting the cytoplasm.

Simpson *et al.* (1963) described a modified French press (French and Milner, 1965) for the disruption of microbial cells. A large percentage of the cells of *Lactobacillus plantarum, Chlorella vulgaris,* and *Rhodotorula* species were ruptured by the sudden release of hydrostatic pressures up to 1700 atm. To the extent that the solution is supersaturated with gas, bubble formation may contribute to the rupture of microbial cells, but sheering action and cavitation phenomena are believed to be largely responsible for the effectiveness of the French press. Milner *et al.* (1950) used such a press to disperse chlorophyll by fragmenting chloroplasts. Yeast cells and dense suspensions of *E. coli* were also disintegrated by the press, resulting in the liberation of endoenzymes and other cellular constituents.

III. Pressure Tolerance of Microorganisms

The pressure required to kill bacteria and allied microorganisms or to inactivate their enzymes is a function of temperature, chemical composition of the medium, duration of compression, and other factors. Different species, varieties, and individual cells differ greatly in their pressure tolerance (ZoBell, 1964; Morita, 1967, 1969b). The rate of compression or decompression may be of great importance when either (a) pressure shock, (b) appreciable adiabatic heating or cooling, or (c) compressed gas is involved (see Section II, E).

A. Pressure Shock

Very few observations have been made on the susceptibility of bacteria to pressure shock. Evidently most stock cultures are not injured by being compressed to 1000 atm within 1 or 2 minutes in nongaseous nutrient media and then immediately decompressed at about the same rate to 1 atm. ZoBell (1964) has reported the compression and decompression of *E. coli* in this manner at 25°C ten successive times within 1 hour without any detectable diminution in the number of viable cells. Several other stock cultures, including deep sea, shallow water, and terrestrial species, have been subjected to such treatment without any evidence of injury.

A good many bacteria have been grown in nutrient media after being hauled from ocean depths exceeding 7000 m and then compressed within a few minutes to *in situ* pressures (700 to 1000 atm). One the other hand, there is evidence that many more deep sea bacteria failed to survive such treatment, probably due to pressure shock. Indeed, Seki and Robinson (1969) have observed adverse effects of pressure shock on bacteria taken from the sea floor at depths of only 400 m (pressure about 40 atm). This was demonstrated by comparing the growth of bacteria (as indicated by ^{14}C uptake) in water samples treated on the sea floor and left there for several days with the growth of bacteria in similar samples which were hauled to the surface and then repressurized for incubation at *in situ* temperature and pressure. The rate of pressure change was less than 10 atm/min.

Pressure changes, particularly decompression, exceeding 50 to 100 atm/sec usually injure microbial cells even in gas-free media and may damage certain apparatus. No stock cultures have been sterilized by such rapid decompression, but many of the cells are killed. When cultures are decompressed more rapidly than 50 atm/sec, piston stoppers may be blown out, glass or plastic culture tubes may be broken, and gauges may be damaged. When working at pressures appreciably higher than 1000 atm near 0°C, adiabatic cooling during rapid decompression may result in ice formation in cells and apparatus with harmful effects. The compression and decompression of gas-free culture media to conditions extant in the biosphere [temperature up to 95°C (Brock, 1967) and pressures up to 1200 atm] may result in temperature changes as much as 7°C (ZoBell, 1959). By applying and releasing pressure by increments of no more than 10 atm/sec, and by having the pressure vessels submerged in a constant temperature water bath, temperature changes in the system being tested could be restricted to less than 2°C and this for only 1 or 2 minutes.

B. Pressure Death Times

Cultures of *E. coli* and *Staphylococcus aureus* examined by Roger (1895) were not sterilized in 10 minutes at 2900 atm at 25° to 30°C, although most of the cells were killed. Cultures of a *Streptococcus* species were sterilized in 10 minutes at 1935 atm but not at 970 atm. The latter conditions killed vegetative cells and some spores of *Bacillus anthracis*. A few spores of *B. anthracis* which survived for 10 minutes at 2900 atm had their virulence attenuated by such treatment. By compressing cultures to 2900 atm at a rate of about 100 atm/min, Roger found that the temperature was increased only 5.3°C. Compressing the cultures to 970 atm several successive times was found to be much more injurious than holding the cultures at this pressure for the same length of time. Decompressing gas-free cultures from 2900 atm to atmospheric conditions in 10 seconds did not appear to be injurious. However, partial gas pressures, especially oxygen and carbon dioxide much in excess of saturation at 1 atm, were highly lethal at pressures less than 100 atm.

The injurious effects of high pressure was observed by Chlopin and Tammann (1903) to be directly proportional to the duration of compression. Whereas only two cultures (*Pseudomonas aeruginosa* and *Vibrio cholera*), among the fourteen species tested, were sterilized within 12 hours at 1935 atm, all except the spore former *B. anthracis* were sterilized in 96 hours by such compression.

According to Hite *et al.* (1914), cultures of bacteria now known as *Serratia marcescens*, *Streptococcus lactis*, *Pseudomonas fluorescens*, and

Aerobacter aerogenes were sterilized at 20° to 25°C within 5 minutes at 5780 to 6800 atm, 10 minutes at 3400 to 4080 atm, or 60 minutes at 2040 to 3060 atm. Pressures of 5780 to 6800 atm killed vegetative cells of *Bacillus subtilis* in 10 minutes, but failed to sterilize cultures, ostensibly due to the greater pressure tolerance of spores. The cultures were sterilized somewhat more rapidly either at 5°C or at 40° to 50°C than at 20° to 25°C. The authors concluded, though, that a pressure of 2040 atm at 50°C for 2 or 3 hours could not be relied upon to prevent the microbial spoilage of canned fruits or vegetables. Pressures of 4080 to 5440 atm for 10 minutes were required to sterilize cultures of *Salmonella typhi* and *Corynebacterium diphtheriae*. The pressure death times for the yeasts *Saccharomyces cerevisiae* and *S. albicans* were 5740 atm for 5 minutes, 3740 to 4080 atm for 10 minutes, or 2040 to 2380 atm for 60 minutes. From 90 to 99.9% of the individual bacteria and yeast cells were killed in much less time than was required for the sterilization of the cultures.

Most but not all spores of *B. subtilis* were found by Larson *et al.* (1918) to be killed in 14 hours by a pressure of 12,000 atm. Heavy cell suspensions of *Salmonella typhi, E. coli, Mycobacterium tuberculosis, Pseudomonas aeruginosa, Staphylococcus* sp., and *Streptococcus* sp. were sterilized in 14 hours at 6000 atm, but not in 90 minutes nor at 3000 atm in 14 hours. All of the nonspore formers were killed in 90 to 150 minutes when suspensions were compressed to 50 atm with CO_2. Prolonged compression as well as its sudden release were believed to be injurious.

Although most *B. subtilis* spores were killed in 45 minutes at 17,600 atm, Basset and Macheboeuf (1932) reported that a few survived such treatment. All of the nonspore-forming bacterial cultures which they examined, including *Serratia marcescens* and *Staphylococcus aureus*, still contained a few viable cells after being compressed to 4000 atm for 45 minutes.

The following tabulation, adapted from Luyet (1937a), shows the death rate of the yeast *Saccharomyces cerevisiae* at 20°C at different pressures in terms of the percentage of cells killed after different periods of time:

	% Killed in (min):			
Pressure (atm)	1	2	4	8
5000	20	31	48	83
5500	42	62	78	97
6000	68	74	91	>99
6500	83	92	>99	>99

According to Luyet (1937b), death was preceded by the gradual flocculation or coagulation of the cell contents. The dead cells were shrunken, but their membranes remained intact.

A pressure of only 540 atm killed many bacteria in samples of soil from flower pots and in deep sea samples within 24 hours, although a small percentage survived for 24 hours at pressures up to 1680 atm (Kriss, 1963). When transferred to nutrient medium and incubated at 1 atm, the survivors reproduced much slower than bacteria from the mud and soil samples which had not been compressed.

The number of viable bacteria in raw milk was found by Timson and Short (1965) to be reduced about tenfold in 30 minutes by a pressure of 2000 atm at 35°C. The viable bacteria were reduced from an average of 10^6/ml to 10^2/ml in 30 minutes at 5000 atm. A few bacterial spores survived for 30 minutes at 10,000 atm.

After reaching the maximum stationary phase of growth (about 9×10^8 viable cells per milliliter of nutrient media), *E. coli,* like most microorganisms, slowly dies. Increasing the pressure accelerates the rate of which *E. coli* dies (ZoBell and Cobet, 1962). When compressed to 1000 atm at 30°C, the number of viable cells decreased to about 2×10^7/ml in 6 hours, to 3×10^5/ml in 12 hours, to 6×10^3/ml in 18 hours, to 1×10^2/ml in 24 hours, and to only 4/ml in 30 hours. None survived compression to 1000 atm for 36 hours, at which time controls held at 1 atm contained 4×10^6 viable *E. coli* per milliliter. At 40°C such cultures were found to be sterilized by compression to 1000 atm within 12 hours and within 124 hours at 20°C.

None of the viruses examined by Basset *et al.* (1935) was entirely inactivated in 30 minutes at 2000 atm, but herpes and yellow fever viruses were inactivated at 3000 atm. Rabies virus was not inactivated in 30 minutes at 3000 atm, but was destroyed at 5000 atm. An encephalomyelitis virus withstood 6000 atm, but was inactivated at 7000 atm in 30 minutes.

Phages for polylysogenic strains of *Actinomyces levoris* were not injured in 30 minutes at 28°C by compression to 200 atm, but Rautenshtein and Muradov (1966) noted a marked inactivation of the phages at 500 atm. Less than 0.01% of them survived compression to 700 atm. On the other hand, phages for *Actinomyces olivaceus* were not inactivated in 30 minutes at 28°C by compression to 700 atm.

Escherichia coli phages T2, T4D, and T5 were inactivated in 5 minutes at 37°C by pressures ranging from 1000 to 2000 atm (Rutberg, 1964). Other observations outlined below show that the pressure tolerance of bacteriophages and viruses is influenced by the temperature of exposure (R. A. C. Foster *et al.*, 1949; Johnson *et al.*, 1948a).

C. Temperature Influences Barotolerance

Ordinarily, duration of compression and temperature of exposure are the two most important conditions which determine the pressure tolerance of microorganisms. Temperature effects depend upon the normal temperature tolerance and optimum growth temperature of any given species.

1. *Subnormal Temperatures*

The death rate of microorganisms at increased pressures is usually slower at temperatures which are appreciably lower than the organism's optimum for growth at 1 atm. For example, *E. coli* normally reproduces most rapidly at 37° to 42°C. Cultures in the maximum stationary phase of growth, containing about 9×10^8 viable cells per milliliter, were disinfected in 12 hours when compressed to 1000 atm at 40°C as compared with 36 hours at 30°C and 124 hours at 20°C (ZoBell and Cobet, 1964). Similarly, Johnson and Lewin (1946a) reported the rate of disinfection of *E. coli* cultures by increased pressure was considerably more rapid at 37°C than at 22.5°C, the rate at 29.1°C being intermediate between the other two.

Compression to 2900 atm was found by Chlopin and Tammann (1903) to disinfect suspensions of *E. coli, Serratia marcescens, Bacillus anthracis,* and a yeast more rapidly at 37°C than at 0°C. On the other hand, *Staphyloccoccus aureus* and *Sarcina rosea* appeared to be killed more rapidly at 0°C that at 37°C. This may have been due to the effects of ice formation at subzero temperatures, because reportedly the medium was cooled 4°C during decompression from 2900 atm to atmospheric conditions. Anomalous results are often obtained when microbial samples are decompressed around 0°C.

ZoBell and associates have observed that barophilic as well as barophobic bacteria in deep sea samples generally survive longer at refrigeration temperatures (3° to 5°C) when pressurized than at 1 atm. Ordinarily, storage temperatures lower than 3°C are avoided in order to eliminate complications caused by ice formation when samples compressed to 1000 atm or more are decompressed. Injurious ice formation can be avoided by compressing cells to high pressures while the cells are being frozen, according to Persidsky and Leef (1969). They reported that *E. coli* cells were not adversely affected by fast freezing to −40°C when pressure was gradually increased to 2000 atm. Simultaneous fast freezing to −40°C and compression to 1500 to 2000 atm was recommended for the *cryopreservation* of bacterial cultures, animal tissues, and organs for transplantation.

2. *High Temperature*

Increased pressure generally retards the death rate of microorganisms at temperatures that are normally lethal at 1 atm, although the retarded death rate may be higher than in cultures similarly compressed at lower temperatures. For example, Johnson and Lewin (1946a) found that at 46.9°C, *E. coli* cultures were disinfected only 26% as rapidly at 400 atm as at 1 atm. Conversely, compression to 400 atm resulted in a fourfold increase in the disinfection rate of *E. coli* cultures at 22.5°C, as shown in the following tabulation:

Temperatures (°C)	Relative death rates (%) at: 1 atm	400 atm	$R_{400}/R_{1\,atm}$
46.9	100	26.6	0.26
37.0	0.23	2.00	8.69
22.5	0.13	0.53	4.07

Stock spore suspensions of *B. subtilis* (about 80,000 viable spores per milliliter of buffered salt solution) were disinfected by holding them for about 1 hour at 93.6°C (Johnson and ZoBell, 1949a). When compressed to 600 atm, it took more than 4 hours to kill all of the spores at 93.6°C.

Increased pressures were found by R. A. C. Foster *et al.* (1949) to retard the thermal destruction of *E. coli* phages. Whereas about 98% of the infective centers of phage T1 were destroyed in nutrient broth at 66°C in 30 minutes at 1 atm, it took 58 minutes at 680 atm for a like percentage of infective centers to be destroyed at 66°C. Phages T2 and T5 likewise survived about twice as long at 66°C when compressed to 680 atm as at 1 atm. The thermal denaturation of tobacco mosaic virus was observed by Johnson *et al.* (1948a) to be retarded by pressures of 400 to 680 atm. At 68.8°C the virus protein was denatured. After 26 minutes at 72.5°C, 90% of the virus protein was denatured at 1 atm as compared with only 22% at 680 atm.

Whereas *Staphylococcus aureus* antitoxin was 85% denatured at 65°C in 30 minutes at 1 atm, at 680 atm a like percentage was denatured in 48 minutes (Johnson and Wright, 1946). The luminescent system of *Photobacterium phosphoreum*, whose normal optimum for growth is 21°C, was 83% destroyed in 5 minutes at 34°C at 1 atm, whereas at 330 atm only 18% of the system was destroyed in 5 minutes (Johnson and Eyring, 1948). At 55°C about the same amount (90%) of the thermostable *B. subtilis* α-amylase was inactivated in 28 days at 1 atm as in 42 days at 1000 atm (ZoBell and Hittle, 1969).

Probably the highest temperature at which microbial growth has been observed is 104°C, made possible by thermal destruction being counteracted by high pressure. ZoBell (1958) reported finding thermophilic sulfate-reducing bacteria, believed to be *Desulfovibrio* species, in cores from oil and sulfur wells at depths exceeding 12,000 feet where the *in situ* temperature was 60° to 105°C and the hydrostatic pressure was as high as 400 atm. Many cultures were grown for several transfers in the laboratory at 65°C at 1 atm and at 85°C at 400 to 600 atm. By gradually increasing the temperature and pressure, a few cultures were induced to grow at 95°C at 1000 atm. One culture was induced to grow at 1000 atm in an ethylene glycol bath at 104°C. Under these conditions of cultivation, sulfate was reduced to sulfide and growth was indicated by hundred- to thousandfold increases in the number of cells, as indicated by direct microscopic determinations. After a day or two at 104°C at 1000 atm, all of the bacteria succumbed.

D. Chemical Composition of Medium

Whereas the ability of bacteria to grow at increased hydrostatic pressure seems to be enhanced by the completeness of the medium (essential amino acids, vitamins, etc.), their lethal pressure points are generally lower in nonnutrient salt solutions than in rich nutrient media. By the same token, bacteria are generally less barotolerant in the early logarithmic phase of growth than during the stationary, dormant, or logarithmic death phase. These circumstances suggest that death near the threshold of pressure tolerance results from faulty metabolism or disordered syntheses. At much higher pressures, death probably results from the irreversible inactivation of essential enzymes or the denaturation of proteins. In both situations, the pressure death point of bacteria is influenced by a great variety of inorganic and organic substances as well as by pH and temperature, as has been pointed out by Johnson *et al.* (1942b, 1945, 1954).

1. *Some Effects of Ions*

Unpublished observations of the author and his students indicate that the adverse effects of increased hydrostatic pressure on the growth and survival of bacteria are accentuated by unfavorable osmotic pressures.

In either nutrient media or in mineral salts solutions, the tolerance of most barophobic bacteria for hypertonicity is lessened by increased hydrostatic pressures. This suggests that increased hydrostatic pressures interfere with the selective permeability of cell walls. The effects of various ions at increased hydrostatic pressures has received relatively little attention. The ionization of most weak electrolytes is increased by increased hydrostatic pressure (Gibson and Loeffler, 1941; Owen and Brinkley, 1941; Hamann, 1957, 1964; Distèche, 1962; Weale, 1967), but probably not enough to have much effect on the osmotic pressure. However, moderate pressures may have biologically significant effects on the hydrogen-ion concentration or pH of solutions containing weak electrolytes. For example, at 0°C, normal sea water having a pH of 8.10 at 1 atm has a pH value of 7.87 at 1100 atm (Distèche and Distèche, 1967; Park, 1966). The extent to which increased pressure changes the pH of solutions depends upon the electrolytes present and the temperature. According to Owen and Brinkley (1941), at 0°C the ionization constant of water is about three times higher at 1000 atm than at 1 atm. Ionization constants of water at 25°C and of some other substances in pure water at 25°C are higher at 1000 atm than at 1 atm by the following number of times:

$H_2O \rightarrow H^+ + CH^-$	2.36
$CH_3COOH \rightarrow H^+ + CH_3COO^-$	1.41
$H_2CO_3 \rightarrow H^+ + HCO_3^-$	3.2
$HCO_3^- \rightarrow H^+ + CO_3^{2-}$	2.65
$CaSO_4 \rightarrow Ca^{2+} + SO_4^{2-}$	5.80
$CaCO_3 \rightarrow Ca^{2+} + CO_3^{2-}$	8.10

More recently, Hamann (1963) experimentally determined the K_w for water at 1000 atm to be 2.14 as compared with a K_w value of 1.00 at 1 atm.

The maximum hydrostatic pressure at which *Vibrio marinus* grew has been shown by Palmer and Albright (1969) to be a function of the salinity of sea water medium, particularly its NaCl content. The effects of salinity on the pressure tolerance of *V. marinus* was markedly influenced by the temperature of incubation.

Hydrostatic pressures up to 680 atm retarded the thermal destruction of *E. coli* phages T1, T2, and T5 in 0.5% NaCl solution at 66°C (R. A. C. Foster *et al.*, 1949). The addition of 0.005 *M* $MgCl_2$ substantially decreased the rate of destruction or inactivation of these phages at increased pressure. Conversely, the addition of 0.005 *M* phosphate increased the rate of thermal destruction of these phages or decreased the counteracting effects of pressure.

In the absence of Ca^{++}, *B. subtilis* α-amylase was inactivated by a pressure of about 500 atm (Meyagawa, 1965). But in the presence of 0.01 *M* Ca^{++}, α-amylase was not appreciably inactivated by pressures of less than 3000 atm. Such pressures were found to favor the reactivation of inactivated α-amylase in the presence of calcium ion. Suzuki and Kitamura (1963) also observed that calcium ions markedly counteracted the pressure inactivation of *B. subtilis* α-amylase. The rate of pressure inactivation of this enzyme was found to depend strongly on the pH of the solution, α-amylase being inactivated much more rapidly by high pressure at pH 4.5 than at pH 8.0.

The pH of the solution at increased pressures influences biochemical reaction rates. For example, Johnson *et al.* (1945) observed the following relative intensities of *Photobacterium phosphoreum* luminescence at 17.5°C at 400 atm, the values being based on normal activity at pH 7.0 at 1 atm being 100:

pH:	4.63	5.10	6.92	8.04
Luminescence:	15	30	95	120

According to Johnson *et al.* (1945), the pH affects the denaturation of enzymes at high pressure, but the volume change accompanying ionization is largely responsible for the effects of pressure in increasing the energy of activation.

Whether any given biochemical reaction rate at increased pressure is increased by higher pH values (as in the example given above), decreased, or unchanged by pH depends upon the system or species, the temperature, the pH range, and other factors. Increased pressures up to 760 atm were found by Johnson *et al.* (1948b) to protect yeast invertase from urethan inhibition more at pH 7.0 than at pH 4.5, the normal optimum for activity at 1 atm in the absence of urethan.

2. *Effects of Pressure on Action of Narcotics*

Luminescence in buffered suspensions of *P. phosphoreum* was found by Johnson *et al.* (1942a) to be reduced about 50% by the addition of 0.78 *M* urethan or 0.05 *M* chloroform at 17° to 18°C. Inhibition by these two narcotics was completely abolished by pressures of 270 to 475 atm. Such pressures were also found to counteract inhibition of bacterial luminescence caused by ethanol, ethyl ether, and procaine. On the other hand, pressure had little effect on inhibition caused by the addition of sulfanilamide, *p*-aminobenzoic acid, chloral hydrate, or barbital.

Similarly, Johnson and Eyring (1948) reported that low concentrations of chloroform, ethyl ether, ethyl carbamate, phenyl carbamate, and novacaine, which normally inhibit *P. phosphoreum* luminescence, were appreciably less inhibitory at increased pressures up to about 400 atm between 17° and 18°C. Low cencentrations of *p*-aminobenzoic acid, sulfanilamide, and chloral hydrate were reported to be somewhat more inhibitory at such increased pressures than at 1 atm.

Whereas the addition of 0.6 *M* ethanol caused a 50% diminution in the intensity of *P. phosphoreum* luminescence at 1 atm at 17.5°C, at 400 atm in the presence of this or lower concentrations of ethanol as well as in its absence, the intensity of luminescence was about 10% less at 400 atm than at 1 atm. The inhibiting effects of higher concentrations of ethanol up to 1.5 *M* were significantly less at increased pressures up to 476 atm than at 1 atm. The inhibiting effects of ethanol on luminescence at increased pressure were greater at 29°C than at 17.5°C (Johnson *et al.*, 1945).

At 37°C, 0.009 *M* quinine caused a 40% decrease in the number of viable *E. coli* cells in nutrient medium within 3 hours at 1 atm. At increased pressures up to about 200 atm, the rate of disinfection of *E. coli* cultures by this concentration of quinine was found by Johnson and Lewin (1946b) to be considerably slower than at 1 atm; but higher pressures, i.e., 300 to 400 atm, accelerated the disinfection of *E. coli* cultures by quinine. Similar inhibiting and accelerating effects were observed at 29°C. At 22°C, however, increased pressure at all levels tested accelerated the disinfection of *E. coli* cultures by quinine.

Within the range of 16° to 45°C at the optimum pH 4.5, the activity of yeast invertase was inhibited by 0.5 *M* urethan somewhat less at increased pressures up to 680 atm than at 1 atm (Johnson *et al.*, 1948b). In the following tabulation are shown the relative amounts of sucrose hydrolyzed by invertase in 120 minutes at 35°C under the stated conditions, based on sucrose hydrolysis in the absence of urethan at 1 atm being 100:

Pressure (atm):	1	170	510
Controls (no urethan):	100	113	123
0.5 *M* urethan:	53	75	90
Control urethan ratio:	0.53	0.66	0.87

Similarly, increased pressure was found by R. A. C. Foster *et al.* (1949) to counteract the destructive action of urethan on *E. coli* phage T5. The relative number of T5 infective centers present in the filtrates after 30 minutes at 68°C at different pressures are shown in the following tabulations:

Pressure (atm):	1	135	270	71
Control (no urethan):	12	36	52	60
0.1 *M* urethan:	0.7	9	27	475
Control urethan ratio:	0.058	0.250	0.519	0.845

The disinfection of *B. subtilis* spores in phosphate-buffered NaCl solution at 93.6°C was found by Johnson and ZoBell (1949b) to be appreciably accelerated by the presence of urethan. The lethal effects of urethan were largely counteracted by a pressure of 600 atm, as illustrated by the data in the following tabulation showing the relative number of viable spores after 60 minutes exposure:

	Urethan			
Pressure	None	0.01 *M*	0.1 *M*	1.0 *M*
Control (1 atm)	100	70	20	0
600 atm	20,000	19,000	13,000	7,400

IV. Microbial Growth at Increased Pressure

A. Highest Pressure Permitting Growth

Microbial species differ greatly from each other in their ability to grow at increased pressure. The temperature of incubation, the chemical composition of the medium, gas tension, and other factors help to determine the pressure level which retards or arrests the growth of microorganisms. The growth of virtually all microbial species examined has been found to be retarded or completely inhibited by pressures 50 to 500 atm higher than the pressure of their normal habitat. This generalization applies to barophilic as well as to barophobic species both marine and terrestrial.

The highest pressure at which sustained reproduction of bacteria has been observed is 1400 atm (ZoBell, 1964), at which pressure the rate of reproduction was very slow and the cultures could be kept alive for only a few transfers. Significantly, only anaerobes have been cultivated at such high pressures, probably because of unresolved difficulties of supplying free oxygen in optimal concentrations to aerobes in closed systems. Although bacterial cells grown at lower pressures have been shown to catalyze certain biochemical reactions at pressures up to 1800 atm, e.g., nitrate reduction (ZoBell and Budge, 1965) and starch hydrolysis (ZoBell and Hittle, 1969), none of these bacteria reproduced at pressures higher than 1200 atm. Indeed, relatively few species have been cultivated at pressures higher than 800 atm, although large bacterial populations appear to be growing on the deep sea floor at pressures ranging from 900 to 1150 atm (ZoBell and Morita, 1956, 1957, 1959; ZoBell, 1968). Most of the barophilic bacteria which have been cultivated at such *in situ* pressures reproduced more rapidly at lower pressures.

Kriss (1963) has reported the cultivation of deep sea bacteria at no pressure higher than 850 atm. Deep sea mud and soil samples were not sterilized by being held at pressure up to 1680 atm for 2 days at 26°C, but under these conditions from 90 to 99% of the bacteria were killed. Kriss *et al.* (1958) isolated from deep sea samples numerous barotolerant bacteria which reproduced at 400 atm. Only a small percentage of these grew feebly at 800 atm. By incubating cultures in nutrient media at pressure increasing by increments of 50 atm with each successive transfer, Chumak *et al.* (1968) induced certain barotolerant bacteria to grow fairly well at pressures up to 800 atm.

ZoBell and Johnson (1949) tested 45 stock cultures of common bacteria and yeasts for their ability to reproduce at increased pressures. When incubated at 30°C, all except three reproduced at 300 atm, but not as rapidly as at 1 atm, and all except eight reproduced to some extent at 400 atm. Two thirds of them failed to reproduce at 500 atm. Only three of the 45 reproduced at 600 atm at 30°C, but 10 reproduced at 600 atm at 40°C.

All of the 63 species of marine bacteria examined by Oppenheimer and ZoBell (1952) grew better at 1 atm than at 200 atm or higher despite the fact that some were isolated from depths of a few thousand meters. Among the 28 that failed to grow at 400 atm, 11 were killed after 8 days of compression at this pressure and 23 of the 56 species that failed to grow at 600 atm were killed by this pressure.

Fischer (1967) demonstrated the occurrence of from 9×10^3 to 1×10^9 viable aerobic bacteria per gram of mud collected from the sea floor at depths of 1600 to 6400 m. Most of these bacteria grew readily at near 0°C at 1 atm but not at 400 to 600 atm. However, when incubated at 19°C, many grew at *in situ* pressures, i.e., 200 to 600 atm.

The majority of the 450 cultures collected from the floor of the Indian Ocean were found by Mishustina (1968) to develop at 400 atm. Both growth and spore formation were observed in the peloscope (Perfil'ev and Gabe, 1964) at 400 atm. Mishustina reported that while investigating the "high pressure tolerance of bacteria from muds of the Kuril-Kamchatka Deep," Chumak, Stupakova, Zheleznikova, and Kaptereva collected more than 900 cultures from depths of 2000 to 9500 m. About half of the cultures increased twentyfold or more in biomass when incubated in nutrient medium at 400 atm. Many of the cultures increased significantly in biomass at 700 to 800 atm, but not as much as at lower pressures. Barotolerant bacteria which grew at pressures up to 500 atm were found in the gut of many benthic mud-eating animals.

Six bacterial species studied by Boatman (1967) grew slowly at 400 atm at 21°C, but only one of the six grew at pressures as high as 280 atm at 4°C.

B. Obligate Barophiles

Barophilic bacteria which grow preferentially or only at increased pressures appear to be very rare. Out of hundreds of enrichment cultures from deep sea samples (7000 to 10,400 m) which grew at 700 to 1000 atm, only a few have yielded bacteria that could be kept growing at these pressures for more than two or three transfers. Cultures of sulfate-reducing bacteria recovered from the Weber Deep in the Indian Ocean (depth, 7250 m) were kept growing in nutrient media at 700 atm at 3° to 5°C for several years during which time they failed to grow at 1 atm (ZoBell and Morita, 1957). Similarly, subcultures of barophilic sulfate-reducing bacteria recovered from the Philippine Trench (depth about 10,000 m) grew for a few transfers at 1000 atm but not at 1 atm. Reproduction of these barophiles was retarded at 1200 atm and arrested at 1500 atm. Growth at 1400 atm was very meager.

According to Morita and Albright (1965), the obligately psychrophilic *Vibrio marinus* grew better during 24 hours at 100 to 200 atm than at 1 atm or 300 atm at 3°C. At 15°C, it grew better at 100, 200, and 300 atm than at 1 atm or 500 atm. At 800 atm, it was killed within 24 hours. Kriss *et al.* (1967) worked with a marine *Pseudomonas sp. 8113,* which formed more biomass in nutrient medium at 300 atm than at 1 atm. A larger percentage of the cells were filamentous and the filaments were longer at higher pressures up to 575 atm, above which neither growth nor reproduction occurred.

ZoBell and Budge (1965) examined 30 species of marine facultative anaerobes for their ability to grow in nutrient media at 25°C at different pressures, including eight species which grew at 1000 atm. Only four of the eight barophilic species grew better at 100 or 200 atm than at 1 atm. None grew as well at 400 atm or higher as at pressures lower than 400 atm. The average amount of biomass produced by the growth of all 30 species was only 91.2% as much at 100 atm as at 1 atm. As shown by the following tabulation, the four preferential barophiles produced only

6.2% more biomass at 100 atm than at 1 atm but only 29.3% as much at 1000 atm as at 1 atm:

Pressure (atm):	1	100	200	300	400	600	800	1000
Average biomass of 30 species:	100	91.2	81.4	65.3	34.7	16.9	8.6	6.4
22 barophobic species:	100	83.7	74.4	48.1	29.1	5.7	0	0
8 barophilic species:	100	98.3	90.5	84.6	65.3	47.5	32.5	24.1
4 preferential barophiles:	100	106.2	103.8	95.5	71.2	56.8	41.7	29.3

C. Growth Inhibition at Increased Pressure

Near the threshold of pressure tolerance, bacterial growth, as indicated by biomass formation, may not be retarded as much as reproduction, as indicated by the number of cells. This is because the cells of many species grow in size, particularly length, without dividing (see Section II, A). Growth, as indicated by protein synthesis by *E. coli* at 37°C, was found by Yayanos (1967) to occur at pressures up to 640 atm, whereas 490 atm was the highest pressure at which *E. coli* reproduced, i.e., increased in the number of colony-forming units.

Not only is the rate of bacterial reproduction retarded by increased pressure, moderately high pressures prolong the lag period. ZoBell and Cobet (1962) observed that whereas at 1 atm at 30°C the number of viable cells of *E. coli* doubled within an hour after small inocula in the logarithmic growth phase were transferred to nutrient media. Such doubling occurred only after several hours at moderate pressures. The average lag periods at different pressures are shown in the following tabulation:

Pressure (atm):	1	300	400	425	450	475	500	525
Lag period (hr):	ca. 1	1–2	2–3	7–9	24	78	174	—

Neither growth nor reproduction occurred at 525 atm, at which pressure all cells were killed within 180 hours. The prolongation of the lag period of mixed cultures of marine bacteria for from 3 to 11 days by high pressure has been reported by Kriss (1963).

The prolongation of the lag period, the retardation of growth, or the death of bacteria at high pressures may result from the leakage of metabolites through cell walls (Berger, 1959; Britten and McClure, 1962), the inactivation of essential enzyme systems (see Section V), depressed ionization of weak electrolytes, unfavorable molecular volume changes, or certain other causes. In certain cases, stifling conditions created by confining bacteria in closed systems for compression may be more injurious than high pressure per se. For example, in many experiments, either insufficient dissolved oxygen to satisfy the requirements of strict aerobes or the toxicity of hyperbaric oxygenation at increased pressures (ZoBell and Hittle, 1967) may have been responsible for the death or the failure of bacteria to grow normally. Another indirect effect of increased pressure is the accumulation of carbon dioxide, resulting in injurious concentrations of carbonic acid.

V. Biochemical Reactions at Increased Pressure

Depending largely on molecular volume changes, the rates of various biochemical reactions may either be retarded or accelerated by increased pressure. However, compression to 300–3000 atm at ordinary temperatures inhibits or arrests most biochemical reactions, the upper limits depending on the nature of the organism and the barotolerence of its enzyme systems and other essential proteins.

Some Reactions Affected by Pressure

1. *Fermentation Reactions*

Beer yeasts observed by Regnard (1884) failed to ferment sugars in 7 hours at 600 atm, although the yeasts were not killed in 1 hour by compression to 1000 atm. Certes (1884a) confirmed this observation. He attributed the inhibition of fermentation at increased pressure to unfavorable gas tension, especially carbon dioxide. The kinds of bacteria that developed and their decomposition products in vegetable infusions were found by Certes (1884b) to be quite different in controls incubated at atmospheric pressure and at 350 to 500 atm. After 42 days incubation at 350 to 500 atm, there was no evidence of putrefaction in sea water enriched with organic matter, although putrefaction took place readily in controls at 1 atm. Similarly, Regnard (1891) reported the occurrence of little or no putrefaction of proteinaceous material by sewage bacteria in 21 days at 650 atm. He also reported that milk did not sour in 12 days at 700 atm. He elaborated upon the inability of beer yeasts to ferment sugar at 600 atm. According to Certes and Cochin (1884), yeasts slowly fermented sugar at 300 to 400 atm, but relatively little alcohol production or cell growth occurred, reportedly due to the accumulation of carbonic acid in injurious concentrations. Juices pressed from yeast cells were shown by Büchner (1897) to catalyze the alcohol fermentation of sugar at pressures up to 400 to 500 atm. According to Chlopin and Tammann (1903), the growth and fermentation of sugars by yeasts were suppressed by pressures of only 300 atm, although some of the cells survived for a few hours in nutrient medium at 3000 atm.

Chumak (1959) worked with a marine bacteria designated Kriss No. 187 which utilized about twice as much glucose per cell at 550 atm as at 1 atm, although its growth was retarded by increased pressure. No gas was formed at 550 atm, but more acid was formed at increased pressure than at 1 atm. This organism was unable to grow or to ferment glucose at pressures higher than 750 atm at 28°C. Similarly, *Pseudomonas desmolyticum* was found by Chumak *et al.* (1964) to produce exceptionally large amounts of low molecular weight acids during the fermentation of glucose at increased pressures up to 500 atm. At this pressure an average of 140 mg of formic acid was formed per gram of glucose consumed as compared with 75 mg at 1 atm. However, *P. desmolyticum* utilized glucose only one-half as rapidly at 500 atm as at 1 atm at 28°C (Chumak and Blokhina, 1964). Only one-fifteenth as much CO_2 was produced by this barotolerant bacterium at 500 atm as at 1 atm.

The average amounts of CO_2 and other gases shown in the following tabulation were found by Koyama (1955) to result from the microbial decomposition of organic matter per 100 gm of lake mud during 9 days incubation at 35°C under anaerobic conditions at different pressures:

Pressure (atm):	1	100	200	360	450
CO_2 (ml):	20.2	19.0	20.6	21.1	24.5
CH_4 (ml):	7.89	6.17	6.35	6.37	4.95
H_2 (ml):	0.51	0.28	0.42	0.09	0.04
N_2 (ml):	6.95	4.25	5.96	5.90	7.56

With all mud samples tested the mixed microflora produced less methane and hydrogen at increased pressures, but there was no consistent difference in the production of nitrogen or CO_2 at different pressures.

2. *Bacterial Luminescence*

The classical studies of Johnson *et al.* (1942a,b, 1945, 1954) on the kinetics of bacterial bioluminescence treat the effects of hydrostatic pressure on the activity and stability of luciferin–luciferase systems under different conditions, including temperature, pH, osmotic pressure, and the chemical composition of the medium. Although luminescence was found to be accelerated by pressures of 50 to 500 atm at temperatures a few degrees above the optimum for bacterial growth, below the optimum temperature such pressures retarded luminescence (Brown *et al.*, 1942). According to Johnson (1957), luminescence caused by cell-free extracts as well as by living cells of *Photobacterium phosphoreum* is retarded by moderate pressures much more at 0° to 20°C than at 25° to 35°C. The thermal destruction of the luminescent system of *P. phosphoreum*, whose normal optimum temperature is 21°C, was shown by Johnson and Eyring (1948) to be retarded by moderate pressures up to 330 atm.

3. *Activity and Stability of Dehydrogenases*

Seeking an explanation of the mechanism whereby increased pressures inhibit microbial metabolism, Morita (1957b) investigated the inactivation of various dehydrogenases of the tricarboxylic acid (TCA) cycle. Compression of washed cell suspensions of *E. coli* at 27°C for brief periods of time resulted in a marked diminution of their formate, malate, and succinate dehydrogenase activity. The relative amounts of each substrate oxidized at 1 atm by whole cell dehydrogenase systems after being pressurized for 15 minutes at the stated pressure are shown in the following tabulation:

Pressure (atm):	1	200	600	1000
Formate oxidized:	100	97	81	21
Malate oxidized:	100	98	74	13
Succinate oxidized:	100	88	52	9

Although the dehydrogenase systems were inactivated by deep sea pressures, resulting in decreased activity, such compression appeared to have little or no effect on the absolute chemical reaction rates. Indeed, the rapid rates of inactivation of certain dehydrogenases during compression made it extremely difficult to assess the effects of increased pressure on the absolute reaction rate. For example, about two-thirds of the succinic dehydrogenase system of *E. coli* was found by Morita and ZoBell (1956) to be inactivated in 3 hours at 30°C when compressed to 600 atm. Inasmuch as the foregoing experiments were conducted under strictly anaerobic conditions, it is not likely that the dehydrogenase systems were destroyed by oxidation. Barron (1955) demonstrated a proportional relationship between the concentration of free oxygen at increased pressures and the oxidation of various biological systems containing the sulfhydryl group, including cysteine, glutathione, coenzyme A, and dehydrogenases. Haugaard (1946, 1955) has demonstrated the inactivation of various dehydrogenases and other enzymes containing sulfhydryl groups by hyperbaric oxygen (see Section VI).

Increased pressure was found by Hill and Morita (1964) to cause a decrease in the mitochondrial dehydrogenase activity of the fungus *Allomyces macrogynus* on α-ketoglutaric, succinic, oxalosuccinic, and isocitric acids. All four dehydrogenases were inactivated at 26.5°C by compression for 72 hours at 600 atm. The fungus was killed by such compression, leading the authors to attribute death to the interruption of the vital TCA energy yielding mechanism.

Various dehydrogenases of the same species differ in their barotolerance, and greater differences occur in the barotolerance of different microbial strains and species. An extreme example is a malic dehydrogenase from the thermophilic *Bacillus stearothermophilus* which was found by Morita and Haight (1962) to be active at 101°C and 1300 atm. Several different cultures of barophilic bacteria isolated from the Philippine Trench (ZoBell and Morita, 1959) were able to oxidize succinic acid while reducing methylene blue in the absence of free oxygen at pressures up to 1000 atm (ZoBell, 1964). Such barophiles have yet to be tested for the pressure tolerance of their dehydrogenases.

4. *Bacterial Phosphatases*

Morita and Howe (1957) noted considerable difference in the effects of deep sea pressures on the phosphatase activity of different species of marine bacteria. The quantities of phosphate, expressed as μM PO_4, shown in the following tabulation, were liberated from *p*-nitrophenyl phosphate by washed cell suspensions of various species in 1 hour at 30°C:

Pressure (atm):	1	200	600	1000
Bacillus borborokoites:	0.058	0.048	0.078	0.076
Flavobacterium marinotypicum:	2.320	3.240	2.600	1.760
Micrococcus aquivivus:	0.010	0.008	0.010	0.007
Pseudomonas azotogena:	0.110	0.093	0.124	0.520
Pseudomonas perfectomarinus:	0.040	0.118	0.044	0.054
Pseudomonas xanthochrus:	0.370	0.320	0.370	0.350

The confidence level in these values, based on quadruplicate determinations, was stated to be ±5%.

Berger (see ZoBell, 1960) made similar observations on the rate of hydrolysis of *p*-nitrophenyl phosphate by the *in vitro* phosphatases of *Bacillus abysseus* and *B. borborokoites*, using *p*-nitrophenol formation as the criterion of activity at 1, 300, 600, and 1000 atm and at 4°, 12°, 22°, and 30°C. Although the phosphatases of both organisms exhibited some activity at 4°C at 1000 atm, the rate of the reaction was retarded by compression and refrigeration almost as a straight-line function of the log of these two parameters.

Morita and Mathemeier (1964) investigated a partially purified pyrophosphatase from *B. stearothermophilus*, which catalyzed the hydrolysis of pyrophosphate with the liberation of phosphate at 90°C at pressures up to 1500 atm. Optimum activity occurred between 500 and 700 atm at 90°C. At 105°C the enzyme was active at pressures up to 1700 atm, this being another example of the counteracting effect of increased pressure on thermal denaturation. At 1 atm the enzyme catalyzed the hydrolysis of pyrophosphate at temperatures up to 95°C.

5. *Activity and Stability of Amylases*

The rate of starch hydrolysis by more than a hundred species of marine bacteria was found by ZoBell and Hittle (1969) to be roughly proportional to their rates of growth at different pressures. A few species of starch hydrolyzers were able to grow at pressures as high as 1200 atm, but the growth of the vast majority was arrested by pressures between 300 and 600 atm. However, whole cultures grown at 1 atm, filtrates therefrom, and washed cells were able to hydrolyze starch at pressures up to 1200 atm. Despite the slow inactivation of bacterial amylases at such pressures, the rate of starch hydrolysis was more rapid under certain conditions. The rate of starch hydrolysis catalyzed by a purified α-amylase from *Bacillus subtilis* was accelerated by deep sea pressure, particularly in the temperature range of 25° to 60°C. Such acceleration is predicted by theory since starch hydrolysis is a reaction that takes place with a volume decrease. According to Laidler (1955), the ΔV^* values for the hydrolysis of starch range from −22 to −27 cm^3/mole.

Whereas deep sea pressures promote the inactivation of α-amylase at temperatures lower than 35°C, this enzyme was more stable at 1000 atm than at 1 atm in the temperature range of 45° to 60°C (ZoBell and Hittle, 1969). Similarly, Suzuki and Kitamura (1963) observed that moderate pressures retarded the thermal inactivation of α-amylase.

Basset and Macheboeuf (1932) and Macheboeuf *et al.* (1933) are often misquoted as having shown that various amylases are stable at pressures of 10,000 to 15,000 atm. Actually, these workers reported that certain amylases retained some activity after being subjected to such pressures for 30 to 45 minutes at room temperature, although the activity was substantially reduced. Although more stable than most enzyme systems, amylases slowly lose their activity under most conditions. ZoBell and Hittle (1969) could detect no diminution in the activity of bacterial α-amylase after several hours of compression at 1000 atm at any temperature between 4° and 35°C, but after a few weeks the enzyme had lost more than half of its activity in calcium acetate phosphate buffer solution at pH 7.2. Inactivation occurred somewhat more rapidly at 1000 atm than at 1 atm. The relative (%) activity of α-amylase after different periods of storage in buffer solutions at different pressure and temperatures was as shown in the following tabulation:

Time (days):	0	7	14	28	42	56
4°C at 1 atm:	100	93.7	86.0	77.6	73.4	69.9
4°C at 1000 atm:	100	90.4	81.7	72.1	65.8	61.2
25°C at 1 atm:	100	82.3	73.4	58.3	48.6	42.4
25°C at 1000 atm:	100	80.1	70.8	54.5	43.6	37.2

When held at 55°C, α-amylase lost more than 99% of its activity in 35 days at 1 atm or in 49 days at 1000 atm. This is another example of the counteracting effect of increased pressure on the thermal inactivation of an enzyme. Interestingly, Meyagawa (1965) observed that in the absence of calcium ions α-amylase was slowly inactived by a pressure of about 500 atm, but in the presence of calcium ions, α-amylase was not rapidly inactivated by pressures of less than 3000 atm.

6. *Effects of Pressure on Other Enzymatic Reactions*

At 10° to 25°C, the rate of hydrolysis of phenylglycoside by the thermolabile glycosidase from a *Streptomyces* species was observed by Berger (1958) to be decreased by increased pressures ranging from 250 to 1000 atm. Such pressures were progressively less inhibitory with increasing temperatures between 30° and 50°C, ostensibly due to the counteracting effect of pressure on thermal denaturation. At 1000 atm the glycosidase lost little of its activity during an hour at 50°C, a temperature which at 1 atm resulted in complete denaturation of the enzyme in about 5 minutes.

Urease catalyzes the hydrolysis of urea with the formation of ammonia and carbon dioxide. The rate of this reaction is substantially retarded by deep sea pressures (ZoBell, 1964). The relative amounts of ammonia formed from urea by some bacterial ureases (in washed cell suspensions) and by a purified urease from the jack bean *Canavalia ensiformis* at different pressures under otherwise comparable conditions during 3 hours at 28°C are shown in the following tabulation:

Pressure (atm):	1	250	500	750	1000
Achromobacter stationis:	100	89.8	61.3	45.4	33.5
Micrococcus euryhalis:	100	87.4	63.4	49.3	48.7
Micrococcus sedimenteus:	100	91.8	64.7	46.3	34.2
Canavalia ensiformis:	100	66.2	35.2	29.3	19.3

New evidence indicates that the retarded rate of urea hydrolysis is largely due to the inactivation of ureases by compression. Most ureases are highly unstable under most conditions, and prolonged compression seems to promote their inactivation at room temperature.

Within the range of the optimum temperature (37° to 44°C) for the growth of *E. coli*, its aspartase activity, as indicated by the deamination of aspartic acid by washed cell suspensions, was found by Haight and Morita (1962) to decrease progressively with increasing pressures up to 1000 atm. At temperatures between 46° and 56°C, pressure up to 1000 atm protected the aspartase from thermal denaturation, provided the substrate was present. In the absence of the substrate, aspartase was rapidly inactivated by increased pressures at all temperatures tested.

The serine deaminase activity of *Vibrio marinus*, an obligate psychrophile which failed to grow at 25°C, was increased by pressures of 200 to 300 atm, according to Albright and Morita (1965). Whereas the deamination of L-serine was inhibited by temperatures higher than 20°C, the partially purified gelatinase of this organism was most active at 40°C (Weimer and Morita, 1968). Between 200 and 300 atm at 15°C, its gelatinase activity was suppressed by pressure.

At 30°C and pH 7.04, a yeast invertase studied by Eyring *et al.* (1946) hydrolyzed sucrose about 33% more rapidly at 476 atm than at 1 atm. A pressure of 476 atm resulted in a 58% increase in the rate of sucrose hydrolysis at pH 7.5 and a 38% increase at pH 1.5. No decrease in the rate of hydrolysis under the influence of pressure up to 680 atm was observed at temperatures up to 30°C, but higher temperatures and pressures tended to inactivate invertase, thereby resulting in decreased sucrose hydrolysis.

All 30 species of marine bacteria tested by ZoBell and Budge (1965) reduced nitrate at the highest pressure at which they were able to grow, the upper limit of growth ranging from 300 to 1200 atm. Washed cells (grown at 1 atm) of all 30 species reduced nitrate to nitrite at pressures as great as 1000 atm, at which pressure the rate was an average of only about one-fifth as fast as at 1 atm, owing to the gradual inactivation of the nitrate reductase system. There was some activity at 1400 atm at 10°C, but the nitrate-reducing enzyme system was irreversibly inactivated within 24 hours at 1800 atm at 10°C. The rate of enzyme inactivation was progressively less rapid at higher temperatures within the range of 21° to 40°C.

Gelatin liquefaction by actively growing cultures of *Bacterium candicans* was observed by Kriss *et al.* (1969b) to be much weaker at 500 atm than at 1 atm. The ability of this barophilic form to produce gelatinase was blocked at increased pressure, although it reproduced almost as rapidly at 500 atm as at 1 atm at 27°C. Filtrates from the cultures grown at 1 atm liquefied gelatin readily at 500 atm, thereby demonstrating that its gelatinase was not inactivated and that the reaction was not inhibited by increased pressure. The viscosity of the gelatin was increased by increased pressure, but this did not affect its liquefaction by gelatin.

7. *Pigment Production*

Compressing cultures now known as *Corynebacterium pseudotuberculosis, Sarcina rosea, Serratia marcescens,* and *Staphylococcus aureus* for a few hours to 2900 atm at 30°C resulted in marked diminution of pigment formation by surviving cells when incubated in nutrient medium at 1 atm (Chlopin and Tammann, 1903). Normal chromogenesis was regained by these cultures only after several days of cultivation under normal conditions.

The normally red *Serratia marinorubra* commonly produces color variants: red > orange > yellow > cream > white (Palmer, 1961). When cultivated at increased pressure, this frequency of color variants is reversed, there being very few red- or orange-colored cultures formed at 300 atm. Whereas broth cultures soon become pink or red under normal conditions of cultivation, cultures which developed more slowly at 300 or 400 atm were colorless or white. When incubated for a day or two at 1 atm, the decompressed cultures formed more biomass and gradually became pink or red. Decompressed cultures streaked on nutrient agar gave rise mainly to pink or red colonies under normal conditions of incubation, indicating that the temporary loss of chromogenesis was physiological and not genetic. Various stress conditions, particularly unfavorable oxygen tensions, suppress chromogenesis in bacteria. According to Williams *et al.* (1965), *Serratia marcescens* produces more red pigment (prodigiosin) at suboptimum temperatures because of the abnormal heat sensitivity of the enzyme which catalyzes the synthesis of the pigment.

VI. Effects of Free Oxygen at Increased Hydrostatic Pressure

Dating from Bert's report (1875) that at 15 to 44 atm, compressed air preserved raw meat and eggs for several days at room temperature, there have been numerous observations on the biological effects of hyperoxygenation. Many bacteria, yeasts, and molds in various foods were found by Bert (1878) to be killed or to have their growth retarded by hyperoxygenation. He attributed the adverse effects of compressed air or oxygen to the increased concentrations of oxygen and not to the increased pressure. In his review of the literature, Bean (1945) pointed out that many microbial species are adversely affected by increased concentrations of oxygen, the maximum tolerated by most aerobic organisms being between 10 and 50 atm of air or an oxygen partial pressure of between 2 and 10 atm, roughly equivalent to a dissolved oxygen concentration of 15 to 75 mg/liter at 25°C. Of course, it is common knowledge that strict anaerobes tolerate little or no oxygen and microaerophilic species grow best or only at reduced oxygen tensions, say 0.05 to 2 mg/liter.

Adverse Effects of Hyperbaric Oxygenation

In his historically significant observations on the effects of high pressure on bacteria, Roger (1895) recognized the adverse effects of increased partial pressures of air, oxygen, and carbon dioxide, but it remained for ZoBell and Hittle (1967) to demonstrate that the adverse effects of hyperbaric oxygen are augmented by increased hydrostatic pressure. They showed that cultures of several species of aerobic bacteria were sterilized within a day or two by compression to 100 atm in media having a dissolved oxygen content of 35 mg/liter, although the bacteria grew readily at 100 atm in similar media having an initial oxygen content of 7 mg/liter or at 1 atm in media having an initial oxygen content of 35 mg/liter. Hydrostatic pressures between 10 and 25 atm suppressed the growth or killed certain aerobic bacteria in nutrient media having an oxygen content of 35 mg/liter.

The toxicity of oxygen for *Streptococcus faecalis*, a facultative anaerobe, was found by Fenn and Marquis (1968) to be increased by hydrostatic pressure and also by increased pressures of various gases. Under microaerophilic conditions, the growth of *S. faecalis* was inhibited by 40 atm or less of xenon, nitrous oxide, argon, or nitrogen, in decreasing order of their toxicity. Helium appeared to be impotent at pressures up to 40 atm. About 90% of the cells in cultures were killed within 16 hours by exposure to 10 atm of oxygen.

Among the many mechanisms whereby hyperbaric oxygenation (also commonly called high pressure oxygen or HPO) is injurious to organisms, is by the destruction of oxidation–reduction systems containing the sulfhydryl group, notably dehydrogenases, oxidases, certain coenzymes, and glutathione. HPO at 7 atm inhibited the activity of several dehydrogenases examined by Haugaard (1946). The activity of pyruvic oxidase and ketoglutaric oxidase were also inhibited by HPO at 7 atm. The greatest inhibition occurred with enzymes containing essential —SH groups, leading Haugaard to believe that such groups were oxidized by HPO. Haugaard (1955) has documented the inactivation of numerous dehydrogenases and certain coenzymes by HPO at 2 to 7 atm.

Finding that the rate of oxidation of cysteine, glutathionine, coenzyme A, alcohol dehydrogenase, and certain other biological systems containing the —SH group was directly proportional to the oxygen pressure between 17 and 35 atm convinced Barron (1955) that the —SH group was involved. Similarly, a progressive decrease in succinic dehydrogenase activity with increasing HPO was observed by Hall and Sanders (1966).

The maximum resistance of *B. subtilis* and *E. coli* to HPO was observed by Gifford (1968) to occur immediately following cell division when catalase activity was greatest in synchronous cultures, leading to the belief that HPO suppressed catalase synthesis. HPO was found by Young (1968) to inhibit leucine uptake by cells of *Pseudomonas saccharophila,* although HPO appeared to have little direct effect on the incorporation of leucine into protein.

An abnormally high oxygen tension seemed to counteract cytochrome peroxidase formation by *Pseudomonas fluorescens,* according to Lenhoff and Kaplan (1953). At increased oxygen tensions, *P. fluorescens* produced less pigment and exhibited less peroxidase and catalase activity.

The formation of injurious peroxides or free oxygen radicals by high hydrostatic pressure from oxygen has been reported by Hedén and Malmborg (1961). Enns *et al.* (1965) have directed attention to the increased excitability of free oxygen at high pressure as another mechanism whereby HPO may be injurious.

VII. Microbial Life at Reduced Pressures

Microorganisms may be subjected to substantially reduced pressures in extraterrestrial environments or experimentally in the laboratory. Despite the reduced oxygen tension which results, a good many species of aerobic bacteria, yeasts, molds, and higher organisms can grow at pressures as low as 0.01 atm (7.6 Torr), roughly equivalent to the atmospheric pressure at an elevation of 31,000 m above sea level. According to Siegel *et al.* (1965), the vital activities of several varieties of insects appeared to be unimpaired for at least 3 days at 0.1 atm and some were not immobilized until the atmospheric pressure was reduced to < 0.01 atm.

At atmospheric pressures appreciably lower than 0.1 atm, oxygen tension becomes a limiting factor for strict aerobes. Anaerobic bacteria, however, have been cultivated in vessels evacuated to 0.1 Torr (or about

0.00013 atm). Sustained growth of most species is retarded by pressures lower than 1 Torr, presumably due to unfavorable partial pressures of volatile substances, including water.

A good many microbial species survive at ultrahigh vacuum. Portner *et al.* (1961) exposed *Bacillus subtilis, Aspergillus fumigatus,* and *Mycobacterium smegmatis* to pressures of 10^{-9} to 10^{-10} Torr for 5 days at ambient temperature. Appreciable kill occurred only with *M. smegmatis.* Spores of *Bacillus megaterium, B. stearothermophilus, B. subtilis, Clostridium sporogenes,* and *Aspergillus niger* were found by Davis *et al.* (1963) to be viable after 4 to 5 days exposure at 10^{-8} to 10^{-10} Torr. There was no significant difference in the survival of these five organisms at 25° and −190°C in vacuum. The survival of numerous *B. subtilis, A. niger,* and *Mycobacterium phlei* cells after 137 days at 25°C at 10^{-10} Torr has been reported by Geiger *et al.* (1965). According to Imshenetskii *et al.* (1964), there was no death of bacteria or fungi on filter paper within 3 days at −23° when *Bacillus mycoides, B. mesentericus, B. subtilis, B. simplex, Sarcina flava, Mycobacterium rubrum, Penicillium variablis, Aspergillus fumigatus,* and *A. flava* were exposed to a vacuum of 10^{-9} Torr. *Streptococcus faecalis* and *Staphylococcus aureus* were found by Silverman and Beecher (1967) to survive for 5 days at 10^{-9} Torr at temperatures lower than 30°C, but at 45° to 75°C all were killed.

VIII. Mutagenic Effects of High Pressures

Subjecting a conidial suspension of *Neurospora crassa* in the presence of nitrogen mustard to high pressure was observed by McElroy and de la Haba (1949) to depress the number of morphological mutants obtained. The reduction of the mutation rate was found to be directly proportional to the pressure, e.g., reductions of 10% at 135 atm, 30 to 35% at 340 atm, and 45 to 50% at 600 atm. On the other hand, increased pressure caused an increase in the number of biochemical mutants obtained with nitrogen mustard. The mutation rates of single colony isolates of *Serratia marinorubra* to streptomycin resistance were found by Palmer (1961) to be about three times higher at 300 atm than at 1 atm, except in the case of one strain in which the mutation rate was significantly lower at increased pressure.

The marine copepod *Tigriopus californicus* survived for an hour or two at 500 atm. Some specimens tolerated pressures up to 650 atm. When eggs of this copepod were subjected to a pressure of 500 atm for an hour or two, the development of males was found by Vacquier (1962) to be suppressed much more than females. The exact mechanism of pressure-induced sex conversion is not known, but Vacquier and Belser (1965) postulate that pressure may alter the genes responsible for sex determination.

In considering the various ways in which high pressure may affect the synthesis and properties of DNA, Hedén (1964) assumed that the double helix is preserved under pressure and that in such a structure the bases are partly protected from contact with the solvent: "In a single strand, they would not be similarly protected, which would explain the higher rate of breakdown." According to Hedén, high pressure opposes single-stranding in DNA at certain temperatures. Although they did not look for mutagenic effects, ZoBell and Cobet (1964) observed the suppression of DNA synthesis by *E. coli* at increased pressures.

A high percentage of mutants permanently lacking chlorophyll and with altered carotenoids resulted from subjecting photosynthetic cultures of *Euglena gracilis* for a few minutes to pressures of 500 to 1000 atm (Gross, 1965). Pressures up to 100 atm had no obvious effects on *E. gracilis* in 2 hours, but the cells were immobilized after 30 minutes at 500 atm. All cells grown in darkness were killed within 3 hours by compression to 1000 atm, but some cells of cultures grown in light tolerated 1000 atm for 3 hours.

ACKNOWLEDGMENT

By conventional criteria the contributions of Jean S. ZoBell to the experimental and field work as well as to the preparation of this and numerous other reports on barobiology during the last twenty years would be acknowledged by listing her as a coauthor. Doing so, however, would complicate bibliographies and reduce other coauthors to *et al.* status when quoted in the text. Both of us appreciate the ideas and encouragement we have received from Frank H. Johnson throughout the years. L. R. Berger, A. B. Cobet, K. M. Budge, D. W. Lear, R. Y. Morita, C. L. Oppenheimer, and F. E. Palmer have helped in various ways to make this report possible. Among the many who have aided in the development of high pressure apparatus are T. R. Folsom, G. W. Harvey, and J. M. Snodgrass. The work has been supported in part by the American Petroleum Institute (Research Project 43A), the Office of Naval Research (NR 103-020), and the National Science Foundation (GB-4240). This paper is a contribution from the Scripps Institution of Oceanography, University of California, San Diego.

REFERENCES

Albright, L. J. (1968). The effect of temperature and hydrostatic pressure on protein, ribonucleic acid and deoxyribonucleic acid synthesis by *Vibrio marinus,* an obligate psychrophile. Ph.D. Thesis, Oregon State University, Corvallis, Oregon.

Albright, L. J., and Morita, R. Y. (1965). Temperature-hydrostatic pressure effects on deamination of L-serine by *Vibrio marinus,* an obligate psychrophile. *Bacteriol. Proc.* **65**, 26.

Albright, L. J., and Morita, R. Y. (1968). Effect of hydrostatic pressure on synthesis of protein, ribonucleic acid, and deoxyribonucleic acid by the psychrophilic marine bacterium, *Vibrio marinus*. *Limnol. Oceanog.* **13**, 637–643.

Aronson, M. H. (1967). "Pressure Handbook." Rimbach Publ., Pittsburgh, Pennsylvania.

Barron, E. S. (1955). Oxidation of some oxidation-reduction systems by oxygen at high pressures. *Arch. Biochem. Biophys.* **59**, 502–510.

Basset, J., and Macheboeuf, M. A. (1932). Etude sur les effets biologiques des ultrapressions: Résistance des bactéries, des diastases et des toxins aux pressions très élevées. *Compt. Rend.* **195**, 1431–1433.

Basset, J., Nicolau, S., and Macheboeuf, M. A. (1935). L'action de l'ultrapression sur l'activité pathogène de quelques virus. *Compt. Rend.* **200**, 1882–1884.

Bean, J. W. (1945). Effects of oxygen at increased pressure. *Physiol. Rev.* **25**, 1–147.

Berger, L. R. (1958). Some effects of pressure on phenylglycosidase. *Biochim. Biophys. Acta* **30**, 522–529.

Berger, L. R. (1959). The effect of hydrostatic pressure on cell wall formation. *Bacteriol. Proc.* **59**, 129.

Berrah, G., and Konetzka, W. A. (1962). Selective and reversible inhibition of the synthesis of bacterial deoxyribonucleic acid by phenethyl alcohol. *J. Bacteriol.* **83**, 738–744.

Bert, P. (1875). Influence de l'air comprimé sur les fermentations. *Compt. Rend.* **80**, 1579–1582.

Bert, P. (1878). "La Pression Barométrique." Masson, Paris; also English translation by M. A. and F. A. Hitchcok, "Bert's Barometric Pressure: Researches in Experimental Physiology." College Book Co., Columbus, Ohio, 1943.

Boatman, E. S. (1967). The effects of hydrostatic pressure on the structure of marine bacteria. Ph.D. Dissertation, University of Washington, Seattle, Washington.

Bradley, R. S. (1963). "High Pressure Physics and Chemistry," Vols. 1 and 2. Academic Press, New York.

Britten, R. J., and McClure, F. T. (1962). The amino acid pool in *Escherichia coli*. *Bacteriol. Rev.* **26**, 292–335.

Brock, T. D. (1967). Life at high temperatures. *Science* **158**, 1012–1019.

Brown, D. E., Johnson, F. H., and Marsland, D. A. (1942). The pressure-temperature relations of bacterial luminescence. *J. Cellular Comp. Physiol.* **20**, 151–168.

Büchner, E. (1897). Alkoholische Gärung ohne Hefezellen. *Ber. Deut. Chem. Ges.* **30**, 117–124.

Byrne, J., and Marsland, D. (1965). Pressure-temperature effects on the form-stability and movement of *Euglena gracilis var. Z*. *J. Cellular Comp. Physiol.* **65**, 277–284.

Certes, A. (1884a). Note relative à l'action des hautes pressions sur la vitalité des micro-organismes d'eau douce et d'eau de mer. *Compt. Rend. Soc. Biol.* **36**, 220–222.

Certes, A. (1884b). De l'action des hautes pressions sur les phénomènes de la putréfaction et sur la vitalité des micro-organismes d'eau douce et d'eau de mer *Compt. Rend.* **99**, 385–388.

Certes, A., and Cochin, D. (1884). Action des hautes pressions sur la vitalité de la levure et sur les phénomènes de la fermentation. *Compt. Rend. Soc. Biol.* **36**, 639–640.

Chlopin, G. W., and Tammann, G. (1903). Über den Einfluss hoher Drucke auf Mikroorganismen. *Z. Hyg. Infektionskrankh.* **45**, 171–204.

Chumak, M. D. (1959). The effect of high pressure on the rate of glucose utilization by barotolerant bacteria. *Dokl.—Biol. Sci. Sect.* (*English Transl.*) **126**, 524–526.

Chumak, M. D., and Blokhina, T. P. (1964). Influence of high pressure on the accumulation of organic acids during glucose fermentation of barotolerant bacteria. *Microbiology* (*USSR*) (*English Transl.*) **33**, 200–204.

Chumak, M. D., Tarasova, N. V., and Blokhina, T. P. (1964). Qualitative composition of organic acids formed in the fermentation of glucose by barotolerant bacteria. *Microbiology* (*USSR*) (*English Transl.*) **33**, 510–513.

Chumak, M. D., Blokhina, T. P., and Kriss, A. E. (1968). The effect of prolonged cultivation of barotolerant bacteria under high pressure on their ability to grow in these conditions. (*Dokl.—Biol. Sci. Sect.* (*English Transl.*) **181**, 462–464.

Clouston, J. G., and Wills, P. A. (1969). Initiation of germination and inactivation *of Bacillus pumilus* spores by hydrostatic pressure. *J. Bacteriol.* **97**, 684–690.

Comings, E. W. (1956). "High Pressure Technology." McGraw-Hill, New York.

Davis, N. S., Silverman, G. J., and Keller, W. H. (1963). Combined effects of ultrahigh vacuum and temperature on the viability of some spores and soil organisms. *Appl. Microbiol.* **11**, 202–210.

Distèche, A. (1962). Electrochemical measurements at high pressure. *J. Electrochem. Soc.* **109**, 1084–1092.

Distèche, A., and Distèche, S. (1967). The effect of pressure on the dissociation of carbonic acid from measurements with buffered glass electrode cells. *J. Electrochem. Soc.* **114**, 330–340.

Duguid, J. P., and Wilkinson, J. F. (1961). Environmentally induced changes in bacterial morphology. *In* "Microbial Reaction to Environment" (G. G. Meynell and H. Gooder, eds.), pp. 69–99. Cambridge Univ. Press, London and New York.

Enns, T., Scholander, P. F., and Bradstreet, E. D. (1965). Effect of hydrostatic pressure on gases dissolved in water. *J. Phys. Chem.* **69**, 389–391.

Eyring, H., Johnson, F. H., and Gensler, R. L. (1946). Pressure and reactivity of proteins, with particular reference to invertase. *J. Phys. Chem.* **50**, 453–464.

Fenn, W. O., and Marquis, R. E. (1968). Growth of *Streptococcus faecalis* under high hydrostatic pressure and high partial pressure of inert gases. *J. Gen. Physiol.* **52**, 810–824.

Fischer, E. C. (1967). "Deep Ocean Microbiological Studies," Proj. 9400–72, Progr. Rept. 1. U.S. Naval Appl. Sci. Lab.

Fisher, R. L., and Hess, H. H. (1963). Trenches. *In* "The Sea" (M. N. Hill, ed.), Vol. 3, pp. 411–436. Wiley (Interscience), New York.

Foster, J. W., Cowan, R. M., and Maag, T. A. (1962). Rupture of bacteria by explosive decompression. *J. Bacteriol.* **83**, 330–334.

Foster, R. A. C., Johnson, F. H., and Miller, V. (1949). The influence of hydrostatic pressure and urethane on the thermal inactivation of bacteriophage. *J. Gen. Physiol.* **33**, 1–16.

Fraser, D. (1951). Bursting bacteria by release of gas pressure. *Nature* **167**, 33–34.

French, C. S., and Milner, H. W. (1955). Disintegration of bacteria and small particles by high-pressure extrusion. *Methods Enzymol.* **1**, 64–67.

Geiger, P. J., Jaffe, L. D., and Mamikunian, G. (1965). Biological contamination of planets. *In* "Current Aspects of Exobiology" (G. Mamikunian and M. H. Briggs, ed.), pp. 283–322. Calif. Inst. Technol., Pasadena, California.

Gibson, R. E., and Loeffler, O. H. (1941). The effect of pressure on acidity in water solutions. *Trans. Am. Geophys. Union* **22**, 503.

Gifford, G. D. (1968). Toxicity of hyperbaric oxygen to bacteria in relation to the cell cycle and catalase synthesis. *J. Gen. Microbiol.* **52**, 375–379.

Gross, J. A. (1965). Pressure-induced color mutation of *Euglena gracilis*. *Science* **147**, 741–742.

Haight, R. D., and Morita, R. Y. (1962). Interaction between the paramenters of hydrostatic pressure and temperature on aspartase of *Escherichia coli*. *J. Bacteriol.* **83**, 112–120.

Hall, I. H., and Sanders, A. P. (1966). Effects of hyperbaric oxygenation on metabolism. III. Succinic dehydrogenase, acid phosphatase, catherpsin and soluble nitrogen. *Proc. Soc. Exptl. Biol. Med.* **121**, 1203–1206.

Hamann, S. D. (1957). "Physico-Chemical Effects of Pressure." Academic Press, New York.

Hamann, S. D. (1963). The ionization of water at high pressure. *J. Phys. Chem.* **67**, 2233–2235.

Hamann, S. D. (1964). High pressure chemistry. *Ann. Rev. Phys. Chem.* **15**, 349–370.

Haugaard, N. (1946). Oxygen poisoning. XI. The relation between inactivation of enzymes by oxygen and essential sulfhydryl groups. *J. Biol. Chem.* **164**, 265–270.

Haugaard, N. (1955). Effect of high oxygen tension upon enzymes. *Natl. Acad. Sci.—Natl. Res. Council, Publ.* **377**, 8–12.

Hedén, C.-G. (1964). Effects of hydrostatic pressure on microbial systems. *Bacteriol. Rev.* **28**, 14–29.

Hedén, C.-G., and Malmborg, A. S. (1961). Aeration under pressure and the question of free radicals. *Sci. Rept. Ist Super. Sanita* **1**, 213–221.

Hill, E. P., and Morita, R. Y. (1964). Dehydrogenase activity under hydrostatic pressure by isolated mitochondria obtained from *Allomyces macrogynus*. *Limnol. Oceanog.* **9**, 243–248.

Hite, B. H., Giddings, N. J., and Weakley, C. E. (1914). The effect of pressure on certain micro-organisms encountered in the preservation of fruits and vegetables. *West Va., Univ. Agr. Expt. Sta., Bull.* **146**, 1–67.

Imshenetskii, A. A., Bogrov, N., and Lysenko, S. (1964). Resistance of microorganisms to high vacuum. *Dokl.—Biol. Sci. Sect.* (*English Transl.*) **154**, 91–92.

Johnson, F. H. (1957). The action of pressure and temperature. *In* "Microbial Ecology" (R. E. O. Williams and C. C. Spicer, eds.), pp. 134–167. Cambridge Univ. Press, London and New York.

Johnson, F. H., and Eyring, H. (1948). The fundamental action of pressure, temperature, and drugs on enzymes, as revealed by bacterial luminescence. *Ann. N.Y. Acad. Sci.* **49**, 376–396.

Johnson, F. H., and Lewin, I. (1946a). The disinfection of *E. coli* in relation to temperature, hydrostatic pressure and quinine. *J. Cellular Comp. Physiol.* **28**, 23–45.

Johnson, F. H., and Lewin, I. (1946b). The influence of pressure, temperature, and quinine on the rates of growth and disinfection of *E. coli* in the logarithmic growth phase. *J. Cellular Comp. Physiol.* **28**, 77–97.

Johnson, F. H., and Wright, G. C. (1946). Influence of hydrostatic pressure on the denaturation of *Staphylococcus* antitoxin at 65°C. *Proc. Natl. Acad. Sci. U.S.* **32**, 21–25.

Johnson, F. H., and ZoBell, C. E. (1949a). The retardation of thermal disinfection of *Bacillus subtilis* spores by hydrostatic pressure. *J. Bacteriol.* **57**, 353–358.

Johnson, F. H., and ZoBell, C. E. (1949b). The acceleration of spore disinfection by urethan and its retardation by hydrostatic pressure. *J. Bacteriol.* **57**, 359–362.

Johnson, F. H., Brown, D. E., and Marsland, D. A. (1942a). Pressure reversal of the action of certain narcotics. *J. Cellular Comp. Physiol.* **20**, 269–276.

Johnson, F. H., Eyring, H., and Williams, R. W. (1942b). The nature of enzyme inhibitions in bacterial luminescence: Sulfanilamide, urethane, temperature and pressure. *J. Cellular Comp. Physiol.* **20**, 247–268.

Johnson, F. H., Eyring, H., Steblay, R., Chaplin, H., Huber, C., and Gherardi, G. (1945). The nature and control of reactions in bioluminescence with special reference to the mechanism of reversible and irreversible inhibitions by hydrogen and hydroxyl ions, temperature, pressure, alcohol, urethane, and sulfanilamide in bacteria. *J. Gen. Physiol.* **28**, 463–537.

Johnson, F. H., Baylor, M. B., and Fraser, D. (1948a). The thermal denaturation of tobacco mosaic virus in relation to hydrostatic pressure. *Arch. Biochem.* **19**, 237–245.

Johnson, F. H., Kauzmann, W. J., and Gensler, R. L. (1948b). The urethan inhibition of invertase activity in relation to hydrostatic pressure. *Arch. Biochem.* **19**, 229–236.

Johnson, F. H., Eyring, H., and Polissar, M. J. (1954). "The Kinetic Basis of Molecular Biology." Wiley & Sons, Inc., New York. 874 pp.

Katchmann, B. S., Spoerl, E., and Smith, H. E. (1955). Effects of cell division inhibition on phosphorus metabolism of *Escherichia coli*. *Science* **121**, 97–98.

Kitching, J. A. (1957). Effects of high hydrostatic pressure on the activity of flagellates and ciliates. *J. Exptl. Biol.* **34**, 494–510.

Kitching. J. A., and Pease, D. C. (1939). The liquefaction of the tentacles of suctorian protozoa at high hydrostatic pressures. *J. Cellular Comp. Physiol.* **14**, 410–412.

Knight-Jones, E. W., and Morgan, E. (1966). Responses of marine animals to changes in hydrostatic pressure. *Oceanog. Marine Biol. Ann. Rev.* **4**, 267–299.

Koyama, T. (1955). Gaseous metabolism of lake muds and paddy soils. *J. Earth Sci., Nagoya Univ.* **3**, 65–76.

Kriss, A. E. (1963). "Marine Microbiology (Deep Sea)" (translated by J. M. Shewan and Z. Kabata). Oliver & Boyd, London and New York.

Kriss, A. E., Biryuzova, V. I., and Abyzov. S. S. (1958). Microorganisms propagating under high pressure (in Russian). *Izv. Akad. Nauk. SSSR, Ser. Biol.*, No. 6, 677–689.

Kriss, A. E., Chumak, M. D., Stupakova, T. P., and Kirikova, N. N. (1967). Glucose metabolism in barotolerant bacteria cultivated under high hydrostatic pressure. *Microbiology (USSR) (English Transl.)* **36**, 41–46.

Kriss, A. E., Mitskevich, I. N., and Cherni, N. E. (1969a). Changes in ultrastructure and chemical composition of bacterial cells under the influence of high hydrostatic pressure. *Microbiology (USSR) (English Transl.)* **38**, 108–113.

Kriss, A. E., Mishustina, I. E., and Zhelezikova, V. A. (1969b). Proteolytic activity of a barotolerant strain of *Bacterium candicans* var. I growing at high pressure (in Russian). *Microbiologiya* **38**, 233–237.

Laidler, K. J. (1955). Some kinetics and mechanistic aspects of hydrolytic enzyme action. *Discussions Faraday Soc.* **20**, 83–96.

Landau, J. V. (1966). Protein and nucleic acid synthesis in *Escherichia coli*: Pressure and temperature effects. *Science* **153**, 1273–1274.

Landau, J. V. (1967). Induction, transcription and translation in *Escherichia coli:* A hydrostatic pressure study. *Biochim. Biophys. Acta* **149**, 506–512.

Landau, J. V., and Thibodeau, L. (1962). The micromorphology of *Amoeba proteus* during pressure-induced changes in the sol–gel cycle. *Exptl. Cell Res.* **27**, 591–594.

Larson, W. P., Hartzell, T. B., and Diehl, H. S. (1918). The effect of high pressure on bacteria. *J. Infect. Diseases* **22**, 271–279.

Lenhoff, H. M., and Kaplan, N. O. (1953). A cytochrome peroxidase from *Pseudomonas fluorescens. Nature* **172**, 730–731.

Luyet, B. (1937a). Sur le mécanisme de la mort cellulaire par les hautes pressions; l'intensité et la durée des pressions léthales pour le levure. *Compt. Rend.* **204**, 1214–1215.

Luyet, B. (1937b). Sur le mécanisme de la mort cellulaire par les hautes pressions; modifications cytologiques accompagnant la mort chez la levure. *Compt. Rend.* **204**, 1506–1508.

McElroy, W. D., and de la Haba, G. (1949). Effect of pressure on induction of mutation by nitrogen mustard. *Science* **110**, 640–642.

Macheboeuf, M., Basset, J., and Levy, G. (1933). Influence des pressions très élevées sur les diastases. *Ann. Physiol.* **9**, 713–722.

Marsland, D. A. (1950). The mechanisms of cell division; temperature-pressure experiments on the cleaving eggs of *Arbacia punctulata. J. Cellular Comp. Physiol.* 36, 205–227.

Marsland, D. A., and Brown, D. E. S. (1936). Amoeboid movement at high hydrostatic pressure. *J. Cellular Comp. Physiol.* **8**, 167–178.

Meyagawa, K. (1965). Reactivation of pressure-inactivated α-amylase of *Bacillus subtilis* under moderate pressure. *Arch. Biochem. Biophys.* **110**, 432–437.

Milner, H. W., Lawrence, N. S., and French, C. S. (1950). Colloidal dispersion of chloroplast material. *Science* **111**, 633–634.

Mishustina, I. E. (1968). Meeting of the geological and water microbiology section of the Moscow Branch of the All-Union Microbiological Society. *Microbiology (USSR) (English Transl.)* **37**, 806–807.

Morita, R. Y. (1957a). Ammonia production from various substrates by previously pressurized cells of *Escherichia coli. J. Bacteriol.* **74**, 231–233.

Morita, R. Y. (1957b). Effect of hydrostatic pressure on succinic, formic, and malic dehydrogenases in *Escherichia coli. J. Bacteriol.* **74**, 251–255.

Morita, R. Y. (1965). The physical environment for fungal growth. 2. Hydrostatic pressure. *In* "The Fungi" (G. C. Ainsworth and A. S. Sussman, eds.), Vol. 1, pp. 551–557. Academic Press, New York.

Morita, R. Y. (1967). Effects of hydrostatic pressure on marine microorganisms. *Oceanog. Marine Biol. Ann. Rev.* **5**, 187–203.

Morita, R. Y. (1969a). Applications of hydrostatic pressure to microbial cultures. *In* "Methods in Microbiology" (D. W. Ribbons and J. R. Norris, eds). Academic Press, New York (in press).

Morita, R. Y. (1969b). Pressure effects on bacteria, fungi and blue green algae. *In* "Marine Biology—Environmental Factors" (O. Kinne, ed.). Wiley (Interscience), New York (in press).

Morita, R. Y., and Albright, L. J. (1965). Cell yield of *Vibrio marinus*, an obligate psychrophile, at low temperature. *Can. J. Microbiol.* **11**, 221–227.

Morita, R. Y., and Haight, R. D. (1962). Malic dehydrogenase activity at 101 C under hydrostatic pressure. *J. Bacteriol.* **83**, 1341–1346.

Morita, R. Y., and Howe, R. A. (1957). Phosphatase activity by marine bacteria under hydrostatic pressure. *Deep-Sea Res.* **4**, 254–258.

Morita, R. Y., and Mathemeier, P. F. (1964). Temperature-hydrostatic pressure studies on partially purified inorganic pyrophosphatase activity. *J. Bacteriol.* **88**, 1667–1671.

Morita, R. Y., and ZoBell, C. E. (1956). Effect of hydrostatic pressure on the succinic dehydrogenase system in *Escherichia coli. J. Bacteriol.* **71**, 668–672.

Oppenheimer, C. H., and ZoBell, C. E. (1952). The growth and viability of sixty-three species of marine bacteria as influenced by hydrostatic pressure. *J. Marine Res.* **11**, 10–18.

Owen, B. B., and Brinkley, S. R. (1941). Calculation of the effect of pressure upon ionic equilibria in pure water and in salt solutions. *Chem. Rev.* **29**, 461–473.

Palmer, D. S., and Albright, L. J. (1969). Salinity effects upon maximum hydrostatic pressure for growth of the obligate psychrophile *Vibrio marinus.* (in press).

Palmer, F. E. (1961). The effect of moderate hydrostatic pressures on the mutation rate of *Serratia marinorubra* to streptomycin resistance. M. S. Thesis, University of California, La Jolla, California.

Park, K. (1966). Deep-sea pH. *Science* **154**, 1540–1541.

Perfil'ev, B. V., and Gabe, D. R. (1964). Study by the method of microbial landscape of bacteria which accumulate manganese and iron in bottom sediments (in Russian). *In* "The Role of Microorganisms in the Formation of Iron-manganese Lake Ores" (M. S. Gurevich, ed.), pp. 16–53. Izd. "Nauka," Moscow-Leningrad.

Persidsky, M. D., and Leef, J. (1969). Cryopreservation under ultrahyperbaric conditions. *Abst. Soc. Cryobiol.* 6th Meeting, C-48.

Pollard, E. C., and Weller, P. K. (1966). The effect of hydrostatic pressure on the synthetic processes in bacteria. *Biochim. Biophys. Acta* **112**, 573–580.

Portner, D. N., Spiner, D. R., Hoffman, R. K., and Phillips, C. R. (1961). Effect of ultrahigh vacuum on viability of microorganisms. *Science* **134**, 2047.

Rautenshtein, Ya. I., and Muradov, M. (1966). Effect of high hydrostatic pressure on some lysogenic strains of actinomycetes and on their moderate phages. *Microbiology (USSR) (English Transl.)* **35**, 571–576.

Regnard, P. (1884). Recherches expérimentales sur l'influence des très hautes pressions sur les organismes vivants. *Compt. Rend.* **98**, 745–747.

Regnard, P. (1891). "Recherches expérimentales sur les conditions physiques de la vie dans les eaux." Masson, Paris.

Roger, H. (1895). Action des hautes pressions sur quelques bacteries. *Arch. Physiol. Normale Pathol.* **7**, 12–17.

Rutberg, L. (1964). On the effects of high hydrostatic pressure on bacteria and bacteriophage. *Acta Pathol. Microbiol. Scand.* **61**, 91–97.

Seki, H., and Robinson, D. G. (1969). Effect of decompression on activity of microorganisms in seawater. *Intern. Rev. Ges. Hydrobiol.* **54** (in press).

Siegel, S. M., Renwick, G., Daly, O., Giumarro, C., Davis, G., and Halper, L. (1965). The survival capabilities and the performance of earth organisms in simulated extraterrestrial environments. *In* "Current Aspects of Exobiology" (G. Mamikunian and M. H. Briggs, eds.), pp. 119–178. Calif. Inst. Technol., Pasadena, California.

Silverman, G. J., and Beecher, N. (1967). Survival of cocci after exposure to ultra-high vacuum at different temperatures. *Appl. Microbiol.* **15**, 665–667.

Silverman, G. J., and Davis, N. S. (1964). Exposure of microorganisms to simulated extraterrestrial space ecology. *Life Sci. Space Res.* **3**, 372–384.

Simpson, K. L., Wilson, A. W., Burton, E., Nakayama, T. O. M., and Chichester, C. O. (1963). Modified French press for the disruption of microorganisms. *J. Bacteriol.* **86**, 1126–1127.

Suzuki, K., and Kitamura, K. (1963). Inactivation of enzyme under high pressure. Studies on the kinetics of inactivation of α-amylase of *Bacillus subtilis* under high pressure. *J. Biochem. (Tokyo)* **54**, 214–219.

Timson, W. J., and Short, A. J. (1965). Resistance of microorganisms to hydrostatic pressure. *Biotechnol. Bioeng.* **7**, 139–159.

Vacquier, V. (1962). Hydrostatic pressure has a selective effect on the copepod *Tigriopus*. *Science* **135**, 724–726.

Vacquier, V. D., and Belser, W. L. (1965). Sex conversion induced by hydrostatic pressure in the marine copepod *Tigriopus californicus*. *Science* **150**, 1619–1621.

Weale, K. E. (1967). "Chemical Reactions at High Pressures." Spon, London.

Weimer, M. S., and Morita, R. Y. (1968). Effect of hydrostatic pressure and temperature on gelatinase produced by an obligately psychrophilic marine vibrio. *Bacteriol. Proc.* **68**, 34.

Williams, R. P., Goldschmidt, M. E., and Gott, C. L. (1965). Inhibition by temperature of the terminal step in biosynthesis of prodigiosin. *Biochem. Biophys. Res. Commun.* **19**, 177–181.

Wolff, T. (1960). The hadal community, an introduction. *Deep-Sea Res.* **6**, 95–124.

Yayanos, A. A. (1967). A study of the effects of hydrostatic pressure on macromolecular synthesis and thiamineless death in *Escherichia coli*, and the compression of some solutions of biological molecules. Ph.D. Dissertation, Pennsylvania State University, University Park, Pennsylvania.

Young, H. L. (1968). Uptake and incorporation of exogenous leucine in bacterial cells under high oxygen tension. *Nature* **219**, 1068–1069.

Zamenhof, S., de Giovanni, R., and Rich, K. (1956). *Escherichia coli* containing unnatural pyrimidines in its deoxyribonucleic acid. *J. Bacteriol.* **71**, 60–69.

Zimmerman, A. M. (1963). Incorporation of ^{3}H-thymidine in the eggs of *Arbacia punctulata*—A pressure study. *Exptl. Cell Res.* **31**, 39–51.

Zimmerman, A. M., and Marsland, D. (1964). Cell division: Effects of pressure on the mitotic mechanisms of marine eggs (*Arbacia punctulata*). *Exptl. Cell Res.* **35**, 293–302.

Zimmerman, A. M., and Silberman, L. (1965). Cell division: The effects of hydrostatic pressure on the cleavage schedule in *Arbacia punctulata*. *Exptl. Cell. Res.* **38**, 454–464.

Zimmerman, A. M., and Silberman, L. (1967). Studies on incorporation of ^{3}H-thymidine in *Arbacia* eggs under hydrostatic pressure. *Exptl. Cell Res.* **46**, 469–476.

ZoBell, C. E. (1958). Ecology of sulfate reducing bacteria. *Producers Monthly* **22**, 12–29.

ZoBell, C. E. (1959). Thermal changes accompanying the compression of aqueous solutions to deep-sea conditions. *Limnol. Oceanog.* **4**, 463–471.

ZoBell, C. E. (1960). "Physiological Activities of Marine Bacteria with Special Reference of Increased Pressure," Progr. Rep., pp. 25–26. Office of Naval Res., Washington, D.C.

ZoBell, C. E. (1964) Hydrostatic pressure as a factor affecting the activities of marine microbes. In "Recent Researches in the Fields of Hydrosphere, Atmosphere and Nuclear Geochemistry" (Y. Miyake and T. Koyama, eds.), pp. 83–116. Maruzen Co., Tokyo

ZoBell, C. E. (1968). Bacterial life in the deep sea. *Bull. Misaki Marine Biol. Inst., Kyoto Univ.* **12**, 77–96.

ZoBell, C. E., and Budge, K. M. (1965). Nitrate reduction by marine bacteria at increased hydrostatic pressures. *Limnol. Oceanog.* **10**, 207–214.

ZoBell, C. E, and Cobet, A. B. (1962). Growth, reproduction, and death rates of *Escherichia coli* at increased hydrostatic pressures. *J. Bacteriol.* **84**, 1228–1236.

ZoBell, C. E., and Cobet, A. B. (1964). Filament formation by *Escherichia coli* at increased hydrostatic pressures. *J. Bacteriol.* **87**, 710–719.

ZoBell, C. E., and Hittle, L. L. (1967). Some effects of hyperbaric oxygenation on bacteria at increased hydrostatic pressures. *Can. J. Microbiol.* **13**, 1311–1319.

ZoBell, C. E., and Hittle, L. L. (1969). Deep-sea pressure effects on starch hydrolysis by marine bacteria. *J. Oceanog. Soc. Japan* **25**, 36–47.

ZoBell, C. E., and Johnson, F. H. (1949). The influence of hydrostatic pressure on the growth and viability of terrestrial and marine bacteria. *J. Bacteriol.* **57**, 179–189.

ZoBell, C. E., and Morita, R. Y. (1956). Bacteria in the deep sea. *In* "The Galathea Deep Sea Expedition" (A. F. Bruun, ed.), pp. 202–210. Macmillan, New York.

ZoBell, C. E., and Morita, R. Y. (1957). Barophilic bacteria in some deep sea sediments. *J. Bacteriol.* **73**, 563–568.

ZoBell, C. E., and Morita, R. Y. (1959). Deep-sea bacteria. *Galathea Report, Copenhagen* **1**, 139–154.

ZoBell, C. E., and Oppenheimer, C. H. (1950). Some effects of hydrostatic pressure on the multiplication and morphology of marine bacteria. *J. Bacteriol.* **60**, 771–781.

CHAPTER 5

JAPANESE STUDIES ON HYDROSTATIC PRESSURE

Tetuhide H. Murakami

I. Introduction

The other chapters in this monograph are concerned mainly with reviewing broad areas of work that relates to the effects of pressure on biological activity. The aims of this chapter are somewhat different. The following account,[1] although not complete, purposes to report some interesting findings of high pressure studies conducted by Japanese investigators over the past two decades.

The areas discussed will include the effects of pressure on (a) cell division and cell proliferation, (b) nerve and muscle, (c) permeability of

[1] *Editor's note*: This survey of Japanese literature is intended to serve as an introduction to extensive Japanese high pressure research. The vitality and diversity of the Japanese workers is shown in this report. Most of the work in this review is published in the Japanese language. Since no monograph on high pressure research would be complete without recognition of the Japanese contribution, it is hoped that this review will reveal to the reader pertinent Japanese literature on high pressure.

the plasma membrane, (d) oxygen consumption, and (e) protein denaturation and enzyme kinetics.

II. Pressure Effects on Cell Division and Cell Proliferation

Murakami (1960) has reported on the effects of pressure on cell division in fertilized sea urchin eggs (*Temnopleurus toreumaticus* and *Hemincentrotus pulcherrimus*). Fertilized eggs placed under pressures of 4300 psi or greater immediately following fertilization did not form a mitotic apparatus (spindle–aster complex), and cleavage in these cells was prevented. Cells subjected to pressure treatment at the time of incipient furrow formation exhibited a recession of furrow formation; however, at later stages (eggs with well-formed deep furrows), recession of the furrows was not observed. At pressures below 4300 psi, the rate of cleavage was delayed and even at low magnitudes of pressure (1400 psi), there was a slight delay. Murakami (1960) was unable to stabilize the mitotic apparatus to the disorganizing effects of pressure with the use of chemical agents such as H_2O_2, $KMnO_4$, and $KCrO_4$. *Temnopleurus* eggs, which were compressed (4300 psi) at prophase for 10 minutes, divided into four cells—without passing through the two-cell stage—20 minutes after decompression. Although this cleavage pattern was common in *Temnopleurus* eggs, the phenomena was rarely observed in *Hemincentrotus* eggs. The DNA content of the compressed cells, as measured by microspectrophotometry prior to furrowing into four cells, indicated that these cells were polyploid.

In a later study, Murakami (1961) investigated RNA and DNA synthesis during the mitotic cycle in *Temnopleurus* eggs and reported a small RNA peak at metaphase and telophase. Pressures of 4300 psi applied at metaphase for 10 or 30 minutes resulted in an increase in total cellular RNA. Following decompression (10–20 minutes), the RNA content of the cell increased still further with the appearance of multipolar mitosis. Pressure treatment (10 minutes) at telophase also resulted in an increase in total RNA, but following decompression there was no further increase in RNA and multipolar mitosis was absent. In *Temnopleurus* eggs, DNA synthesis occurs between 15 and 25 minutes after fertilization, and division is completed at 40 minutes. When pressure (4300 psi) was initiated at prophase for a duration of 10 minutes, the cells exhibited a rapid synthesis of DNA 10–20 minutes after decompression, often yielding DNA values suggesting octaploidy. The changes in RNA and DNA were closely correlated with multipolar mitosis induced by the pressure treatment.

Yasuda *et al.* (1960) also investigated the effects of hydrostatic pressure on cell division. These investigators observed that fertilized eggs divided directly into an eight-cell stage when pressure (4300 psi) was initiated 30 minutes after insemination for a duration of 1 hour. For further pressure studies on marine eggs, see reviews by Marsland and Zimmerman, Chapters 11 and 10, respectively.

III. Pressure Effects on Nerve and Muscle

Isolated muscle preparations (skeletal, cardiac, and smooth muscle) undergo contractions when subjected to pressures of 4300 psi or greater (Tanbara, 1952; Yasuda, 1959a,b,c; Miki, 1960a,b). Upon release of pressure, recovery is instantaneous. Inherent differences can be demonstrated in these different preparations on compression and decompression. Cardiac muscle, exposed to pressures of 4300–11,400 psi, undergo a twitch contraction on application and release of pressure. Smooth muscle, on the other hand, exhibits a twitch response only upon decompression, whereas with skeletal muscle, the twitch phenomena does not occur (Miki, 1960a).

Skeletal muscle exhibits a decrease in electrical threshold under hydrostatic pressure. Cardiac muscle subjected to relatively low pressures (1400 psi) displays an increased excitability; the rate of beating is also increased, and the refractory period is prolonged about 20%. However, the response to electrical stimuli disappears at pressures above 5700 psi (Yasuda, 1959a,b,c). Longitudinal muscle strips exhibit increased frequency and tonus while placed under relatively low pressures up to 1400 psi. When pressures of 4300–7100 psi are employed, the length of the muscle strip is increased, but shortens with time (Miki, 1960a). The lengthening and subsequent shortening under pressure appear to be of myogenic nature.

Since pressure has a marked effect on the electrical activity of both nerve and muscle, it is difficult to record effects on the action potential from the surface of both nerve and muscle tissues under high pressures. These facts support the theory that there is a simultaneous depolarization of membrane potential on the whole surface of nerves and muscles (Hayasi, 1961a,b). For further references on the electrical activity of muscles and nerves, see Murakami and Zimmerman, Chapter 6, Section I.

IV. Pressure Effects on Permeability

The effect of pressure on the permeability of the plasma membrane of onion epidermal cells (*Allium cepa*) was investigated by Murakami

(1963). Using varying concentrations of electrolytes and nonelectrolytes, he investigated the rate of plasmolysis and deplasmolysis during varying pressures and temperatures. Gelatin blocks were employed as models, and volume changes were measured in various test solutions. The volume of the gelatin blocks increased markedly above that of the controls at 7100 psi in solutions of glycerine, NaCl, urea, and formaldehyde; the volume changes in sucrose solutions were retarded under pressure as compared with atmospheric controls. In *Alluim* cells the time for plasmolysis to occur in hypertonic solutions was increased under high pressure; the time for deplasmolysis, on the other hand, was decreased.

Okada (1954a–e) studying muscle and Yamato (1952a,b) studying erythrocytes also investigated the effect of hydrostatic pressure on cell permeability. Okada reported that the resting current and the injury potential of skeletal muscle was markedly decreased following the application of high pressure (5700 psi). In studying salt current of isolated frog skin he reported the voltage was considerably reduced under pressure. Yamato reported that hemolysis of mammalian blood cells in isotonic salt is increased at 300 atm (4400 psi); however, in the latter study, the pressure chamber was not equipped with windows, hence it was necessary to remove the cells from the pressure chamber in order to observe the pressure effects.

Miyatake (1957), investigating high pressure effects on blood cells, reported that electrical conductivity increases under pressure (14,300 psi) and following decompression. Moreover, the amount of ^{32}P incorporation into erythrocytes increases at pressures between 2900 and 14,300 psi. Additional reports on permeability can be found in the reviews of Landau, Chapter 2, Section V and Zimmerman, Chapter 10, Section II.

V. Effects of Pressure on Oxygen Consumption

Hayasi and Kono (1958) investigated the influence of high hydrostatic pressure on buffered and nonbuffered chemical solutions by means of an improved polarographic apparatus equipped with a stationary platinum electrode. In general, the pH of nonbuffered solutions is more susceptible to the action of high pressure, probably owing to the quantity of dissociable weak electrolytes.

The oxygen consumption of isolated tissue (muscle, heart, kidney, cerebrum) is accelerated under pressures of 1400, 4300, and 7100 psi. Pressures in excess of 14,200 psi resulted in an inhibition of oxygen consumption that was proportional to the magnitude of pressure. Upon decompression, the pressure effects on muscle and cerebrum were reversed. However, heart muscle generally did not recover following decompression (Kono, 1958a,b).

The oxygen consumption of *E. coli* was inhibited in proportion to the strength of pressure (1400–28,400 psi). When the temperature was increased, the inhibitory effect was slightly reduced; however, at atmospheric pressure, the oxygen consumption was not augmented even when the temperature rose above 37°C. Pressures in the range of 1400–4300 psi did not affect the oxygen consumption of lymphocytes obtained from thymus; however, pressures in excess of 7100 psi caused a reduction in oxygen consumption that was proportional to the magnitude of pressure (Kono, 1958a,b). For further studies see ZoBell, Chapter 4, Section VI.

VI. Pressure Inactivation of Enzyme and Protein Denaturation

Biological organisms are seldom subjected to pressures in excess of 16,000 psi; thus most investigators do not usually exceed these values in studying alterations of biological activities. However, there is an extensive area of pressure research which is concerned with the effects of high pressure (in excess of 16,000 psi) on enzyme inactivation and protein denaturation. For the past two decades Chieko and Keizo Suzuki and coworkers have conducted extensive research on the denaturation of protein systems and inactivation of enzymes; their laboratories remain among the most active in the world. They have investigated the pressure denaturation of albumin (C. Suzuki, 1963a,b; C. Suzuki and Suzuki, 1962, 1963; C. Suzuki *et al.*, 1963a,b; K. Suzuki *et al.*, 1963) and hemoglobin (K. Suzuki, 1960; K. Suzuki and Kitamura, 1960a,b), as well as pressure inactivation of α-amylase (K. Suzuki and Kitamura, 1963), trypsin (Miyagawa and Suzuki, 1963a), and chymotrypsin (Miyagawa and Suzuki, 1963b).

Depending upon the magnitude of pressure, protein denaturation may be either retarded or accelerated. K. Suzuki and Kitamura (1960a,b) reported that heat denaturation of hemoglobin is retarded at pressures below 29,400 psi and accelerated at pressures above this magnitude of pressure. Moreover, the reaction of pressure denaturation follows first-order kinetics.

The denaturation reaction of proteins is one of a complex class of reactions, and at present there is no general agreement concerning the definition of the word "denaturation" (C. Suzuki, 1963a). However, it is generally believed that protein denaturation is a phenomena that results from changes in the conformation of the molecule. However, in all probability, different conformational changes will occur depending upon the methods used to denature the proteins. C. Suzuki and Suzuki (1962) have demonstrated that the optical rotation of ovalbumin will vary markedly when the protein is denatured with pressure, heat, or urea. C. Suzuki *et al.* (1963b) reported that maximum stability of ovalbumin to pressure denaturation (86,000–106,700 psi for 5 minutes at 30°C) was found at pH 9.0, and that the rate of the pressure denaturation reaction of ovalbumin is proportional to the square-root of the hydrogen-ion concentration. These investigators also found that sulfate and glucose inhibited pressure denaturation, whereas urea and ethanol accelerated denaturation. Miyagawa and Suzuki (1963a,b) investigated the kinetics of pressure inactivation of trypsin and chymotrypsin. Inactivation of trypsin and chymotrypsin increased with increasing magnitude of pressure. Above a critical pressure, about 114,000 psi for trypsin and 85,300 psi for chymotrypsin, no additional inactivation occurred following a single pressure pulse. Below the critical pressure for each enzyme, if the total duration of compression was the same, repeated compression of the enzyme yielded the same inactivation as that caused by a single compression. However, above the critical pressure, repeated compression caused more inactivation than that caused by a single compression. The process of inactivation of trypsin and chymotrypsin by pressure followed first order kinetics. The thermodynamic quantities of the inactivation process of the enzymes are similar to the protein denaturation of ovalbumin and hemoglobin, except that the ΔH (enthalpy) for the proteins is negative, whereas positive values are obtained for the enzymes.

Murakami (1960), investigating the RNase activity obtained from beef pancreas, reported that pressures of 7100, 14,200, and 21,300 psi increase the RNase activity as measured in a specially designed spectrophotometer-pressure apparatus.

The action of succinic dehydrogenase was accelerated at 4300 psi; however, pressures in excess of 11,400 psi inhibited the activity in proportion to the magnitude of applied pressure (Tokumoto, 1962). Murakami (1958) found that magnesium-activated muscle ATPase and freshly prepared myosin ATPase were inhibited under pressure (14,300 psi).

REFERENCES

Hayasi, K. (1961a). Biological action of high hydrostatic pressure. *Nisshin Igaku* **48**, 292–305 (in Japanese).

Hayasi, K. (1961b). Reizwirkunkung hoher hydrostatischer Drucke und ihre Eigentümlichkeiten. *J. Physiol. Soc. Japan* **23**, 527–546 (in Japanese).

Hayasi, K., and Kono, I. (1958). Study on the influence of high hydrostatic pressure on solutions by the use of an improved stationary platinum electrode polarography. *Japan. J. Physiol.* **8**, 246–253.

Hayasi, K., Yasuda, H., Murakami, T. H., and Miki, H. (1958). On the observations with high pressure apparatus containing two windows. *Kagaku* (*Tokyo*) **28**, 365–336 (in Japanese).

Kono, I. (1958a). Oxygraphic studies on the influence of high hydrostatic pressure upon the oxygen consumption of tissues. Part I. Studies on the oxygraphic measurement of dissolved oxygen in solutions under normal and high hydrostatic pressures. *Okayama Igakkai Zasshi* **70**, 4521–4533 (in Japanese).

Kono, I. (1958b). Oxygraphic studies on the influence of high hydrostatic pressure upon the oxygen consumption of tissues. Part II. Influence of high hydrostatic pressure upon the oxygen consumption of tissues. *Okayama Igakkai Zasshi* **70**, 4535–4545 (in Japanese).

Miki, H. (1960a). Effect of high hydrostatic pressure on smooth muscle. Part I. On the longitudinal muscle of frog intestine. *Okayama Igakkai Zasshi* **72**, 1609–1613 (in Japanese).

Miki, H. (1960b). Effect of high hydrostatic pressure on smooth muscle. Part II. Effect of some drugs on the longitudinal muscle of frog intestine. *Okayama Igakkai Zasshi* **72**, 1615–1621 (in Japanese).

Miyagawa, K., and Suzuki, K. (1963a). Pressure inactivation of enzyme: Some kinetic aspects of pressure inactivation of trypsin. *Rev. Phys. Chem. Japan* **32**, 43–50.

Miyagawa, K., and Suzuki, K. (1963b). Pressure inactivation of enzyme: Some kinetic aspects of pressure inactivation of chymotrypsin. *Rev. Phys. Chem. Japan* **32**, 51–56.

Miyatake, T. (1957). Studies on effects of high hydrostatic pressure on blood cells. II. On the exchange of ions in the erythrocyte. *Okayama Igakkai Zasshi* **69**, 461–471 (in Japanese).

Murakami, T. H. (1958). Hydrostatic pressure effects on the adenosinetriphosphatase activities. *Symp. Cellular Chem.* **8**, 71–77 (in Japanese).

Murakami, T. H. (1960). The effects of high hydrostatic pressure on cell division. *Symp. Cellular Chem.* **10**, 233–244 (in Japanese).

Murakami, T. H. (1961). The effects of high hydrostatic pressure on cell division: Synthesis of nucleic acids. *Symp. Cellular Chem.* **11**, 223–233 (in Japanese).

Murakami, T. H. (1963). Effect of high hydrostatic pressure on the permeability of plasma membrane under the various temperature. *Symp. Cellular Chem.* **13**, 147–156 (in Japanese).

Okada, K. (1954a). The effects of high hydrostatic pressure on the permeability of plasma membrane. I. Resting current of muscle. *Okayama Igakkai Zasshi* **66**, 2071–2075 (in Japanese).

Okada, K. (1954b). Physiological studies of hydrostatic high pressure. Supplement. II. On pH found in a neutral red solution. *Okayama Igakkai Zasshi* **66**, 2105–2110 (in Japanese).

Okada, K. (1954c). Effects of hydrostatic high pressure on the permeability of plasma membrane. III. On salt current. *Okayama Igakkai Zasshi* **66**, 2083–2088 (in Japanese).

Okada, K. (1954d). Effects of hydrostatic high pressure on the permeability of plasma membrane. IV. On electric conductivity. *Okayama Igakkai Zasshi* **66**, 2089–2094 (in Japanese).

Okada, K. (1954e). Effects of hydrostatic high pressure on the permeability of plasma membrane. V. On plasmolysis. *Okayama Igakkai Zasshi* **66**, 2095–2099 (in Japanese).

Suzuki, C. (1963a). The denaturation of protein under high pressure. I. The reversal from the pressure denaturation of ovalbumin and horse serum albumin. *Rev. Phys. Chem. Japan* **33**, 85–98.

Suzuki, C. (1963b). The denaturation of protein under high pressure. II. The gelations of ovalbumin solution by pressure and by heat. *Rev. Phys. Chem. Japan* **33**, 99–108.

Suzuki, C., and Suzuki K. (1962). The protein denaturation by high pressure. Changes of optical rotation and susceptibility of enzymic proteolysis with ovalbumin denatured by pressure. *J. Biochem. (Tokyo)* **52**, 67–71.

Suzuki, C., and Suzuki, K. (1963). The gelation of ovalbumin solutions by high pressures. *Arch. Biochem. Biophys.* **102**, 367–372.

Suzuki, C., Kitamura, K., Suzuki, K., and Osugi, J. (1963a). The protein denaturation under high pressure. Horse serum albumin. *Rev. Phys. Chem. Japan* **32**, 30–36.

Suzuki, C., Suzuki, K., Kitamura, K., and Osugi, J. (1963b). The protein denaturation under high pressure. Effects of pH and some substances on the pressure denaturation of ovalbumin solution. *Rev. Phys. Chem. Japan* **32**, 37–42.

Suzuki, K. (1960). Studies on the kinetics of protein denaturation under high pressure. *Rev. Phys. Chem. Japan* **29**, 91–98.

Suzuki, K., and Kitamura, K. (1960a). Denaturation of hemoglobin under high pressure. I. *Rev. Phys. Chem. Japan* **29**, 81–85.

Suzuki, K., and Kitamura, K. (1960b). Denaturation of hemoglobin under high pressure. II. *Rev. Phys. Chem. Japan* **29**, 86–91.

Suzuki, K., and Kitamura, K. (1963). Inactivation of enzyme under high pressure: Studies on the kinetics of inactivation of a-Amylase of *Bacillus subtilis* under high pressure. *J. Biochem. (Tokyo)* **54**, 214–219.

Suzuki, K., Miyosawa, Y., and Suzuki, C. (1963). Protein denaturation by high pressure. Measurements of turbidity of isoelectric ovalbumin and horse serum albumin under high pressure. *Arch. Biochem. Biophys.* **101**, 225–228.

Tanbara, H. (1952). Effect of high pressure on the skeletal muscle. *Okayama Igakkai Zasshi* **64**, 909–950 (in Japanese).

Tokumoto, H. (1962). Effects of high hydrostatic pressure on respiration of liver tissue. *Okayama Igakkai Zasshi* **74**, 639–648 (in Japanese).

Yamato, H. (1952a). Effects of high hydrostatic pressures on the action of erythrocyte. III. On the kalium contents. *Okayama Igakkai Zasshi* **64**, 874–881 (in Japanese).

Yamato, H. (1952b). Effects of high hydrostatic pressure on the action of erythrocyte. V. On the lysis. *Okayama Igakkai Zasshi* **64**, 888–900 (in Japanese).

Yasuda, H. (1959a). Effects of high pressure on the cardiac muscle. Part I. On the isolated frog heart. *Okayama Igakkai Zasshi* **71**, 5849–5855 (in Japanese).

Yasuda, H. (1959b). Effects of hydrostatic pressure on the cardiac muscle. Part II. On the refractory period. *Okayama Igakkai Zasshi* **71**, 5857–5861 (in Japanese).

Yasuda, H. (1959c). Effects of high hydrostatic pressure on the cardiac muscle. Part III. On the excitability of muscle. *Okayama Igakkai Zasshi* **71**, 5863–5870 (in Japanese).

Yasuda, H., Murakami, T. H., and Miki, H. (1960). Effect of high hydrostatic pressure on cell division. *Okayama Igakkai Zasshi* **72**, 1635–1641 (in Japanese).

CHAPTER 6

A PRESSURE STUDY OF GALVANOTAXIS IN *Tetrahymena*

Tetuhide H. Murakami and Arthur M. Zimmerman

I. Introduction

Among the earliest reports on ciliary activity and movement are the studies of Regnard (1884) and Certes (1884) concerning the effects of pressure on ciliary action. Ebbecke (1935) reported that with increasing pressure, *Paramecium* exhibit decreased movement, an effect which is readily reversed upon decompression. In a wide variety of cells that have cilia or flagella, the effect of hydrostatic pressure is variable and complex. Cells such as *Epistylis plicatilis* and *Euplotes* exhibit increased activity at moderate pressures up to 3000–4000 psi; other protozoa such as *Stentor* show progressively decreased frequency of beating of the membranelles up to 6000 psi (Kitching, 1957). Renewed interest in the effects of pressure on form and activity as well as on the biochemistry of ciliates has in recent years led to further investigations in this field (cf. Kitching, Chapter 7; Zimmerman and Zimmerman, Chapter 8).

It is well established that *Paramecium* is cathodally galvanotactic. A constant electric potential applied to *Paramecium* will cause ciliary reversal at the cathodal end; this results in the animal exhibiting a cathodal galvanotactic response. Cilia reversal also occurs in solutions of monovalent and certain divalent cations; the organisms swim backward in these solutions (Jahn, 1968). In the recent reports of Sleigh (1962), Jahn and Bovee (1967), and Kinosita and Murakami (1967), the response of cells to electrical current has been extensively reviewed.

Pressure is known to exert a marked effect on the electrical activity of muscles and nerves, which modifies the duration, magnitude, and threshold of response (Cattell and Edwards, 1928; Ebbecke and Schaefer, 1935; Spyropoulos, 1957a,b; Hayasi, 1961). It was thought that pressure could also alter the electrical characteristics of the protozoan, *Tetrahymena*. In this connection we should like to report some recent experimental results on the galvanotactic response of *Tetrahymena* under hydrostatic pressure.

II. Preparation of Cells for Study

Log phase cultures of *Tetrahymena pyriformis* strain GL were grown in 2% proteose-peptone supplemented with 0.1% liver extract. Before measuring the galvanotactic response, *Tetrahymena* were washed twice by hand centrifugation with either of two test solutions (Hahnert's solution or Plesner's solution) and resuspended in the same test solution for 30 minutes before they were investigated. Hahnert's test solution (1932) contained 8 mmoles of $CaCl_2$, 10 ml of 6.67×10^{-2} *M* Sørenson's phosphate buffer (pH 6.64), and deionized water to form 1 liter of test solution. Modified Plesner's inorganic test solution (1964) contained 2.75 gm NaCl, 0.25 gm $MgSO_4 \cdot 7H_2O$, 100 ml of 6.67×10^{-2} *M* potassium phosphate buffer (at a pH of 4.95–8.50), and deionized water to form 1 liter of test solution.

Two types of testing chambers were used in the present studies. One experimental chamber was used for measurement of cell accumulation at an electrode (Fig. 1A). Electrical connection was made via a AgCl coated silver wire inserted into the 4% agar in 3.3 *M* KCl, which filled the stem of the glass electrode. Glass electrodes of 300–500-μ diameter were selected. The distance between electrodes was 5 mm. Another chamber, of a capillary type, was used for measurement of response time (Fig. 1B). The internal diameter of the capillary chamber was 1.0 mm, and the distance between electrodes was 20 mm. Either testing chamber

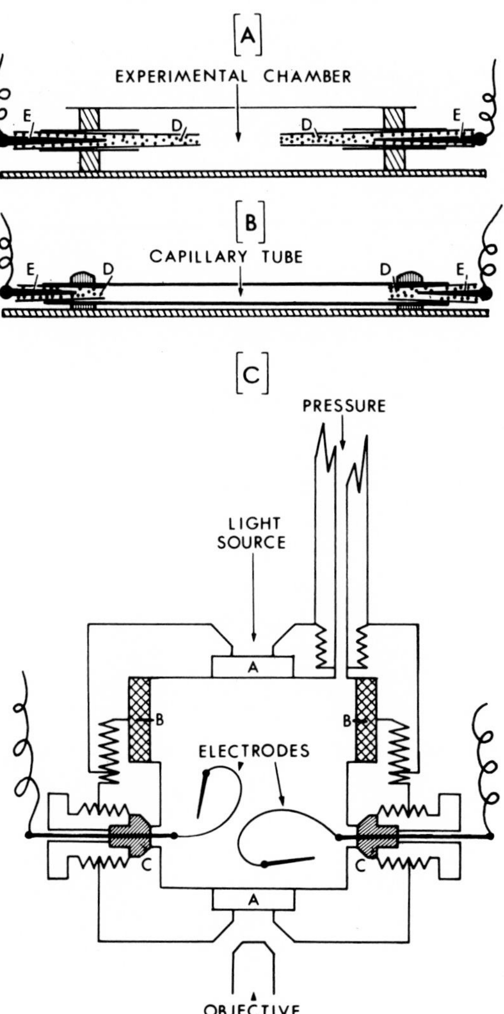

FIG. 1. Schematic drawings of pressure chamber and electrical chambers. [A] Electrical experimental chamber used for observing galvanotaxis was prepared from glass. The glass electrodes were filled with 4% agar in 3.3 *M* KCl, *D;* and a silver chloride-coated silver wire *E* was inserted into the glass electrodes and connected to the stimulator. [B] Capillary type electrical chamber employed for response time studies. Either the experimental chamber or the capillary tube chamber was placed above the window of the pressure chamber; the pressure chamber was filled with light paraffin oil. [C] The pressure chamber was provided with two windows, *A*. Large rubber gaskets *B* provided a tight seal; electrical connections enter from the exterior through ebonite seals *C*.

was placed into the pressure chamber (Fig. 1C), and the entire unit was placed into a controlled temperature housing maintained at 28°C (Marsland, 1950). The microscope–pressure chamber permits observation of the cells in the testing chamber at magnifications up to 600× while subjected to hydrostatic pressure as high as 20,000 psi. A Grass Medical Instruments' model SD-5 stimulator was connected to the pressure chamber.

III. Galvanotaxis at Atmospheric Pressure

When *Tetrahymena* were placed in an electrical field, the cells migrated toward the cathode. As the strength of the voltage was increased from 1 to 3 V, more cells accumulated in the region of the cathode. In order to quantitate the response of the *Tetrahymena* in the electrical field, a scoring system was employed. A grading system of 1 plus (+) to 4 plus (++++) was established which represented the relative number of cells accumulating at an electrode within a specific period of time following the experimental treatment. In Fig. 2, a series of photographs illustrate the scoring system employed in the present studies.

The number of cells which accumulated at the cathode was dependent upon the voltage and the duration of the treatment (see Table I). Thirty seconds after the voltage was increased from 1 to 3 V, there was a progressive increase in the number of cells which accumulated at the cathode. Extending the duration of voltage to 2 minutes resulted in a further increase in the number of cells at the cathode. However, if the voltage was increased from 4 to 10 V, there was no appreciable increase in the number of cells that accumulated at the cathode. When the voltage was returned to zero, the cells immediately moved away from the cathode.

Heat or formalin (2%) fixed cells did not exhibit any migration toward either electrode, even at the highest voltages studied (10 V).

IV. Effect of Pressure on Cathodal Galvanotaxis

It was previously reported that the activity of *Tetrahymena* decreases at high pressures (Zimmerman, 1969). The effects of pressure on motility are dependent upon the magnitude and duration of pressure treatment. In general, pressures up to 8000 psi for a duration of 2 minutes did not cause ciliary arrest. Pressures of 1000–2000 psi (for a duration of 2 minutes) did not seem to affect the coordination of ciliary activity nor the rate of movement of the cells. At 4000 and 6000 psi, movement was generally decreased. At 8000 psi, movement was markedly reduced and the shape of the cell changed to a more rounded form (see Zimmerman and Zimmerman, Chapter 8). The pressure effects were readily reversible, and the cells subjected to 1000–8000 psi immediately recovered their previous rate of movement following decompression.

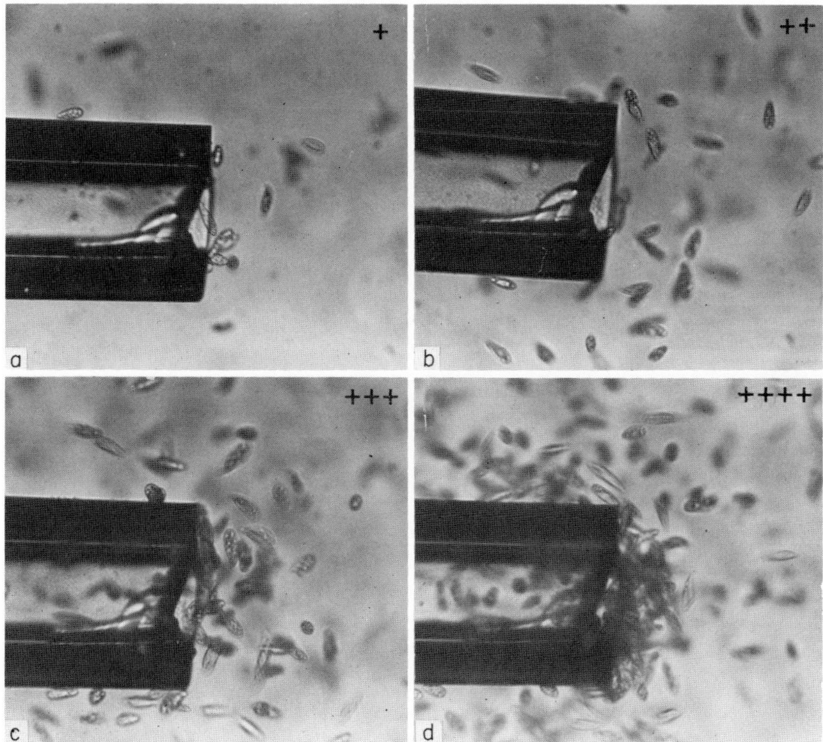

FIG. 2. Standardized scoring for galvanotaxis. A series of photomicrographs illustrating the relative number of cells at an electrode under specified conditions. (a) 1 plus; (b) 2 plus; (c) 3 plus; (d) 4 plus.

The galvanotactic response of *Tetrahymena*, at varying pressures, following a 30-second and 2-minute treatment of 4 V is shown in Fig. 3. At atmospheric pressure, a 3 plus (+++) response was elicited within 30 seconds following the application of 4 V, and after 2 minutes, a 4 plus (++++) response was observed. When the cells were subjected to 1000 psi and then supplied with a 4-V stimulus, a 1 plus (+) response was obtained 30 seconds later. Two minutes after pressure–voltage treatment a 2 plus (++) response was observed. When the voltage was

TABLE I
RESPONSE AT THE CATHODE TO A PRESSURE–VOLTAGE STIMULUS[a]

Volts	Atmospheric pressure		1000 psi		2000 psi		4000 psi		6000 psi		8000 psi	
	30 sec	2 min	30 sec	2 min	30 sec	2 min	30 sec	2 min	30 sec	2 min	30 sec	2 min
1	0	0	0	0	0	0	0	0	0	0	0	0
2	++	+++	+	+	+	+	0	0	0	0	0	0
3	+++	++++	+	+	+	+	+	+	0	0	0	0
4	+++	++++	+	++	+	++	+	+	0	+	0	0
5	+++	++++	++	+++	++	+++	+	++	+	+	0	0
6	+++	++++	++	+++	++	+++	+	++	+	++	0	0
10	+++	++++	+++	+++	++	+++	++	+++	+	++	0	0

[a] The cells were subjected to the desired pressure and the voltage was immediately applied.

further increased to 10 V, while maintaining a pressure of 1000 psi, the response increased to 3 plus (+++), but it was never as strong as that found at atmospheric pressure (++++) with lower voltages (e.g. 2 minutes at 4 V). When the pressure was increased to 6000 psi after 30 seconds at 4 V, a zero response was recorded; however, within 2 minutes, a 1 plus (+) response was observed. At higher pressures, there was no

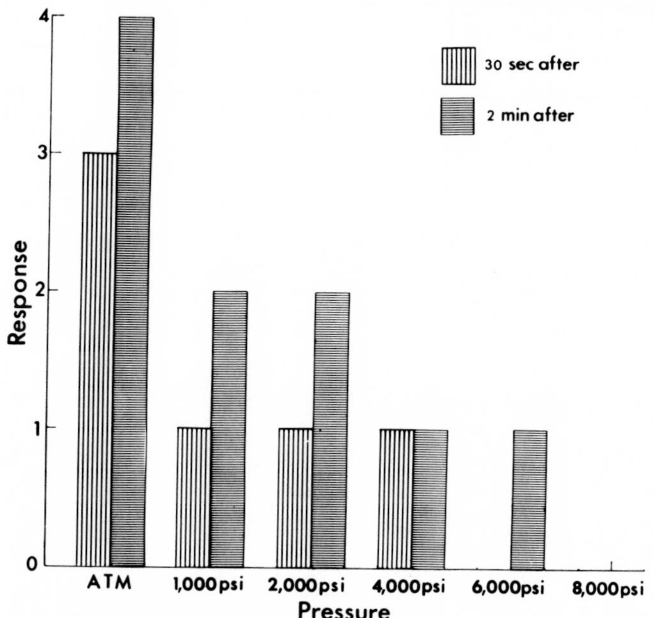

FIG. 3. Response at the cathode following pressure–voltage treatment. The cells were subjected to the desired pressure and 4 V of direct current was immediately applied. The response at 30 seconds and 2 minutes after the application of the voltage is shown.

accumulation of cells at the cathode. The galvanotactic response at varying pressures and voltages is summarized in Table I, which illustrates the relationship between pressure and voltage; as pressure is increased, a greater voltage is required to attain cathodal galvanotaxis, as measured by the relative number of cells which accumulate at the electrode. However, even with maximal voltage (10 V), a 4 plus (++++) response was not attained at the minimal pressure treatment (1000 psi).

In order to establish whether or not decreased galvanotaxis (decreased accumulation at the cathode) was merely a result of decreased activity following compression, the cells were placed in the electrical field for

1–2 minutes and *then* subjected to the experimental pressure. The results from these studies were generally similar to that found in the previous series where the order of the voltage–pressure application was reversed. In Fig. 4, the response to varying pressures at 4 V is shown. Although

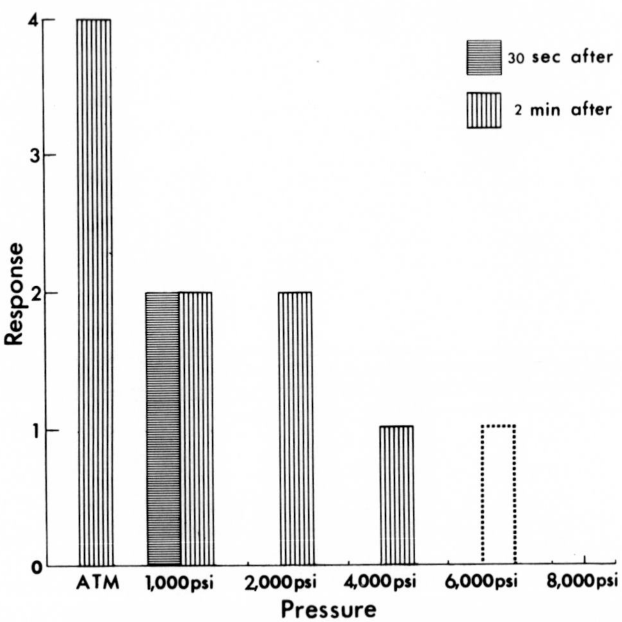

FIG. 4. Response of *Tetrahymena* at the cathode electrode following voltage–pressure treatment. The cells were subjected to 4 V of direct current for 1–2 minutes and while the voltage was maintained the pressure was applied. The response at 30 seconds and 2 minutes after compression is shown. At 6000 psi, the response of 1 plus was variable.

there was a 4 plus response prior to the application of pressure, about 10 seconds after compression the cells began to migrate away from the cathode and then, depending upon the magnitude of pressure, slowly returned. The response at 1000 and 4000 psi was 2 plus (++) and one plus (+), respectively, similar to the results obtained at 4 V in the pressure–voltage series. However, at 6000 psi (at 4 V), the response was variable. At 8000 psi there was no accumulation of cells at the cathode. Since the data in both series of pressure–voltage experiments are similar, it would appear that decreased galvanotaxis, under varying pressures, is not due to a decrease in activity of the cells.

V. Polarity Change under High Pressure

As previously shown (Figs. 3 and 4), *Tetrahymena* exhibited cathodal galvanotaxis at pressures below 8000 psi. However, at 8000 psi, the cells displayed anodal galvanotaxis. The application of 3–10 V (at 8000 psi) resulted in a 1 or 2 plus response at the anode; at lower voltages, galvanotaxis was not observed. Figure 5 shows cells accumulating at the anode at 8000 psi.

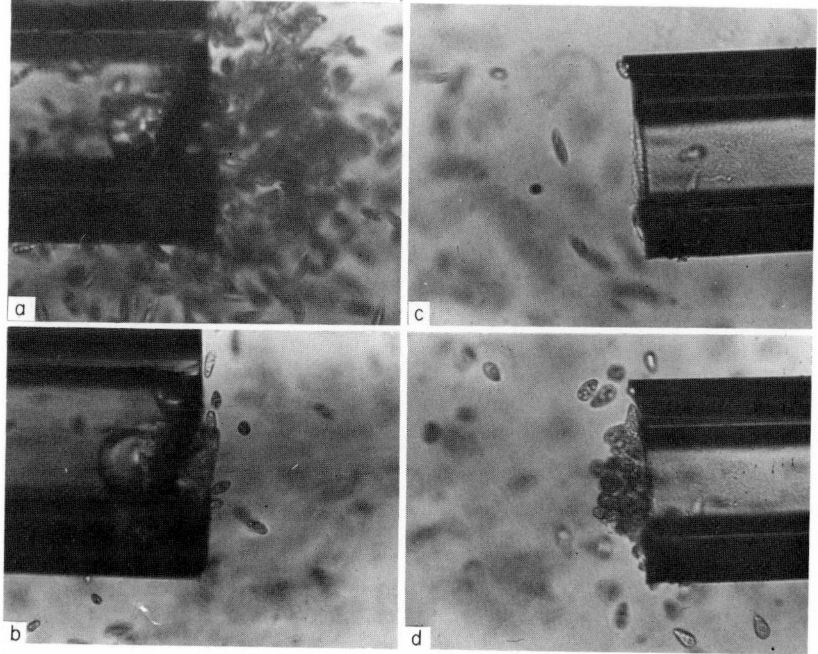

Fig. 5. Photomicrographs of *Tetrahymena* illustrating the response at the electrodes to 4 V of direct current under atmospheric pressure and 8000 psi. (a) The response at the cathode at atmospheric pressure and (b) 10 seconds after the application of 8000 psi. (c) The response at the anode under atmospheric conditions and (d) 2 minutes after the application of 8000 psi.

For a more critical analysis of polarity changes under 8000 psi, the capillary electrical chamber was employed. The advantage of this chamber was that the cells were retained within a relatively confined area, thus permitting observations concerning the rate and direction of movement following the application of a current. When the voltage (4

V) was applied at atmospheric pressure, the cells migrated toward the cathode; after the application of 8000 psi they continued to move toward the cathode for a few seconds and then slowly reversed their direction toward the anode. When pressure was applied prior to voltage, a similar pattern was observed. If the pressure was maintained for less than 10 minutes, the cells migrated toward the cathode immediately following decompression.

VI. pH Studies

In order to study the effect of hydrogen-ion concentration on the migration of the cells in an electrical field, the pH of the testing solutions were varied between 4.95 and 8.50 (in Plesner's inorganic solution) in the capillary chamber. At atmospheric pressure, it was not possible to measure quantitatively any alteration of galvanotaxis when the hydrogen-ion concentration of the test solution was modified. However, it was possible to quantitate the effects of varying hydrogen-ion concentrations at 8000 psi by measuring the time that was required for cells to change direction (response time) when the electrode polarity was reversed. As previously mentioned, at 8000 psi, *Tetrahymena* migrated toward the anode; when the electrode polarity was reversed, the cells continued to move in their original direction for 5–35 seconds and then reversed their movement toward the "new anode." As shown in Fig. 6, the response time of the cells varied at 8000 psi, depending upon the hydrogen-ion concentration. At low pH values the response time exceeded 30 seconds, whereas at the higher pH values (7.5 and 8.5) the response time was less than 10 seconds.

VII. Effects of Selected Chemical Agents

To evaluate further the galvanotaxis of *Tetrahymena*, two chemical agents were studied, KCl and ouabain. Potassium chloride was employed in an attempt to depolarize the cells and thus alter galvanotaxis under pressure. Ouabain, a known inhibitor of potassium- and sodium-activated ATPase, was used to disrupt the electrical characteristics of the cell membrane, possibly through interference with the "sodium pump."

Cells were washed three times with 0.0145 *M* KCl following which their movement became disoriented, frequently showing a backward motion. This activity could continue for as long as 2 minutes. During this time, the cells did not exhibit any galvanotactic response at atmospheric pressure. When the concentration was increased to 0.06 *M* KCl, similar loss of galvanotaxis was initially observed. An increase in the concentration to 0.1 *M* KCl stops activity almost immediately, and very few if any of the cells recover. In 1.0 *M* KCl, there is no activity observed.

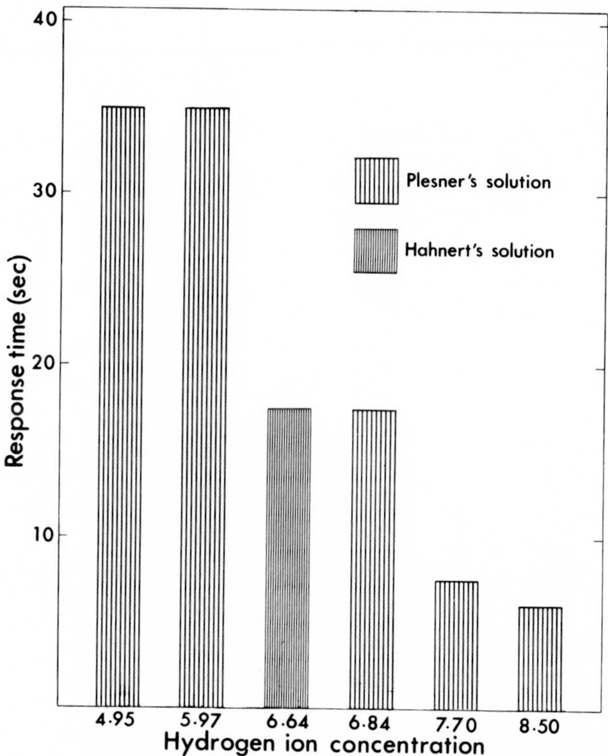

FIG. 6. The effect of hydrogen ion concentration on the response time at 8000 psi.

After 10 minutes immersion in 0.0145 *M* KCl, galvanotaxis returned; however, a pressure of 6000 psi blocked this response. At 30 minutes after immersion, the cells exhibited a cathodal preference even at 6000 psi, and at 8000 psi there was anodal preference. When the capillary chamber was employed, it was evident that the response time was markedly delayed.

Ouabain (strophanthin K) was mixed in Hahnert's solution to a final concentration of 10^{-4} gm/ml. The activity and galvanotaxis of cells in the presence of ouabain, at atmospheric pressure, appeared normal. At 6000 psi (at 4 V) galvanotaxis of the ouabain-treated cells was comparable with that of the nontreated controls. At 8000 psi, the cells formed a halo around the anode, and it appeared that they were milling in this area but did not come into contact with the electrode. It is interesting to note that the galvanotactic response of cells treated with choline chloride (0.1 *M*) was comparable with that of nontreated cells at the various pressures studied.

VIII. Discussion and Summary

These studies show that cathodal galvanotaxis of *Tetrahymena* is progressively weakened with increasing pressure. However, at 8000 psi, the *Tetrahymena* become anodally galvanotactic and they migrate toward the anode.

An explanation of the mechanism of galvanotaxis in ciliate protozoa has been sought after since Ludloff (1895) reported that there is ciliary reversal when *Paramecium* are subjected to an electrical current. One of the most comprehensive theories to explain the Ludloff phenomena has been put forth by Jahn (1961). He has proposed that the organism (*Paramecium*) may be considered to be similar to a core conductor immersed in a volume conductor, and the electric current, which tends to depolarize the membrane, results in ciliary reversal. It is a well-established fact that ciliary reversal can be caused by a variety of chemical agents; monovalent ions and divalent ions such as manganese, thallium, and barium will cause a temporary reversal of all the cilia (cf. Jahn, 1968). There is extensive literature to support the theory that ciliary reversal is correlated with the removal of calcium ions from the cell membrane and that this removal of calcium can result from either an applied electrical potential or from a physical ion exchange mechanism that conforms to the Gibbs-Donnan law (Jahn, 1961, 1962, 1968; Naitoh, 1966, 1968; Naitoh and Yasumasu, 1967). Further evidence which illustrates the important role that calcium plays in ciliary reversal is shown in the studies of Grebecki (1965) and Kuznicki (1966) on *Paramecium.* A gradual reduction in the available calcium (decalcification) results in a sequence of ciliary responses (behavioral responses) that are dependent upon the calcium concentration of the medium (Grebecki, 1965). However, in the absence of calcium in the external medium, ciliary activity returns to normal; it was, therefore, concluded that ciliary reversal was caused by incomplete decalcification. Kuznicki (1966) extended these studies using varying concentrations of different cations. He observed that the types of motor responses appeared in a definite sequence in conjunction with the rise in ion concentration. Ciliary responses could be induced and reversed, depending upon the level of calcium that was used concurrent with the various cations. It is interesting to note that the pressure response of *Spirostomum ambiguum*, which is uncalcified at early growth stages, but at later stages becomes calcified (filled with particles of hydroxyapatite), depends upon its stage of development. At early stages *Spirostomum* is sensitive to pressure, becoming quiescent at 3700 psi and disrupting at 8100 psi; however, mature calcified animals exhibit no changes at 8100 psi and only slight changes at 14,700 psi (Bien, 1967).

Action potentials have been suspected to occur in ciliary activity, but evidence does not support this supposition (Jahn, 1966). In *Paramecium*, the cathodal current causes a ciliary reversal by means of membrane depolarization, but there is no evidence for transmission of this depolarization (Jahn, 1961).

In the present study there was an initial loss of galvanotaxis in 0.0145 *M* KCl, accompanied by a transient backward movement; however, the cells recovered their cathodal galvanotactic response within 10 minutes, and by 30 minutes, galvanotaxis under pressure was similar to the pressurized controls in Hahnert's solution. In 0.1 *M* KCl, both movement and galvanotaxis were blocked. Kinosita (1954) reported that isotonic KCl induced a temporary reversal of ciliary beat in *Opalina* and that the cell membrane potential fell to zero (see Kinosita and Murakami, 1967). Perhaps the transient loss of galvanotaxis in low concentrations of KCl and the permanent loss at high concentrations, in *Tetrahymena*, is due to an effect of KCl on the cell membrane potential.

Although ouabain has been employed extensively to inhibit the potassium- and sodium-activated ATPase of the cell membrane and presumably to affect the sodium pump, it had no visible effect on galvanotaxis of *Tetrahymena* at atmospheric pressure or under varying hydrostatic pressures. It is possible that ouabain does not restrict the permeability of univalent cations in *Tetrahymena*.

The fact that the cilia continued to beat under pressures as high as 8000 psi for the 2-minute period during which the galvanotactic response was observed, has led the present authors to assume that the anodal response is galvanotactic rather than electrophoretic. Additional support for this arises from the observations that heat and formalin fixed cells did not migrate to either electrode. Moreover, in high concentrations of KCl (0.1 or 1.0 *M*), activity of the cells was stopped and the cells did not accumulate at either electrode. If the anodal response at 8000 psi was electrophoretic, one would expect the physically or chemically inactivated cells to migrate toward the anode.

In view of the well-known effect of pressure which results in solation of cytoplasmic gel structure (e.g. in the cortex, cilia, axopodia; cf. Zimmerman and Zimmerman, Chapter 8), it is possible that pressure reduces the structural cohesion (causes a solation) of the fibrous protoplasmic network of the cell, resulting in a disruption of the internal cytoplasmic membranes. Thus, a pressurized cell in an electric field may have a distribution of ions (calcium and potassium) which is different from that found in a nonpressurized cell. Jahn (1966) has analyzed the redistribution of cations in a cell placed in an electrical field and he proposes that the redistribution of ions may be responsible for the observed protoplasmic activity.

The present studies of pressure effects on galvanotaxis in *Tetrahymena* have raised many questions that remain to be answered. For example: Why do the cells, which are cathodally galvanotactic under atmospheric pressure and low hydrostatic pressures, become anodally galvanotactic when exposed to high hydrostatic pressures? What role does calcium play in pressure-induced galvanotaxis reversal? Can this reversal be altered by monovalent or divalent cations? The answers to these questions would provide a better understanding of both the galvanotactic mechanism as well as the role that pressure plays in modifying this mechanism.

ACKNOWLEDGMENT

The experiments from the authors' Laboratory have been carried out with the support of research grants from the National Research Council of Canada. Their assistance is gratefully acknowledged.

REFERENCES

Bien, S. M. (1967). High hydrostatic pressure effects on *Spirostomum ambiguum. Calcified Tissue Res.* **1**, 170–172.

Cattell, McK., and Edwards, D. J. (1928). The energy changes of skeletal muscle accompanying contraction under high pressure. *Am. J. Physiol.* **86**, 371–382.

Certes, A. (1884). Note relative à l'action des hautes pressions sur la vitalité des micro-organismes d'eau douce et d'eau de mer. *Compt. Rend. Soc. Biol.* **36**, 220–222.

Ebbecke, U. (1935). Das Verhalten von Paramecien unter den Einwirkung hohen Druckes. *Arch. Ges. Physiol.* **236**, 658–661.

Ebbecke, U., and Schaefer, H. (1935). Uber den Einfluss hoher Drucke auf den Aktionsstrom von Muskeln und Nerven. *Arch. Ges. Physiol.* **236**, 678–692.

Grebecki, A. (1965). Role of Ca^{++} ions in excitability of protozoan cell. Decalcification, recalcification, and the ciliary reversal in *Paramecium caudatum. Acta Protozool.* (*Warsaw*) **3**, 275–289.

Hahnert, W. F. (1932). A quantitative study of reactions to electricity in *Amoeba proteus. Physiol. Zool.* **5**, 491–526.

Hayasi, K. (1961). Reizwirkung hoher hydrostatischer Drucke und ihre Eigentumlichkeiten. *J. Physiol. Soc. Japan* **23**, 527–546 (in Japanese).

Jahn, T. L. (1961). The mechanism of ciliary movement. I. Ciliary reversal and activation by electric current; the Ludloff phenomenon in terms of core and volume conductors. *J. Protozool.* **8**, 369–380.

Jahn, T. L. (1962). The mechanism of ciliary movement. II. Ion antagonism and ciliary reversal. *J. Cellular Comp. Physiol.* **60**, 217–228.

Jahn, T. L. (1966). Contraction of protoplasm. II. Theory: Anadol vs. cathodal in relation to calcium. *J. Cell Physiol.* **68**, 135–148.

Jahn, T. L. (1968). The mechanism of ciliary movement. III. Theory of suppression of reversal by electrical potential of cilia reversed by barium ions. *J. Cell Physiol.* **70**, 79–90.

Jahn, T. L., and Bovee, E. C. (1967). Motile behaviour of protozoa. *Res. Protozool.* **1**, 41–200.

Kinosita, H. (1954). Electric potentials and ciliary response in *Opalina. J. Fac. Sci., Imp. Univ., Tokyo, Sect. IV* **7**, 1-14.

Kinosita, H., and Murakami, A. (1967). Control of ciliary motion. *Physiol. Rev.* **47**, 53–82.

Kitching, J. A. (1957). Effects of high hydrostatic pressure on the activity of flagellates and ciliates. *J. Exptl. Biol.* **34**, 494–510.

Kuznicki, L. (1966). Role of Ca++ ions in the excitability of protozoan cell. Calcium factor in the ciliary reversal induced by inorganic cations in *Paramecium caudatum. Acta Protozool.* **4**, 241–256.

Ludloff, K. (1895). Untersuchungen uber den galvanotropismus. *Arch. Ges. Physiol.* **59**, 525–554.

Marsland, D. (1950). The mechanisms of cell division; temperature-pressure experiments on cleaving eggs of *Arbacia punctulata. J. Cellular Comp. Physiol.,* **36**, 205–227.

Naitoh, Y. (1966). Reversal response elicited in non-beating cilia of *Paramecium* by membrane depolarization. *Science* **145**, 660–662.

Naitoh, Y. (1968). Ionic control of the reversal response of cilia in *Paramecium. J. Gen. Physiol.* **51**, 85–103.

Naitoh, Y., and Yasumasu, I. (1967). Binding of Ca ions by *Paramecium caudatum. J. Gen. Physiol.* **50**, 1303.

Plesner, P. (1964). Nucleotide metabolism during synchronized cell division in *Tetrahymena pyriformis. Compt. Rend. Trav. Lab. Carlsberg* **34**, 1–76.

Regnard, P. (1884). Note relative à l'action des hautes pressions sur phénomènes vitaux (movement des cils vibratiles, fermention). *Compt. Rend. Soc. Biol.* **36**, 187–188.

Sleigh, M. A. (1962). "The Biology of Cilia and Flagella." Pergamon Press, Oxford.

Spyropoulos, C. S. (1957a). Response of single nerve fibres at different hydrostatic pressures. *Am. J. Physiol.* **189**, 214–218.

Spyropoulos, C. S. (1957b). The effects of hydrostatic pressure upon the normal and narcatised nerve fibre. *J. Gen. Physiol.* **40**, 849–857.

Zimmerman, A. M. (1969). Effects of high pressure on macromolecular synthesis in synchronized *Tetrahymena. In* "The Cell Cycle" (G. M. Padilla, G. H. Whitson, and I. L. Cameron, eds.), pp. 203–225, Academic Press, New York.

CHAPTER 7

SOME EFFECTS OF HIGH PRESSURE ON PROTOZOA

J. A. Kitching

I. Foreword

My association with high pressure is a casual one. It provides an outlet for occasional unpremediated research rather than a planned research program, although in contrast my interest in this field is substantial.

My first contact with high pressure was in the stimulating atmosphere of Newton Harvey's Laboratory at Princeton. D. C. Pease, then a young research worker, had just made himself a pressure vessel and was looking for something to put into it. After the war G. A. Shepherd of Imperial Chemical Industries and C. R. Burch of the University of Bristol led me to a much improved general and optical design. The suctorian, *Discophrya collini*, turned up in material brought in by Edward Livingstone of Bristol from the Botany Pond and was successfully cultured by him to produce several years of work, including unexpected effects of high pressure. Then another visit to the same pond produced the heliozoon, *Actinophrys sol*, also successfully cultured by him, and with it a few more years of work not without interest for biological studies of high pressure. The study of axopodial structure was continued here at Norwich in collaboration with Susan Craggs and Andrew C. Macdonald. Finally, Alister G. Macdonald has achieved an important advance by measuring the respiration of *Tetrahymena pyriformis* W. at high pressure with an oxygen electrode and the growth of populations of the same organism at high pressure by electronic counting. This development points the way to biochemical studies. He has also taken to an advanced stage the design of equipment for recovery of small animals from the deep sea without decompression and their transfer to a continuous flow observation vessel.

II. Introduction

The action of moderately high hydrostatic pressure (ranging up to 6000 psi or about 400 atm) in reducing the structural cohesion of protoplasmic gels and the opposing effects of an increase in temperature have long been known through the work of Marsland and his colleagues. Together with this reduction in "viscosity" goes a retardation or cessation of protoplasmic movements such as ameboid activity or cytokinesis. An association of movement with structural elements within cells, long accepted for muscle, for cilia, and for "myonemes," is now generally suspected in other cases; and the mechanisms—both biochemical and fine structural—which are characteristic of muscle, can be used to provide, or at least to suggest, hypotheses by which other protoplasmic movements may be explained. Thus a sliding of filaments will account for ciliary movement (Sleigh, 1968) and no doubt also for the body and stalk contractions of various ciliates. In some cases, movements appear to be associated closely with microtubules, but it is uncertain to what extent microtubules act merely as an elastic skeleton and to what extent they generate movement, either by microundulation (which itself could be produced by molecular sliding) or by providing a surface against which molecular sliding could take place. The flow of endoplasm in amebas is most easily explained by contraction of the surrounding gelated granular ectoplasm, but it is not clear to what extent such a contraction is dependent on fibrous substructure (such as that described by Nachmias, 1968). There is probably considerable diversity at the fine structural level in the application of fundamentally similar biochemical processes to the generation of movement in Protozoa. All these forms of movement are affected by high pressure, and the effects are complicated. It seems likely that pressure acts directly on the mechanism of movement, but perhaps also on the supply of energy. It is also unclear to what extent the constant supply of energy is necessary for the integrity of the cell machinery upon which movement depends.

Against this background we shall now consider the effects of high pressure on some rather complicated complete single cells. These enquiries may throw some light on the action of high pressure as such; they may help toward an understanding of the physiological mechanisms of these cells and they may ultimately help to define the adaptations necessary for life in the deep sea.

III. The Axopods of Actinophrydea

Axopods of two Heliozoa—*Actinophrys* and *Actinosphaerium* of the order *Actinophrydea*—contain an axial rod or axoneme which consists of a number of parallel (or perhaps nearly parallel) microtubules which in transverse sections present the configuration of a double spiral (Fig 1) (Kitching, 1964; Tilney and Porter, 1965; Kitching and Craggs, 1965). *Actinophrys* is uninucleate, and the axonemes rest internally on the nucleus; *Actinosphaerium* is a larger multinucleate version of the same general body form, and the axonemes reach internally to the nuclear layer. Basal sections of the axonemes of *Actinophrys sol* contain about 100–200 microtubules, and sections of *Actinosphaerium nucleofilum* contain about 500–600; in both the numbers decrease toward the tips. In *Actinosphaerium* the array of microtubules is seen in transverse sections to consist of twelve similar but distinct radial sectors. In favorable transverse sections (A. C. Macdonald and Kitching, 1967) the microtubules are seen to be connected by a system of faint dark bands, and in a few sections these are more clearly defined here and there. These connections take the form of tangential links joining adjacent members of the same spiral, radial links between microtubules arranged in radial lines bounding the sectors, and secondary links connecting microtubules in oblique parallel lines (Fig. 2) branching off one side of the radii. In spite of changes likely to be produced by fixation and subsequent treatment, these links may reasonably be held to represent a related pattern of structure in the original material. In any interpretation of the molecular structure of these microtubules, the 30° angle between radial linkages and between radial and secondary linkages will need to be considered.

The direction of twist of the double spirals is not a matter of chance.

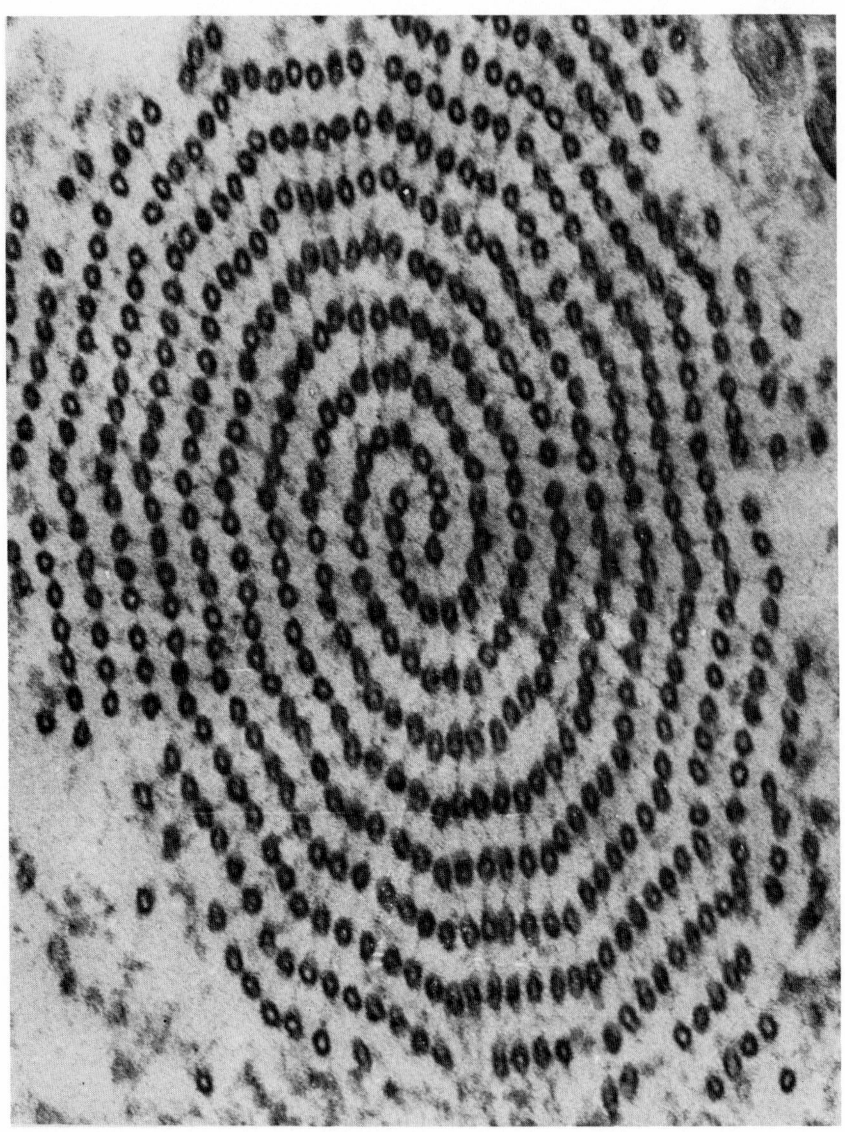

FIG. 1. Transverse section of an axoneme of *Actinosphaerium nucleofilum* cut within the body. Note the dark bands which link microtubules together in accordance with the scheme depicted in Fig. 2. Photograph prepared by Andrew C. Macdonald from negative by Kitching and Craggs.

In any one section taken through the cortical cytoplasm, all of the double spirals were found to turn in the same direction; and in sections of known orientation it was found that to an observer looking at the animal from the outside, and tracing the spirals from the center to the periphery, the track was clockwise (Kitching and Craggs, 1965). This

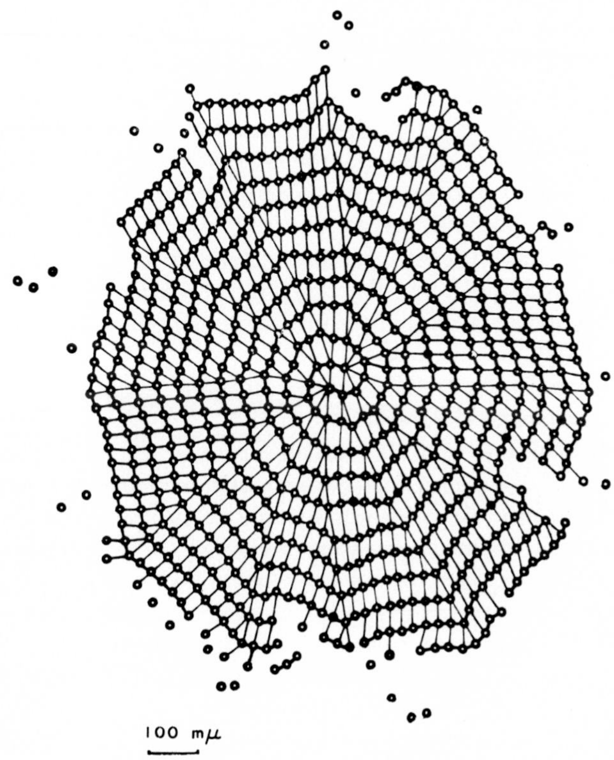

FIG. 2. Interpretation of system of links in axoneme of *Actinosphaerium nucleofilum*. (From A. C. Macdonald and Kitching, 1967.)

results renders it highly improbable that the spirals are equally distributed in either direction, and it is a plausible extrapolation to postulate that they always have the same direction of twist. Inevitably this proposition suggests a comparison with the asymmetrical organization of the arms of the outer fibers of cilia, although the detail is here very different. It suggests some fundamental and inherent asymmetry in the molecular organization of the microtubules.

Both in *Actinophrys sol* (Kitching, 1957b) and in *Actinosphaerium nucleofilum* (Tilney *et al.*, 1966), application of moderate pressures causes the axopods to disappear. They become beaded within a minute or so of the application of pressure and beading is followed by folding and collapse of the axonemes. Release of pressure is followed within about 2 minutes by the protrusion of points which grow out within a further few minutes into new axopods. In *Actinosphaerium* it has been established (Tilney *et al.*, 1966) that a limited recovery will take place even while pressure is maintained.

As in the case of amebas (Landau *et al.*, 1954) and various marine eggs (Marsland and Landau, 1954), temperature and pressure act in opposition. In experiments in which the pressure was raised by steps (each of 1000 psi and 5 minutes duration), the critical pressure for the collapse of all or nearly all of the axopods of *Actinophrys* was 4000–5000 psi at 15°–20°C and 2000–3000 psi at 5°–10°C, with considerable variation. Likewise the critical pressure for the restoration of axopods in stepwise release was lower the lower the temperature. A temperature increment of 10°C counterbalanced a pressure increment of 1200 psi (82 atm) (Kitching, 1957b).

The structural effects of high hydrostatic pressure upon *Actinosphaerium nucleofilum* were investigated by Tilney *et al.* (1966). Fixation was carried out with the organisms held at various appropriate pressures, and the material was subsequently sectioned and examined by electron microscopy. At pressures leading to beading and retraction of the axopods, the microtubules were no longer to be found. When the treatment was such that axopods were reformed or regenerated, microtubules were once more found in the sections. Fine fibrillar material was found in pressurized *Actinosphaerium* at sites where microtubules were believed to have disappeared, both in degenerating axopods and within the cell body. Fibers 10–20 mμ in diameter also appeared in the cell body. The role of these two products of compression has yet to be established, although each may represent breakdown products or constituents of microtubules.

The axonemes of *Actinophrys* and *Actinosphaerium* undoubtedly support the axopods and are thus skeletal. In their instability they resemble the microtubules of division spindles, and we may presume that their instability has a functional significance: the axopods are required to retract readily and subsequently to reconstitute themseves. There may also be some connection between the microtubules and the streaming of lines of cytoplasm in the axopods. The microtubules of cilia, different in their fine structural organization, are presumably much more stable at high pressure; cilia remain extended and in some cases may beat very slowly at 10,000–13,000 psi (Kitching, 1957a). It is evident that microtubules are not all alike (Behnke and Forer, 1967); a comparative study of the effects of high pressure on their stability and on any associated motility should be instructive. For instance, it would be interesting to investigate the effects of high pressures on the microtubules of the axostyles of pyrsonymphid flagellates with their strongly marked hexagonal system of linkage (Grimstone and Cleveland, 1965). These axostyles undergo and possibly generate undulatory movements. (Further studies on the effects of pressure on ultrastructure may be found in the reviews of Marsland, Chapter 11, and Zimmerman and Zimmerman, Chapter 8.)

IV. Effects of Hydrostatic Pressure on Suctoria

Two explanations are possible of the way in which Suctoria "suck" the protoplasm out of their prey: the first, pumping by tentacles, which might drive material along their inner tube, perhaps by microperistalsis; the second, an active expansion of the body surface, which could produce a negative pressure inside the body provided that there was some resistance to passive collapse. Each or both of these two explanations might be correct, and it is also possible that the tentacles might pump and the body surface undergo some process of relaxation that did not result in the creation of a negative pressure.

Suctorian tentacles are generally thought to contain an inner tube through which food material passes from prey to predator. Electron micrographs of *Tokophrya infusionum* (Rudzinska, 1965) show that the wall of the inner tube in this suctorian contains 49 microtubules, with a further 21 microtubules in 7 groups in the surrounding cytoplasmic layer. The corresponding microtubules in *Acineta tuberosa* number 32 and 24 (Bardele and Grell, 1967), and it is supposed for this animal that the inner tube forms a temporary structure during feeding. The feeding tentacles of *Ephelota* have an even more elaborate array of microtubules (Batisse, 1966). When suitable prey swims up against a tentacle of a suctorian, the prey sticks. Opaque "missilelike bodies" stream along the tentacles toward the tip, and are probably responsible for the breakdown of the surface and the immobilization of the prey (Rudzinska, 1967). Attached tentacles very quickly shorten and thicken. Their cuticle must either be elastic or capable of some considerable relaxation or deformation. The role of microtubules remains uncertain. Those lining the inner tube might provide elasticity and other structural properties necessary to permit and to guide the passage of lumpy food material, or might generate peristaltic movements, and it has also been suggested that microtubules are responsible for the shortening of the tentacles (Rudzinska, 1967).

In considering the role of microtubules in the feeding of Suctoria it is useful to draw comparisons with some other cases in which food material is caused to move along a set course through the body. For instance, groups of microtubules lie in the longitudinal ridges in the walls of the esophageal tube of peritrich ciliates (Fauré-Fremiet *et al.*, 1962). Food vacuoles are conducted rapidly along the esophageal tube into the body of a peritrich, and it seems necessary to assume that energy is imparted to them continuously along the tube (Kitching, 1939), possibly by the action of microtubules. The ciliate *Nassula* has an elaborate pharyngeal basket composed of very many microtubules, and observations of algal filaments in the process of being eaten suggest that the motive force for uptake is generated by or in the neighborhood of the microtubules (Tucker, 1968). Nevertheless, we must admit that we do not know what microtubules actually do in any of these instances.

When the suctorian *Discophrya collini* captures prey, and one or more tentacles become attached, there is a change in the shape of the body surface of the predator. Apparently (Kitching, 1952) this change involves an increase in the surface area, usually accompanied by a volume increase, so that the body surface remains rounded. However, the body surface occasionally wrinkles, presumably because the body volume has failed to keep pace with the surface expansion. Additional support was drawn for this conclusion by experiments in which the prey was reduced osmotically, so as to delay or prevent the transfer of any substantial quantity of food material into the *Discophrya*. Under these conditions, a wrinkling regularly developed in the predator when prey was captured, with a simultaneous increase in profile perimeter. On the other hand, Hull (1961) observed a change in the shape of the predator from pyriform to nearly spherical with an increase in body volume, but observed no wrinkling; presumably the increase in body volume must be ascribed to a rapid uptake of prey substance.

Application of high hydrostatic pressure to a nonfeeding *Discophrya* (Kitching, 1954a) causes, within a minute or two but not instantly, an increase in profile perimeter and a severe surface wrinkling, as is very clearly seen in photographs (Figs. 3 and 4). Pressures as low as 2700 psi (183 atm) are effective. Later, or at higher pressures, there is a separation of the protoplasmic surface from the cuticle (Fig. 4). Within a few minutes after release of pressure there is a spreading of the protoplasmic

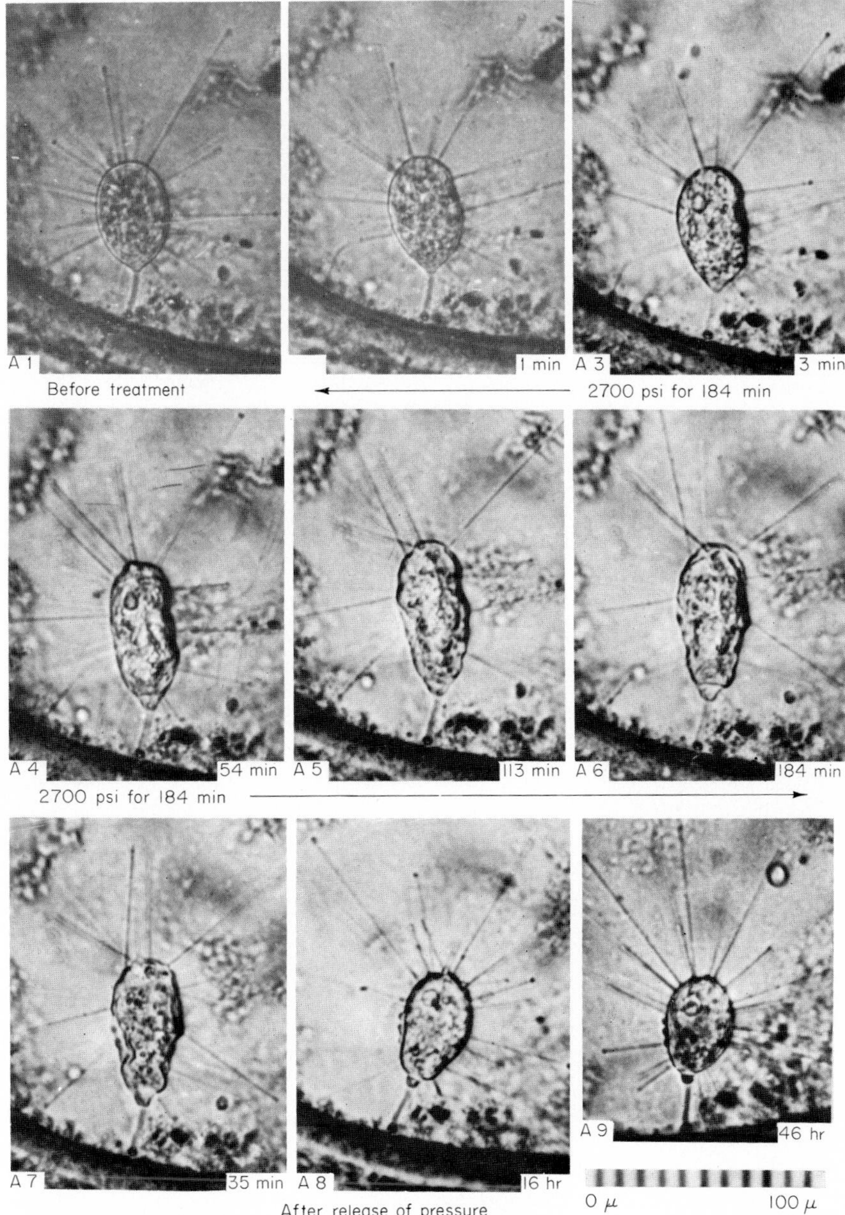

FIG. 3. A series of photographs illustrating the effects of a moderately high pressure (2700 psi) on the suctorian *Discophrya collini*. (From Kitching, 1954a.)

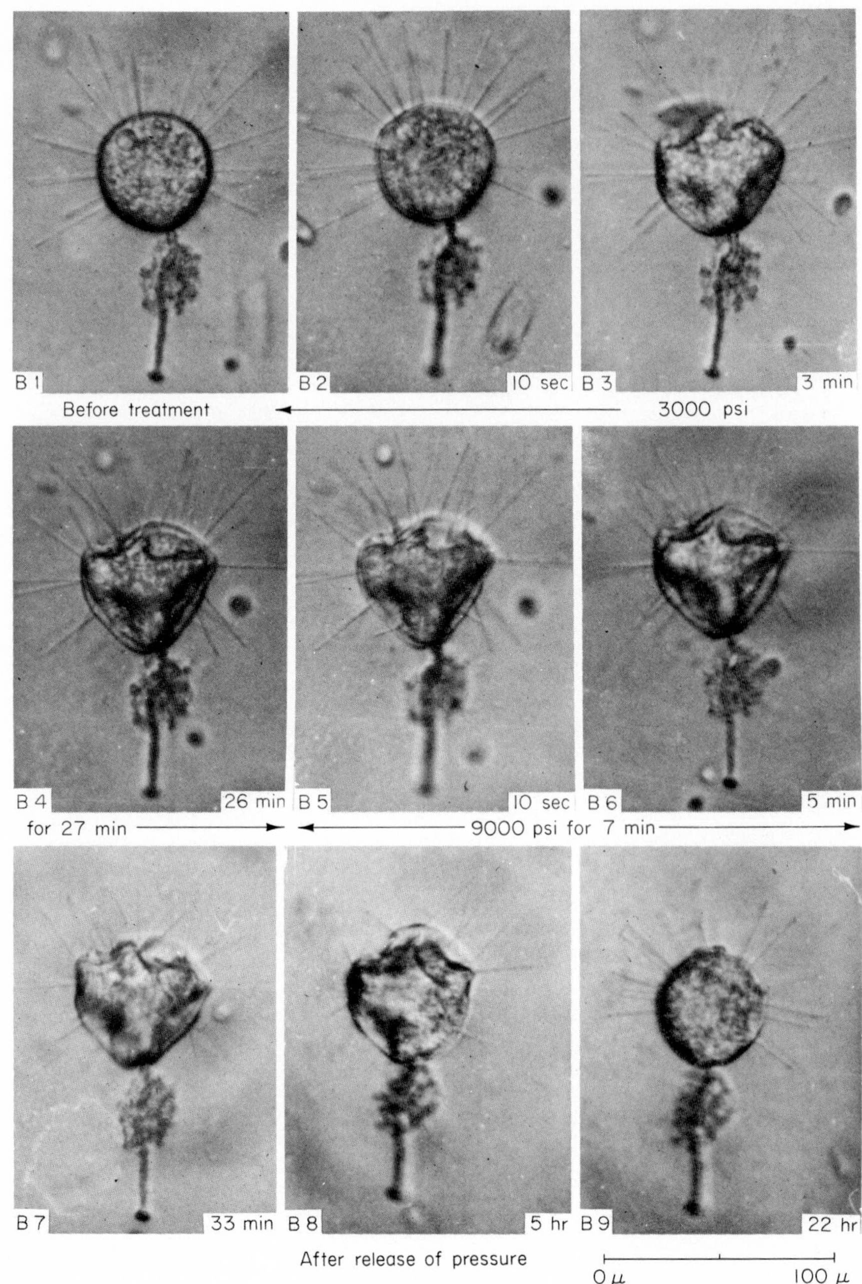

FIG. 4. A series of photographs illustrating the effects of moderate and more severe high pressure (3000 psi and 9000 psi) on the suctorian *Discophrya collini*. (From Kitching, 1954a.)

surface into contact with the cuticle which may become even more creased. The whole body surface with overlying cuticle then gradually becomes smooth and rounded over a period of 24 hours or more.

I attribute the surface wrinkling which follows the application of high hydrostatic pressure to an increase in surface area. It is too extensive and too slow to develop to be caused by compression of the body substance. It might be attributed to a transfer of water from the body to the medium, but in *Discophrya* caused to shrink osmotically there is a shortening of the profile perimeter and a reduction in profile area, whereas this is not true of *Discophrya* subjected to high hydrostatic pressures. Moreover, a loss of water from the protoplasm would probably be accompanied by a reduction in rate of output of the contractile vacuole, and this reduction should be drastic for a shrinkage sufficient to cause the observed wrinkling. Yet at hydrostatic pressures sufficient to cause substantial wrinkling (2500 psi or 169 atm, in Fig. 5), the rate of vacuolar ouput was still only slightly reduced (Kitching, 1954c). The most plausible explanation is that high hydrostatic pressure releases a reaction that involves surface extension and which normally, though less drastically, forms part of the feeding process. The fine structural basis for this reaction (possibly even microtubules?) requires investigation.

At higher pressures, such as 6000 to 10,000 psi (approximately 400–700 atm), other complicating effects intervene: a retreat of the body from the pellicle; a stoppage of the contractile vacuoles; and probably a shrinkage of the body, reflected in the smaller ultimate volume on recovery. This process of recovery, which follows release of pressure, is a slow one and involves a reversal of the various effects already noted. It presumably requires a reorganization of the pellicle to fit the reduced and reconstructed body surface.

Expansion of the body surface upon capture of prey or upon treatment with high hydrostatic pressure may be an active process. According to Hull (1961), it can be initiated by injection of 1–2% ATP. Alternatively, high pressure might break down a structional organization responsible for the maintenance of body form, so as to lead to the observed wrinkling or rounding up and surface expansion. This might develop the 0.2 atm of negative pressure (Hull, 1961) necessary to produce the observed rates of flow in the tentacles.

Experiments on the effects of pressure on feeding *Discophrya* (Kitching, 1956) confirm the view that at relatively low pressures (2700 psi, or 187 atm) the surface expansion occurs with relatively little interference from other complicating factors. Feedings continues and the wrinkled and enlarged protoplasmic surface gradually fills up. At 4000

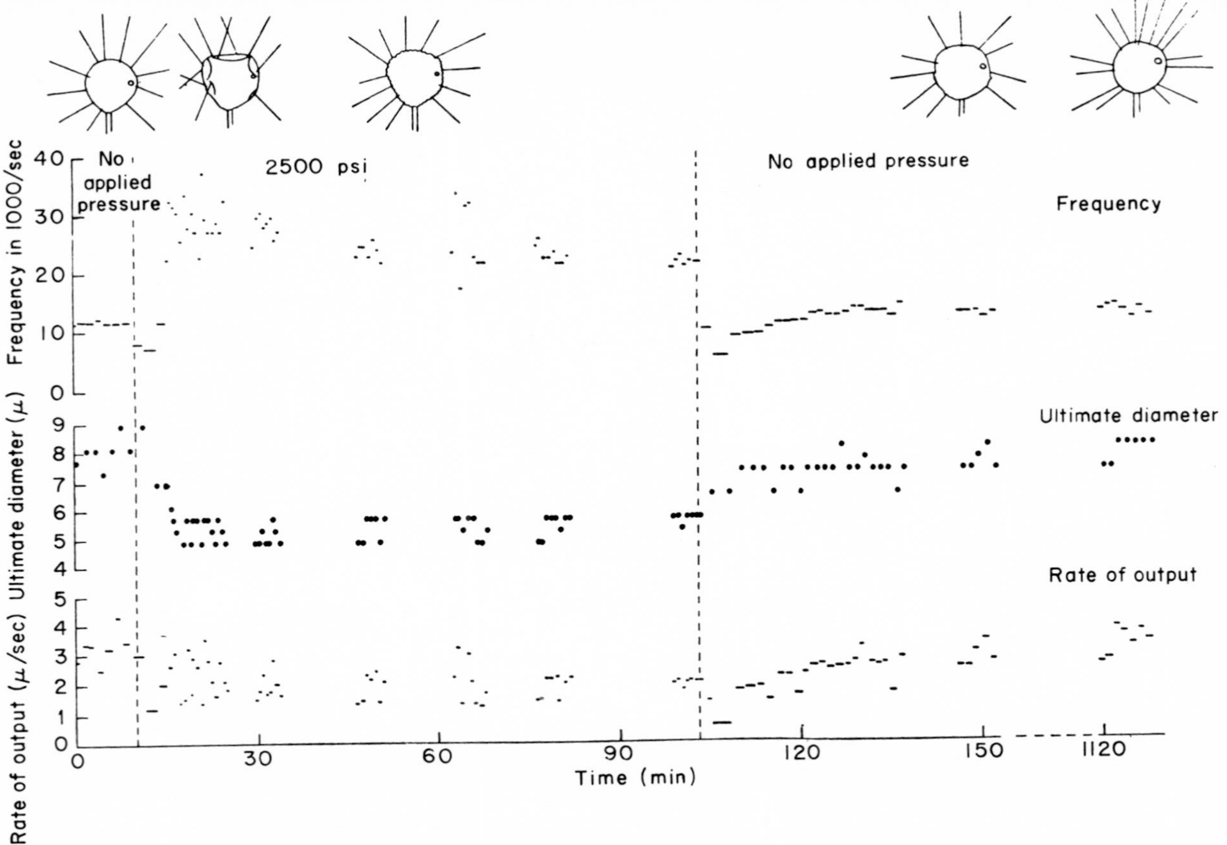

FIG. 5. Effects of a pressure of 2500 psi (170 atm) on the activity of the contractile vacuole of *Discophrya collini*. (From Kitching, 1954c.)

psi (272 atm) or over, feeding is arrested and the feeding tentacles re-extend. If the pressure is released, feeding may be resumed, but with prolonged application of pressure the prey separates from the tentacles of the suctorian. We may suppose that pressures exceeding 4000 psi inhibit a process taking place within the tentacles and at their tips—a process involving thickening and longitudinal contraction of the tentacles and at least the maintenance of an open central channel. There might also be some microperistaltic assistance toward injection of prey substance, but this remains purely speculative and seems a very inadequate explanation of the vacuum cleaner action reported by Hartog (1901) for *Choanophrya infundibulifera.* It is nevertheless quite plausible to suggest that pressure interferes with some action toward which the tentacular microtubules contribute.

V. Effects of Hydrostatic Pressure on the Behavior of Heterotrich Ciliates

It is well known that various heterotrich ciliates react to mild external stimuli, such as contact with a solid object, by making an avoiding reaction, and to noxious stimuli by sudden contraction of the body. This section concerns some observations (Kitching, 1969) on the effects of pressure on these reactions.

In *Stentor* (Randall and Jackson, 1958) and *Spirostomum* (Finley *et al.*, 1964) there are "ectomyonemes" which consist of groups or stacks of microtubules lying close to the body surface and "endomyonemes" which are fibrillar but not microtubular. One of each is associated with each kinety, and the ectomyonemes connect with basal bodies. It is reasonable to suppose that either ectomyonemes or endomyonemes or both are responsible for sudden contraction. The fact that *Blepharisma* has ectomyonemes but no endomyonemes and is not contractile (Kennedy, 1965) suggests that the endomyonemes are responsible. Moreover, it has recently been shown by Bannister and Tatchell (1968) that the parallel rows of microtubules, which are piled together as a stack to form an ectomyoneme, become thinned out to one or two rows in an extended organism. Bannister and Tatchell suggest that the ectomyonemes re-extend the body after contraction—a process which might be achieved by a sliding of one row upon another, thanks to reformation of certain linkages, seen on the electron micrographs, at the head of the rows.

At pressures of 2000–4000 psi and over, *Spirostomum* ceases to make an avoiding reaction when it reaches the edge of the hanging drop.

Instead, its anterior end bends over and it follows the margin of the drop round and round, while the middle of the drop becomes empty of *Spirostomum*. Nevertheless, when *Spirostomum* was tested with a direct current, or by sudden contact with a sodium chloride solution, it was found to be fully able to reverse its cilia at pressure. The threshold voltage for reversal on the cathodal side was either the same or not greatly elevated. Light blows on the anterior end with a hair, which cause reversal at atmospheric pressure, do not cause reversal at the high pressure in question. Only heavy blows, sufficient to kink the body surface, are effective. It therefore seems that there is failure to receive the mechanical information or to transmit it to the rest of the surface ciliature. There is evidence (reviewed by Jahn and Bovee, 1967) to suggest that the movement of calcium is involved in reversal, but it is not yet known whether there is a spreading of some surface change to initiate this over the whole body, or whether a locally stimulated area, acting as a temporary pacemaker, imposes its beat on the remainder of the body surface by mechanical coordination of the cilia. This second explanation has the attraction of simplicity, and could be applied in other cases (Kitching, 1961, for *Opalina*). In any case, we may suspect that the anterior cilia have a sensory function, as it is these which make contact with the stimulating surface or object. Decreased activity might reduce their sensitivity or might prevent them from setting the stroke for the rest of the ciliature. Alternatively, the suggestion that the microtubules of the "ectomyonemes" act to extend the body by sliding on one another offers a possible clue. High pressure might decrease the rigidity of the body and thus impair mechanoreception. It is interesting that at pressures over 6000 psi there is a shortening of the body; it is not known whether some shortening by contracture of endomyonemes may not also be involved. The normal sensitivity of the body-contracting mechanism, whether to direct mechanical stimulation or to electric shocks, is also drastically though reversibly reduced.

VI. Comments on the Mechanism of the Effects of High Pressure

The effects of high hydrostatic pressure on biochemical reactions and equilibria may be directly manifest in changes in some visible form of activity, such as luminescence, or they may act in an even more complicated way through changes in the fine structure of cells, as in the case of the microtubules of axopods or mitotic spindles. The interactions of temperature and deuterium (with increase favoring stability) and of high pressure and colchicine (with increase reducing stability) (Marsland, 1965, 1968; Marsland and Hecht, 1968; Tilney, 1968) has led to the suggestion (cf. Marsland, Chapter 11) that polymerization of microtubular proteins is directly affected by these influences. Allison and Nunn (1968) have proposed that many anesthetics act specifically on microtubular protein, leading to depolymerization. The anesthetic halothane in clinical concentrations rapidly causes a reversible withdrawal of the axopods of *Actinosphaerium*, according to observations, not yet published, by Allison, Nunn, Macdonald, and Kitching. The axopods of *Actinosphaerium* also retract slowly but reversibly when the animal is exposed in a hanging drop to an atmosphere of pure hydrogen. The effects of a lack of oxygen on the fine structure of the axoneme remains to be investigated, but it appears—as might be expected—that respiration is necessary for maintenance of the normal axopodial organization, and we may ask to what extent high pressure interferes with the transfer of energy from ATP to contractile proteins in axopods, or ameboid phasmagel, or beating cilia. This process might provide a common point of susceptibility. Also, under normal circumstances, its regulation could determine the mobility and behavior—such as they are—of Heliozoa as well as of amebas.

The actual results of the application of high pressure upon Protozoa and doubtless on other cells are to be regarded as a mixture of effects. Metabolic dislocation in combination with structural damage leads to changes in rates of biological activities, which are sometimes accelerated at moderate pressures but always depressed at high pressures (Fig. 6), and to abnormalities in the form of these activities, often accompanied by rounding up, withdrawal of pseudopods or axopods, blistering, or prolapse of the body surface. It is not possible to apportion the blame for structural collapse between increased tension on the part of contractile elements and decreased structural cohesion such as might be caused by solation of ectoplasmic gel or disintegration of microtubules or other fibers.

There are a number of instances in which the behavior of ciliates is changed by moderately high pressure—that of *Spirostomum* has been discussed in some detail. Some ciliates (*Colpoda cucullus, Holophrya* sp., *Euplotes* sp.) are provoked into increased activity, whereas others are caused to stop swimming. *Paramecium aurelia* becomes less active at 500 psi, and an increase in activity is noticeable on release from 200 psi (Kitching, 1957a). These and other changes may perhaps be ascribed to the effect of high pressure on the control of ciliary activity rather than on the ciliary mechanism itself. This possibility leads in turn to a consideration of the role of electrical potential changes in controlling ciliary activity and of the possibility—still disputed and unresolved—that microtubular structures may take part in ciliary coordination. Ciliary reversal is associated—at any rate in some cases—with a depolarization of the cell surface. Although the effects of moderate pressures on various properties

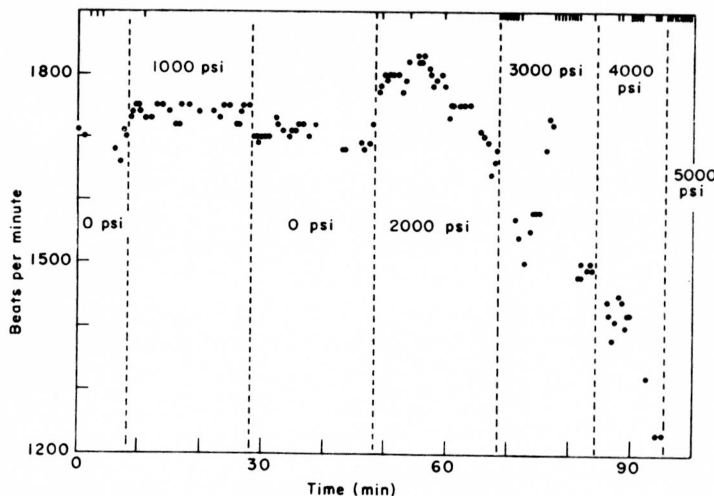

FIG. 6. Effects of hydrostatic pressure on ciliary frequency in the peritrich ciliate *Epistilis plicatilis*. (From Kitching, 1957a.)

of the surface membrane of the squid giant axon are relatively slight, the threshold current of excitation is decreased even at 100 psi, and there is spontaneous excitation at 3000–7000 psi (Spyropoulos, 1957). It is by no means inconceivable that membrane changes induced by high pressure affect the activity of cilia (see also Murakami and Zimmerman, Chapter 6).

Even more speculatively, there is also the possibility that microtubules play a part in interciliary coordination. The anal cirri of *Euplotes* are connected with the system of adoral membranellas, of which they may perhaps be regarded as an extension, by "fibers" which are known to consist of bundles of microtubules (Roth, 1957). Before their fine structure was understood, Taylor (1920) transected these fibers and found that the coordination of adoral membranellas and anal cirri was destroyed. More recently, Okajima and Kinosita (1966) have disputed Taylor's conclusion. They made cine films of the movements of *Euplotes* before and after transection of these fibers, and from their analysis of these records they conclude that changes of the pattern of ciliary beat occur simultaneously in the adoral membranellas and in the anal cirri, even after the transection. On the other hand, Gliddon (1965) simultaneously confirmed Taylor, both in experiments in which the fibers were cut near their anterior ends and in experiments in which they were cauterized by radiofrequency. Although the view that microtubules can conduct has recently found little support (Jahn and Bovee, 1967), there remains the possibility that microtubules coordinate the actions of compound cilia through the constituent basal bodies which they interconnect. The elaborate system of interconnections described by Gliddon (1966) no doubt has morphogenetic significance, but the possibility also remains of coordination by some form of contraction or chemical change. It is possible that the effects of high pressure on coordinated ciliary activity, especially in hypotrichs, may be mediated partly through the microtubular system.

VII. Adaptation

Although the application of high pressures normally leads to an instantaneous change in rate of a biological activity and the release of pressure to a reversal of this change, there are instances in which there occurs some degree of regulation back towards normality. For instance, the frequency of beat of the lateral cilia of *Mytilus edulis* is briefly accelerated by the application of pressure but soon reverts (Pease and Kitching, 1939). Examples of this sort can be interpreted in terms of a transition period between the steady state of metabolic reactions at atmospheric pressure and a new steady state to be established at the experimental pressure, as discussed by Johnson *et al.* (1954). (See, e.g., Chapter 1.) For instance, at the purely metabolic level, acceleration of a reaction late in a series of reactions could initially lead to an acceleration of the terminal activity, but the subsequent depletion of reactants earlier in the series could finally slow the later reaction down again. The process of "adaptation" could involve physical or structural changes rather than, or as well as, a purely metabolic readjustment. To use an example based on a change of temperature rather than of pressure, a sudden rise of temperature over an appropriate range causes a complete cessation of passage of water into the contractile vacuole of *Discophrya;* there is then a slight but rapid swelling of the body, as indicated by the filling up of furrows in the pellicle, which is presumably due to the osmotic uptake of water uncompensated by any equivalent discharge; finally, after several minutes, the contractile vacuole resumes activity at a much greater rate than previously, perhaps in response to the increased protoplasmic hydration (Kitching, 1954b). Several instances of short-term adaptation to high pressure involve structural effects. When *Discophrya* is subjected to a relatively small increase of pressure (2500 psi), the surface wrinkling, already described and interpreted as an expansion, is not permanent; while still at pressure this suctorian undergoes some degree of surface reorganization and tends to resume its former shape (Kitching, 1954a). Similarly, the axopods of *Actinosphaerium* partially recover after they have been caused to retract by the application of moderate pressure (Tilney *et al.*, 1966). Finally, the multiplication of *Tetrahymena* in a culture subjected to moderate pressure shows signs of a comparable phenomenon (A. G. Macdonald, 1967a). Whereas exposure to 100 atm (1470 psi) has little effect on the growth rate of the population, and exposure to 250 atm (3700 psi) immediately and completely prevents cell division, the effects of an intermediate pressure (175 atm, about 2600 psi) are more complicated: division continues for an hour, then stops, but then after several hours is resumed at a lower rate than normal (Fig. 7). The retraction and partial recovery of the axopods of *Actinosphaerium* involves the destruction and reconstruction of microtubules and their reassembly (in smaller numbers) in axonemes. It is not known what fine structural changes are involved in the surface wrinkling and surface reorganization of *Discophrya*. It seems likely that the arrest of divisions in *Tetrahymena* involves structural proteins and that there is a distinct similarity in all three phenomena. A. G. Macdonald (1967b), using Zeuthen's interpretation of effects of inhibitors, has suggested that high pressure delays precursor reactions necessary for the formation of substances or structures which actually carry out cell division: if these precursors have already accumulated sufficiently, or can accumulate after some interval of time, division takes place. It is also interesting that high pressure decreases the oxygen consumption of *Tetrahymena* (Fig. 8), reversibly if the pressure is not too high (A. G. Macdonald, 1965). The rate of oxygen consumption is reduced to about 85% of the normal at 250 atm (3700 psi) and to about a third of the normal at pressures ranging from 550 to 1000 atm (8100 to 14700 psi) in short experiments (A. G. Macdonald and Gilchrist, 1969a). It is not possible to say to what extent this reduction is a cause or effect of other results of compression. A similar explanation involving metabolic readjustment could be invoked for the axopods of *Actinosphaerium* and the surface structure of *Discophrya*. One might suppose that normally the length of axopods is in some way regulated in accordance with the biochemical and energetic resources available to them and that pressure depresses this normal morphogenetic process. This possibility suggests the need for studies of the effect of high pressure on morphogenesis in other Protozoa. Finally,

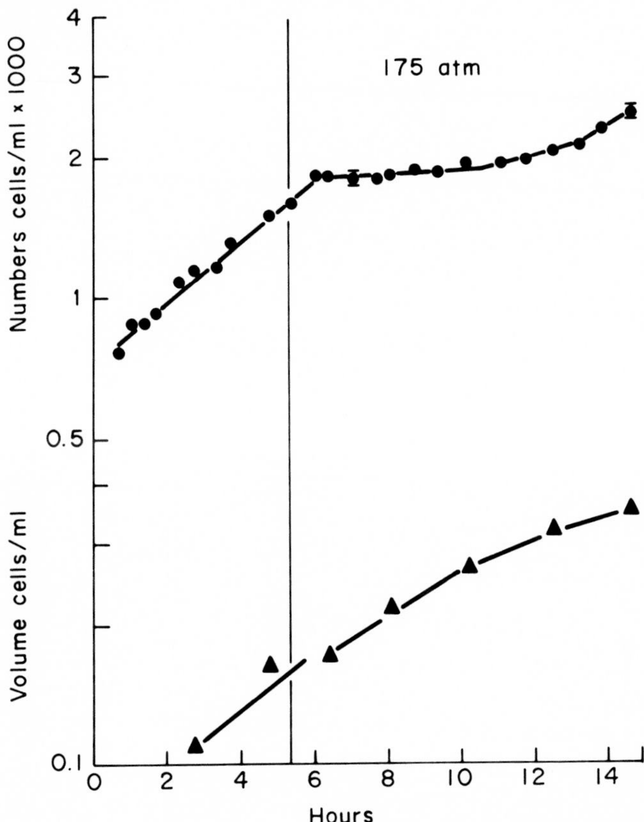

FIG. 7. Growth in numbers and in total cell volume of a population of *Tetrahymena pyriformis* before and during exposure to 175 atm. (From A. G. Macdonald, 1967a).

we may suppose that in *Discophrya* the application of high pressure elicits to the full a reaction normally undergone at a more moderate and controlled rate in response to feeding, which also carries a mechanism of restoration by which the body surface is normally maintained with a snug fit over the underlying body substance. On this interpretation high pressure merely triggers off a physiological response of the animal already built in.

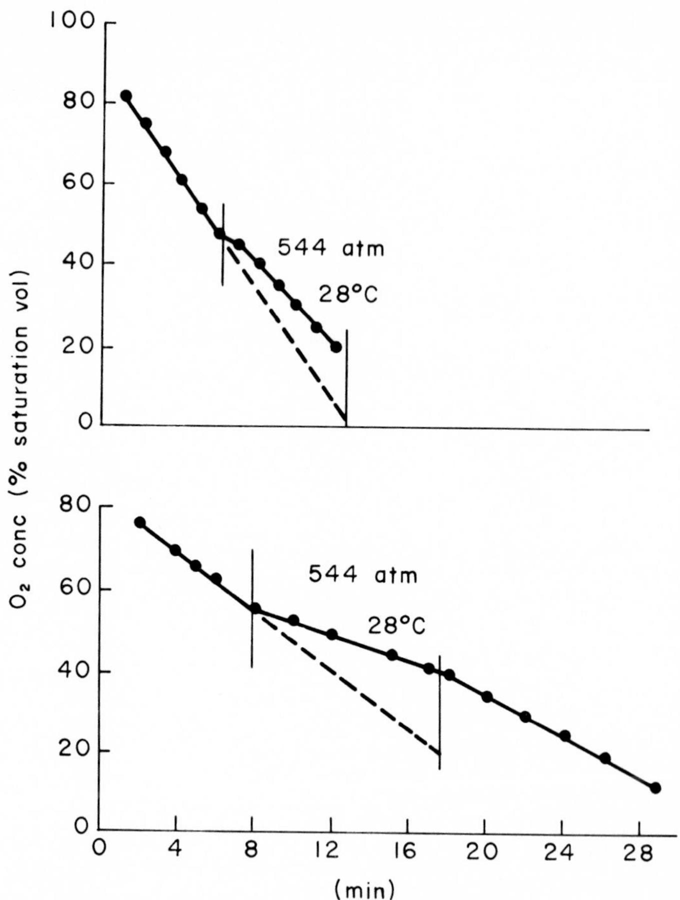

FIG. 8. Results of two experiments showing oxygen consumption of *Tetrahymena pyriformis* subjected to increased hydrostatic pressure. Ordinate—oxygen concentration as percentage of saturation value in equilibrium with air. Abscissa—time in minutes. The first part of each graph shows the oxygen consumption at 1 atm and 25°C, the second part at 544 atm and 28°C, the third part (lower graph only) at 1 atm and 25°C again. (From A. G. Macdonald, 1965).

The possibility that "adaptations" of the kind described above can occur over short periods of time contributes little to the understanding of mechanisms of adaptation to the deep sea, beyond indicating that slow invasion of the deep is not quite as drastic as sudden pressurization. Nevertheless, only the retrieval and culture at pressure of deep sea organisms, as done by ZoBell and Oppenheimer (1950) and Kriss *et al.* (1967) for bacteria, can reveal the biochemical modifications necessary for survival at these pressures. The equipment designed by A. G. Macdonald and Gilchrist (1969a,b), and now brought to an advanced state of development, for retrieval of deep sea plankton and for its transfer to an observation pressure vessel with only minimal change of pressure and temperature throughout, should open the way for such studies.

ACKNOWLEDGMENTS

I am indebted to the editors and publishers of *Nature* and *Experimental Cell Research* and to the Company of Biologists Limited (*Journal of Experimental Biology*) for permission to reproduce figures and plates as stated from these journals.

REFERENCES

Allison, A. C., and Nunn, J. F. (1968). Effects of general anaesthetics on microtubules: A possible mechanism of anaesthesia. *Lancet* **II**, 1326–1329.

Bannister, L. H., and Tatchell, E. C. (1968). Contractility and the fibre systems of *Stentor coeruleus*. *J. Cell Sci.* **3**, 295–308.

Bardele, C. F., and Grell, K. G. (1967). Electronenmikroskopiske Beobachtungen zur Nahrungsaufnahme bei dem Suktor *Acineta tuberosa* Ehrenberg. *Z. Zellforsch. Mikroskop. Anat.* **80**, 108–123.

Batisse, A. (1966). L'ultrastructure des tentacules suceurs d'*Ephelota gemmipara* Hertwig. *Compt. Rend.* **D262**, 771–774.

Behnke, O., and Forer, A. (1967). Evidence for four classes of microtubules in individual cells. *J. Cell Sci.* **2**, 169–192.

Fauré-Fremiet, E., Favard, P., and Carasso, N. (1962). Etude au microscope électronique des ultrastructures d'*Epistylis anastatica* (Cilié Péritriche). *J. Microscopie* **1**, 287–312.

Finley, H. E., Brown, C. A., and Daniel, W. A. (1964). Electron microscopy of the ectoplasm and infraciliature of *Spirostomum ambiguum*. *J. Protozool.* **11**, 264–280.

Gliddon, R. (1965). Ciliary activity and co-ordination in *Euplotes eurystomus*. *Proc. 2nd Intern. Conf. Protozool., London, 1961* p. 246. Excerpta Med. Found., Amsterdam.

Gliddon, R. (1966). Ciliary organelles and associated fibre systems in *Euplotes eurystomus* (Ciliata, Hypotrichida). I. Fine structure. *J. Cell Sci.* **1**, 439–448.

Grimstone, A. V., and Cleveland, L. R. (1965). The fine structure and function of the contractile axostyles of certain flagellates. *J. Cell Biol.* **24**, 387–400.

Hartog, M. (1901). Notes on Suctoria. *Arch. Protistenk.* **1**, 372–374.

Hull, R. W. (1961). Studies on suctorian protozoa: The mechanism of ingestion of prey cytoplasm. *J. Protozool.* **8**, 351–359.

Jahn, T. L., and Bovee, E. C. (1967). Motile behaviour of Protozoa. *Res. Protozool.* **1**, 41–200.

Johnson, F. H., Eyring, H., and Polissar, M. J. (1954). "The Kinetic Basis of Molecular Biology." Wiley, New York.

Kennedy, J. R. (1965). The morphology of *Blepharisma undulans* Stein. *J. Protozool.* **12**, 542–561.

Kitching, J. A. (1939). On the mechanism of movement of food vacuoles in peritrich ciliates. *Arch. Protistenk.* **91**, 78–88.

Kitching, J. A. (1952). Observations on the mechanism of feeding in the suctorian *Podophrya*. *J. Exptl. Biol.* **29**, 255–266.

Kitching, J. A. (1954a). The effects of high hydrostatic pressure on a suctorian. *J. Exptl. Biol.* **31**, 56–57.

Kitching, J. A. (1954b). The physiology of contractile vacuoles. IX. Effects of sudden changes in temperature on the contractile vacuole of a suctorian; with a discussion of the mechanism of contraction. *J. Exptl. Biol.* **31**, 68–75.

Kitching, J. A. (1954c). The physiology of contractile vacuoles. X. Effects of high hydrostatic pressure on the contractile vacuole of a suctorian. *J. Exptl. Biol.* **31**, 76–83.

Kitching, J. A. (1956). Effects of high hydrostatic pressure on a feeding suctorian. *Protoplasma* **46**, 475–480.

Kitching, J. A. (1957a). Effects of high hydrostatic pressure on the activity of flagellates and ciliates. *J. Exptl. Biol.* **34**, 494–510.

Kitching, J. A. (1957b). Effects of high hydrostatic pressures on *Actinophrys sol* (Heliozoa). *J. Exptl. Biol.* **34**, 511–517.

Kitching, J. A. (1961). The physiological basis of behavior in the Protozoa. *In* "The Cell and the Organism" (J. A. Ramsay and V. B. Wigglesworth, eds.), pp. 60–78. Cambridge Univ. Press, London and New York.

Kitching, J. A. (1964). The axopods of the sun animalcule *Actinophrys sol* (Heliozoa). *In* "Primitive Motile Systems in Cell Biology" (R. D. Allen and N. Kamiya, eds.), pp. 445–455. Academic Press, New York.

Kitching, J. A. (1969). Effects of high hydrostatic pressure on the activity and behavior of the ciliate *Spirostomum*. *J. Exptl. Biol.* **51**, 319–324.

Kitching, J. A., and Craggs, S. (1965). The axopodial filaments of the heliozoon *Actinosphaerium nucleofilum*. *Exptl. Cell Res.* **40**, 658–660.

Kriss, A. E., Chumak, M. D., Stupakova, T. P., and Kirikova, N. N. (1967). Glucose metabolism in barotolerant bacteria cultivated under high hydrostatic pressure. *Mikrobiologiya* **36**, 51–57.

Landau, J. V., Zimmerman, A. M., and Marsland, D. A. (1954). Temperature-pressure experiments on *Amoeba proteus*; phasmagel structure in relation to form and movement. *J. Cellular Comp. Physiol.* **44**, 211–232.

Macdonald, A. C., and Kitching, J. A. (1967). Axopodial Filaments of Heliozoa. *Nature* **215**, 99–100.

Macdonald, A. G. (1965). The effect of high hydrostatic pressures on the oxygen consumption of *Tetrahymena pyriformis* W. *Exptl. Cell Res.* **40**, 78–84.

Macdonald, A. G. (1967a). The effect of high hydrostatic pressure on the cell division and growth of *Tetrahymena pyriformis*. *Exptl. Cell Res.* **47**, 569–580.

Macdonald, A. G. (1967b). Delay in the cleavage of *Tetrahymena pyriformis* exposed to high hydrostatic pressure. *J. Cell Physiol.* **70**, 127–130.

Macdonald, A. G., and Gilchrist, I. (1969a). Life in the ocean depths—the physiological problems and equipment for the recovery and study of deep sea animals. *Proc. Conf. Oceanol. Intern., Brighton, Engl., 1969.*

Macdonald, A. G., and Gilchrist, I. (1969b). Recovery of deep sea water at constant pressure. *Nature* **222**, 71–72.

Marsland, D. A. (1965). Partial reversal of the anti-mitotic effects of heavy water by high hydrostatic pressure. An analysis of the first cleavage division in the eggs of *Strongylocentrotus purpuratus*. *Exptl. Cell Res.* 38, 592–603.

Marsland, D. A. (1968). Cell division—enhancement of the anti-mitotic effects of colchicine by low temperature and high pressure in the cleavage eggs of *Lytechinus variegatus*. *Exptl. Cell Res.* **50**, 369–376.

Marsland, D. A., and Hecht, R. (1968). Cell division: Combined anti-mitotic effects of colchicine and heavy water on first cleavage in the eggs of *Arbacia punctulata*. *Exptl. Cell Res.* **51**, 602–608.

Marsland, D. A., and Landau, J. V. (1954). The mechanisms of cytokinesis: Temperature–pressure studies of the cortical gel system in various marine eggs. *J. Exptl. Zool.* **125**, 507–539.

Nachmias, V. T. (1968). Further electron microscope studies on fibrillar organization of the ground cytoplasm of *Chaos chaos*. *J. Cell Biol.* **38**, 40–50.

Okajima, A., and Kinosita, H. (1966). Ciliary activity and co-ordination in *Euplotes eurystomus*. 1. Effects of microdissection of neuromotor fibres. *Comp. Biochem. Physiol.* **19**, 115–130.

Pease, D. C., and Kitching, J. A. (1939). The influence of hydrostatic pressure upon ciliary frequency. *J. Cellular Comp. Physiol.* **14**, 135–142.

Randall, J. T., and Jackson, S. F. (1958). Fine structure and function in *Stentor polymorphus*. *J. Biophys. Biochem. Cytol.* **4**, 807–830.

Roth, L. E. (1957). An electron microscope study of the cytology of the protozoon *Euplotes patella*. *J. Biophys. Biochem. Cytol.* **3**, 985–1000.

Rudzinska, M. A. (1965). The fine structure and function of the tentacle in *Tokophrya infusionum*. *J. Cell Biol.* **25**, 459–477.

Rudzinska, M. A. (1967). Ultrastructure involved in the feeding mechanism of Suctoria. *Trans. N.Y. Acad. Sci. [2]* **29**, 512–525.

Sleigh, M. A. (1968). Patterns of ciliary beating. *Symp. Soc. Exptl. Biol.* **22**, 131–150.

Spyropoulos, C. S. (1957). The effects of hydrostatic pressure upon the normal and narcotised nerve fibre. *J. Gen. Physiol.* **40**, 849–857.

Taylor, C. V. (1920). Demonstration of the function of the neuromotor apparatus in *Euplotes* by the method of microdissection. *Univ. Calif. (Berkeley) Publ. Zool.* **19**, 403–471.

Tilney, L. G. (1968). Studies on the microtubules in Heliozoa. IV. The effect of colchicine on the formation and maintenance of the axopodia and the redevelopment of patterns in *Actinosphaerium nucleofilum* (Barrett). *J. Cell Sci.* **3**, 549–562.

Tilney, L. G., and Porter, K. R. (1965). Studies on microtubules in Heliozoa. I. The fine structure of *Actinosphaerium nucleofilum* (Barrett), with particular reference to the axial rod structure. *Protoplasma* **60**, 317–344.

Tilney, L. G., Hiramoto, Y., and Marsland, D. (1966). Studies on the microtubules in Heliozoa. III. A pressure analysis of the role of these structures in the formation and maintenance of the axopodia of *Actinosphaerium nucleofilum* (Barrett). *J. Cell Biol.* **29**, 77–95.

Tucker, J. B. (1968). Fine structure and function of the cytopharyngeal basket in the ciliate *Nassula*. *J. Cell Sci.* 3, 493–514.

ZoBell, C. E., and Oppenheimer, C. H. (1950). Some effects of hydrostatic pressure in the multiplication and morphology of marine bacteria. *J. Bacteriol.* **60**, 771–781.

CHAPTER 8

BIOSTRUCTURAL, CYTOKINETIC, AND BIOCHEMICAL ASPECTS OF HYDROSTATIC PRESSURE ON PROTOZOA

Selma B. Zimmerman and Arthur M. Zimmerman

I. Introduction

Since the earliest studies of high pressure (Regnard, 1884; Certes, 1884), protozoa have been a favorite cell form in which to study the effects of high hydrostatic pressure. The ease with which protozoa may be cultured, as well as the facility of directly observing them under pressure, have added to their usefulness as experimental organisms.

This chapter deals primarily with high pressure studies on *Amoeba* and *Tetrahymena*, although other protozoa are also discussed. This report is mainly concerned with the effects of pressure on biochemical synthesis, as well as the use of pressure in evaluating the action of physical and chemical agents on gelation reactions. In addition, pressure-induced modifications of cell form and structure, at the levels of the light microscope and electron microscope, are discussed.

Hydrostatic pressure has been used to analyze the structural characteristics of protoplasmic gels and the role that these gels play in form and movement. It is well established that both temperature and pressure exert marked effects on the gelational state of cytoplasmic structures, and that these effects can be evaluated quantitatively. The formation of gelated structures within cells appears to represent an endothermic reaction which is accompanied by a volume increase (cf. Marsland, Chapter 11). Thus, decreasing temperatures and increasing pressures tend to weaken protoplasmic gel structure by causing a shift in the sol–gel equilibria toward the sol state.

Heat-synchronized cultures of *Tetrahymena pyriformis* have been studied for the past 15 years, since Scherbaum and Zeuthen (1954) reported on a method for synchronizing division in these cells. Synchronized *Tetrahymena* provide excellent material for studying cell division and they have been used in our laboratory for investigating the effects of high pressure on cell division. The technique of synchronization consists of subjecting the cells to a total of eight heat shocks at 34°C for 30-minute durations, alternated by 30-minute intervals at 28°C. At 70–75 minutes after the last heat shock (EH), 89–90% of the cells divide synchronously. This method permits the biochemical analysis of large numbers of cells, at comparable stages of development, under hydrostatic pressure or following decompression.

II. Pressure Effects on Morphology

A. Form and Movement

Direct observations of protozoa following compression and decompression at various pressure levels offer insight into the relationship of the protoplasmic gel structure to the morphology, stability, and locomotion of these cells.

1. *Stability of Amoeba under Pressure*

One of the earliest pressure studies which demonstrated that the form and movement of *Amoeba* was directly related to the maintenance of the plasmagel structure was that of Marsland and Brown (1936). Landau and co-workers (1954) conducted experiments with *Amoeba proteus,* at a constant temperature of 25°C at various pressure levels. They reported that at 1000–1500 psi, pressurized amebae appeared similar to the controls, although there were fewer pseudopodia that were frequently monopodal. At 2000–3000 psi, the pseudopodia were shorter with bulbous tips. Movement was reduced, and changes of form occurred slowly. With increasing pressure, the pseudopodia became more unstable and retracted. At 5000–6000 psi there was a slow (5–10 minute) rounding of the amebae into motionless spheres. Where there were well-extended pseudopodia, these quickly collapsed, pinched off from the distal end, and rounded up upon pressure treatment. Following the return to atmospheric pressure from 5000 psi, (within 10–20 seconds), a broad hyaline zone filled with a clear fluid developed between the cell membrane and the granular plasmagel layer, as a result of a generalized contraction of the plasmagel. This was followed by a slow return of protoplasmic streaming and formation of pseudopodia. Within 4–5 minutes, normal locomotion reappeared. In pressure studies with *Amoeba dubia* similar behavior was observed.

Recently, Landau (1965) investigated the "sphering" of *Amoeba proteus* under hydrostatic pressure to see whether the "sphering" was accompanied by a volume increase. Since the collapse of the contractile vacuole occurs following pressure application, he considered that osmoregulation might be disrupted, which could result in a swelling or volume increase of the cell. Volume measurements were made by comparing photographs of cells 15 seconds after application of pressure (8000 psi at 25°C) with those of cells under this pressure for 20 minutes. This analysis revealed a slight decrease in volume that occurred within 2 minutes of the applied pressure. Thus, Landau (1965) concluded that osmotic forces appeared to play no role in "sphering."

Pinocytosis in amebae has been the subject of numerous investigations (cf. Rustad, 1964; Chapman-Andresen, 1962). In experiments employing hydrostatic pressure (Zimmerman and Rustad, 1965), pinocytosis was reversibly blocked within 1 to 2 minutes by a pressure of 3000 psi. Although existing channels regressed at this pressure, some small pseudopodia persisted up to 10 minutes. The channels reappeared within 20 seconds following decompression. Thus, the maintenance of pinocytotic channels appears more sensitive to high pressure than that of the pseudopods. This would indicate that pinocytosis may involve at least two distinct physiologic processes, the formation of small pseudopodia, and the formation of channels.

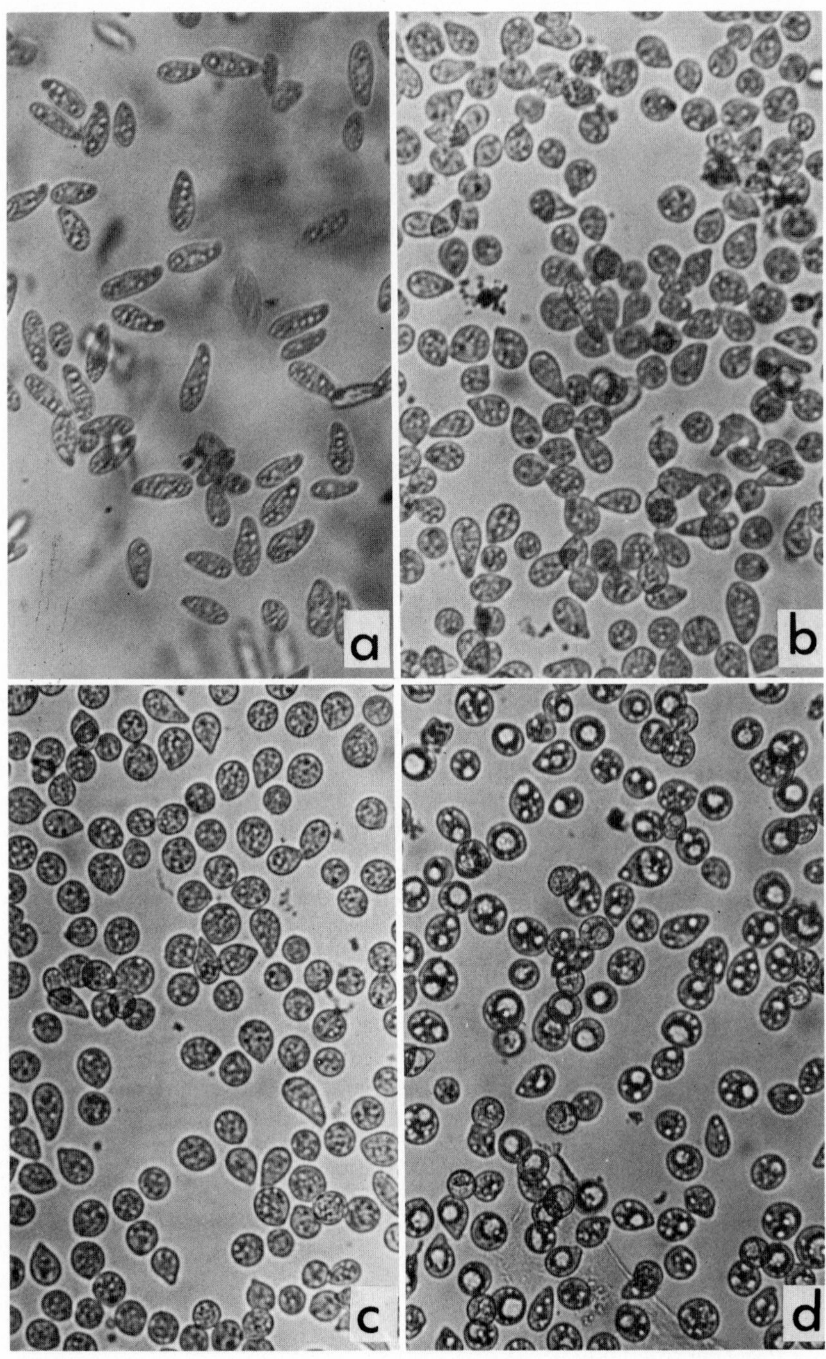

FIG. 1

2. *A Comparison of Pressure Effects on Synchronized and Nonsynchronized Tetrahymena*

Hydrostatic pressure is known to cause a decrease in locomotor activity and a rounding of the cell body in *Tetrahymena pyriformis* (Simpson, 1966; Lowe-Jinde and Zimmerman, 1969). These pressure effects become more evident as the magnitude and duration of pressure treatment are increased. The effect of high pressure on the movement of *Tetrahymena* cells is rapid and readily reversible. The rounding of the *Tetrahymena* cell body following pressurization is not as rapid an effect nor as easily reversed as is the reduction in locomotion. Numerous vacuoles appear in the cell cytoplasm following decompression and may be representative of some damage caused by the pressure treatment.

A comparison of the effects of various levels of pressure on the morphology and locomotion of logarithmically growing cells and heat-synchronized cells (at 10 and 60 minutes after EH) was conducted in our laboratory. Observations revealed that the log cells were more sensitive to high pressure than were the heat-synchronized cells. In addition, recovery of normal shape and movement occurred more slowly in the log cells than in the synchronized cells. At the lowest pressure (2000 psi for 10 minutes) there were no obvious changes in locomotion or appearance of the cells except for a transient increase in the rate of movement (within 1 minute) which was more marked in the log cells. Within 3 minutes the log cells became slightly rounded and the synchronized cells showed a slightly bulbous posterior. Upon decompression (following 10 minutes of pressure) the cells resumed a normal rate of movement; however, within 5–10 minutes, vacuoles were observed in the cytoplasm of both the synchronized and the log cells, the latter retaining a rounded appearance. At 7500 psi, there was a rapid reduction in the rate of locomotion of log and synchronized cells, and within 5 minutes, the majority of log cells ceased their forward spiraling motion, although the cilia of some of these cells continued to beat. The shape of the log cells was modified to that of a tear drop within 2 minutes, whereas the heat-synchronized cells rounded up posteriorly within 5–10 minutes. Vacuoles were observed in the cell

---

FIG. 1. The effects of pressure on a log growth culture of *Tetrahymena pyriformis* GL. The photomicrographs were taken while the cells were in the pressure chamber. (a) Control cells photographed at atmospheric pressure. (b) Cells under 10,000 psi for a duration of 5 minutes. Although cilia continue to beat, movement is markedly decreased. The cells become contracted and many cells are tear shaped. (c) After 10 minutes at 10,000 psi many cells have uncoordinated ciliary activity and there is very little translational movement. (d) Cells photographed 5 minutes after decompression exhibit prominent vacuoles.

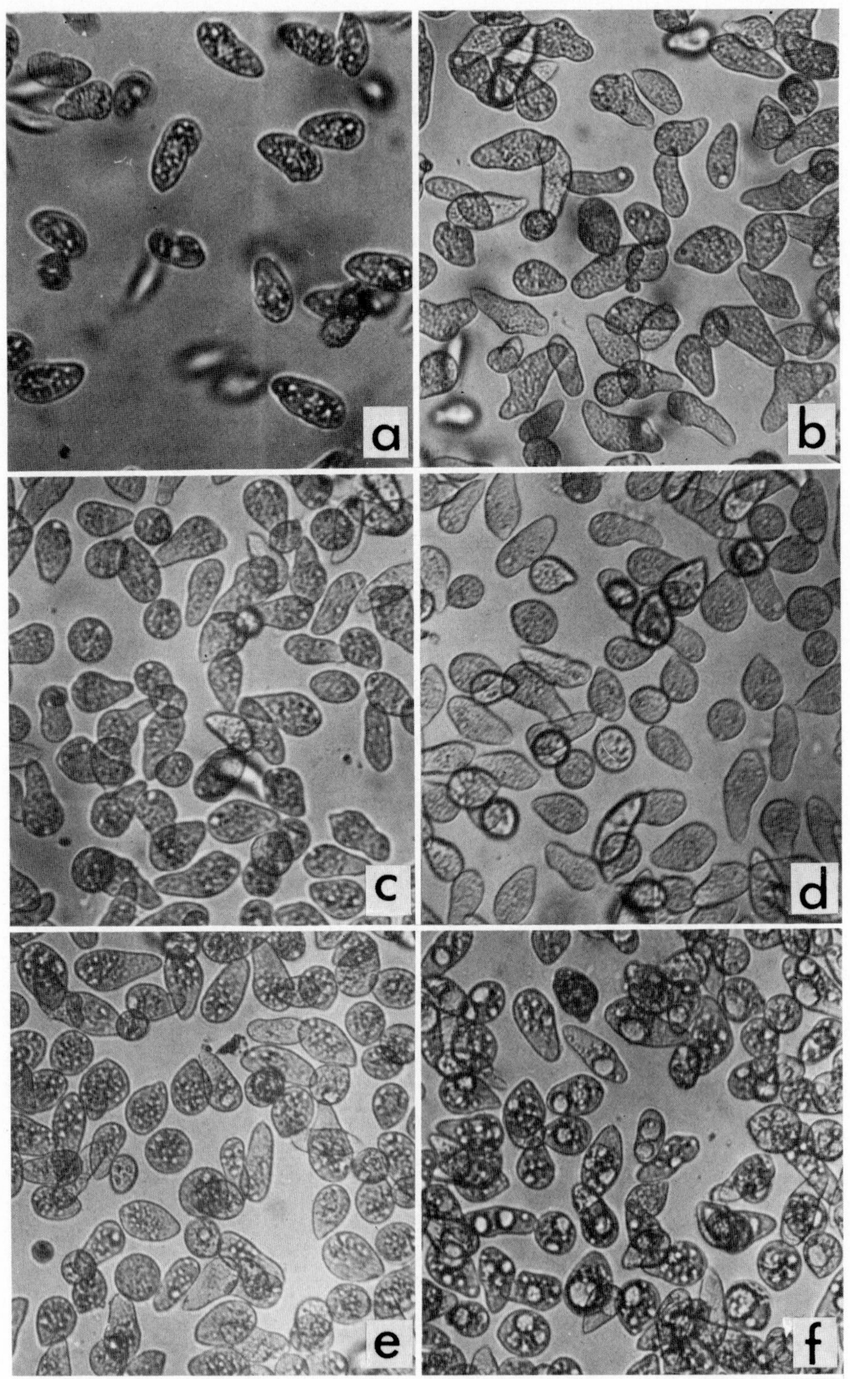

FIG. 2

cytoplasm of log and synchronized cells 5 minutes after pressure release. At 10,000 psi there was an immediate loss in locomotor activity. The majority of log cells halted their movement within 2–5 minutes after compression, whereas many of the synchronized cells ceased to move approximately 5 minutes following pressure initiation. The pressure effect on the shape of the log cells and the synchronized cells was similar to that found at 7500 psi, except that at 10,000 psi there were a larger number of cells that changed their shape. Vacuolization was observed in log and synchronized cells approximately 5–10 minutes after decompression. The effects of pressure on log phase and heat-synchronized cells are shown in Figs. 1 and 2.

3. *Studies in Other Protozoa*

The agency of high hydrostatic pressure has been employed on other protozoa such as *Blepharisma* and *Paramecium* (Auclair and Marsland, 1958; Asterita and Marsland, 1961; Giese, 1968) and *Euglena* (Byrne and Marsland, 1965) in a further analysis of the importance of the peripheral gelated cytoplasm to the form and movement of these cells. Generally these protozoa lose their form stability and become rounded following pressurization; nevertheless, the susceptibility of different species will vary for similar temperature and pressure parameters. Such variation of the resistance to high pressure solation among different protozoa is presumably related to structural differences of the surface membranes of these organisms (cf. Kitching, Chapter 7).

B. Modification of Ultrastructure under Pressure

Landau and Thibodeau (1962) designed a pressure-fixation chamber that permits material to be fixed while under hydrostatic pressure. The vessel consists of two chambers separated by a cover glass that can be broken by a steel ball, allowing fixation of the cells (cf. Landau, Chapter 2). They studied the ultrastructure of *Amoeba proteus* at atmospheric

---

Fig. 2. Effects of pressure on a heat-synchronized culture of *Tetrahymena pyriformis* GL at 60 minutes after EH. (a) Photograph taken in pressure chamber at atmospheric pressure. (b) Cells at 10,000 psi for a duration of 5 minutes; most cells have stopped their movement. (c) Photograph of cells after 10 minutes of 10,000 psi. Many cells accumulate on the lower window of the pressure chamber. Some cells tend to become spherical while others exhibit a bulbous posterior. (d). Photograph of cells 30 seconds after decompression. Immediately upon decompression many cells start to exhibit translational movement. (e) Five minutes after decompression many cells are moving and a large number exhibit vacuoles. (f) Ten minutes after decompression vacuoles become more prominent.

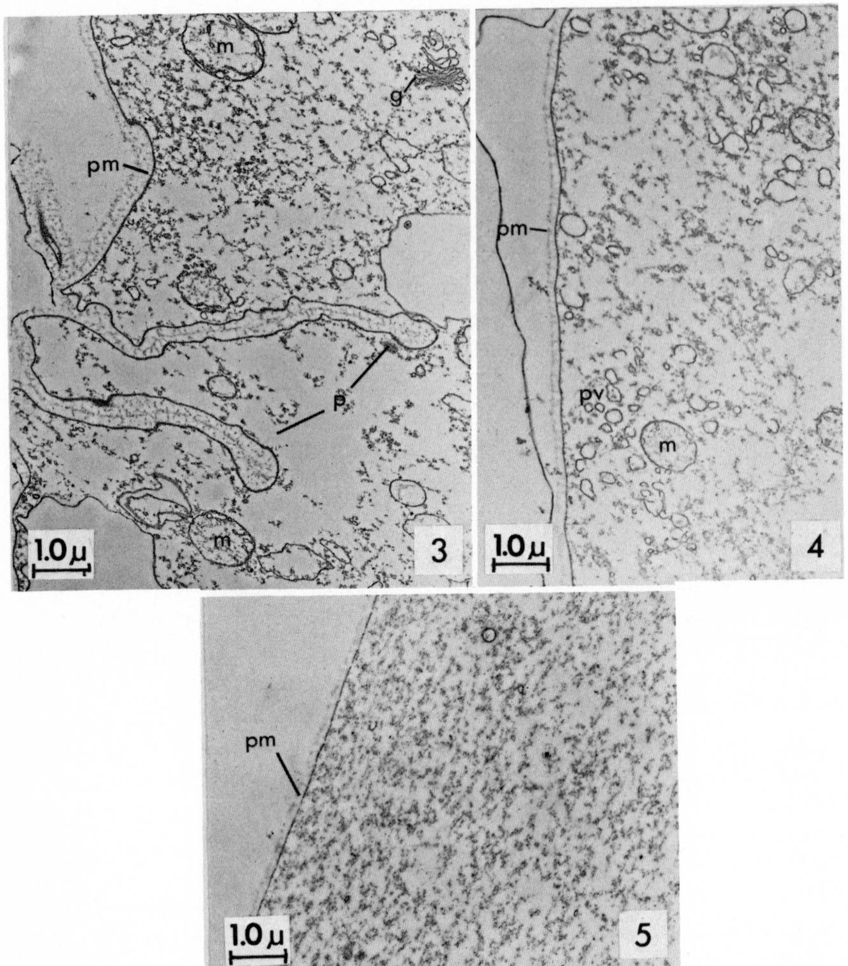

FIG. 3. Electron micrograph of *Amoeba proteus* at atmospheric pressure. Section near the surface illustrates pinocytosis channels. The specimens were fixed in 2% potassium permanganate: p—pinocytosis channels; pm—plasmalemma; m—mitochondrion; g—golgi body. (From Landau and Thibodeau, 1962.)

FIG. 4. Section through *Amoeba* fixed under pressure. Specimen was subjected to 8000 psi for 20 minutes at 20°C and fixed in a pressure chamber designed by Dr. J. Landau. Pinocytosis channels disappear: pv—may represent vesicles following breakdown of pinocytosis channels; pm—plasmalemma; and m—mitochondrion. (From Landau and Thibodeau, 1962.)

FIG. 5. Section near the surface of *Amoeba* fixed 20 seconds after decompression. pm is the plasmalemma. (From Landau and Thibodeau, 1962.)

pressure (Fig. 3) under 8000 psi for a duration of 20 minutes (Fig. 4) and at 20 seconds after decompression (Fig. 5). The control specimens clearly showed Golgi complex, mitochondria, and pinocytotic channels. The pressure-treated cells revealed a loss of Golgi complex and pinocytotic channels. The pressurized plasmalemma membrane was smooth in contour and quite similar to the control specimens except for the absence of pseudopodia. Small vesicles appearing near the plasmalemma were thought to be indicative of the breakdown of the pinocytosis channels. Following decompression (20 seconds) the plasmalemma was unchanged but the hyaline zone appeared to contain no structural components. Tilney *et al.* (1966) investigated the role of microtubules in the heliozoan, *Actinosphaerium nucleofilum*, by means of high pressure analysis. Electron micrographs revealed that hydrostatic pressure caused disintegration of the microtubule system, which was accompanied by an instability of the axopodia (cf. Marsland, Chapter 11; Kitching, Chapter 7).

Kennedy and Zimmerman (unpublished) undertook studies of the effects of high pressure on the fine structure of logarithmically growing cultures of *Tetrahymena pyriformis*. In general, there were marked structural changes at pressures of 7500 and 10,000 psi which were related to the duration of treatment. Alterations in ultrastructure could be seen with pulses as short as 2 minutes at 7500 psi; these alterations became more pronounced with increasing duration and magnitude of pressure.

The cytoplasm of the *Tetrahymena* is covered by a series of membranes (Allen, 1967). The outer membrane of the pellicle covering the cilium is wavy. Under this membrane there are two membranes which cover the basal body cilium complex. The unpaired central ciliary tubules extend from the terminal plate through the cilium (Fig. 6). The proximal portion of the central ciliary tubules, as well as the basal body material, are sensitive to hydrostatic pressure. The unpaired central ciliary tubules, peripheral to the axosome, undergo dissolution following a pressure of 7500 psi for 2 minutes, which is accompanied by some disorganization of the basal cilium (Fig. 7). At an increased duration and magnitude (10,000 psi for 10 minutes) the two central tubules in the area of the terminal plate undergo further dissolution, and the granular mass that is associated with the basal body is either missing or frequently displaced to one side (Fig. 8). At the lower pressures there appears to be a definite swelling of the granular mass associated with the basal bodies.

The longitudinal microtubules are found below the cell surface as a series of ten to seventeen discrete filaments situated between the inner pellicular membrane and the amorphous material lining this membrane. They appear as an electron transparent core surrounded by an electron dense cylinder. These microtubules, which extend to the anterior end of the animal, have an outside diameter of 240 Å and are separated by a space of about 80 Å. (cf. Allen, 1967; Pitelka, 1961). The microtubules may be seen in cross section (Fig. 9) and in longitudinal section (Fig. 10).

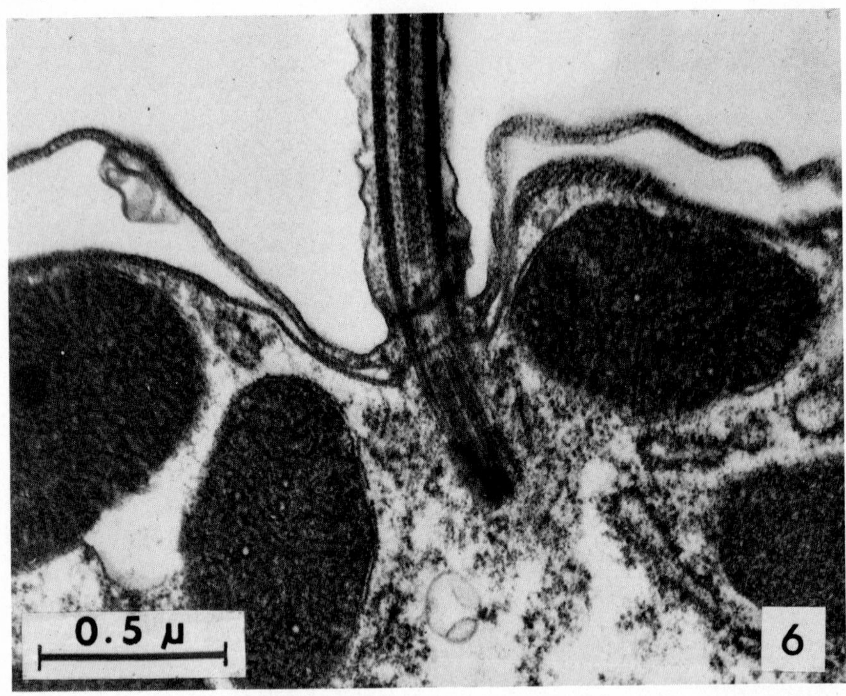

FIG. 6. Longitudinal section through surface of *Tetrahymena* at atmospheric pressure. Section through basal body cilium complex shows the outer membrane of the pellicle covering the cilium. The inner membranes of the cilium are continuous with the basal body. Log phase cultures were fixed in 3% glutaraldehyde solution buffered with 0.1 *M* cacodylate, pH 7.2. Cells were mixed with the fixative 1:1 (v/v). Following 20 minutes in glutaraldehyde the cells were washed in 0.1 *M* cacodylate buffer containing 10% sucrose and postfixed in 1% osmium for 20 minutes prior to embedding. Magnification 51,600 ×. (Kennedy and Zimmerman, 1968.)

FIG. 7. Section through surface of *Tetrahymena* subjected to 7500 psi for a duration of 2 minutes. Cells were fixed under pressure in a Landau designed pressure chamber. Longitudinal section through cilium shows dissolution of inner fibers and some disorganization of the basal body. Magnification 51,600 ×. (Kennedy and Zimmerman, 1968.)

FIG. 8. Section through surface of *Tetrahymena*. The cells were subjected to 10,000 psi for a duration of 10 minutes and fixed under pressure. Further disorganizing effects of pressure are visible in the cytoplasm and in the cilia. Magnification 51,600 ×. (Kennedy and Zimmerman, 1968.)

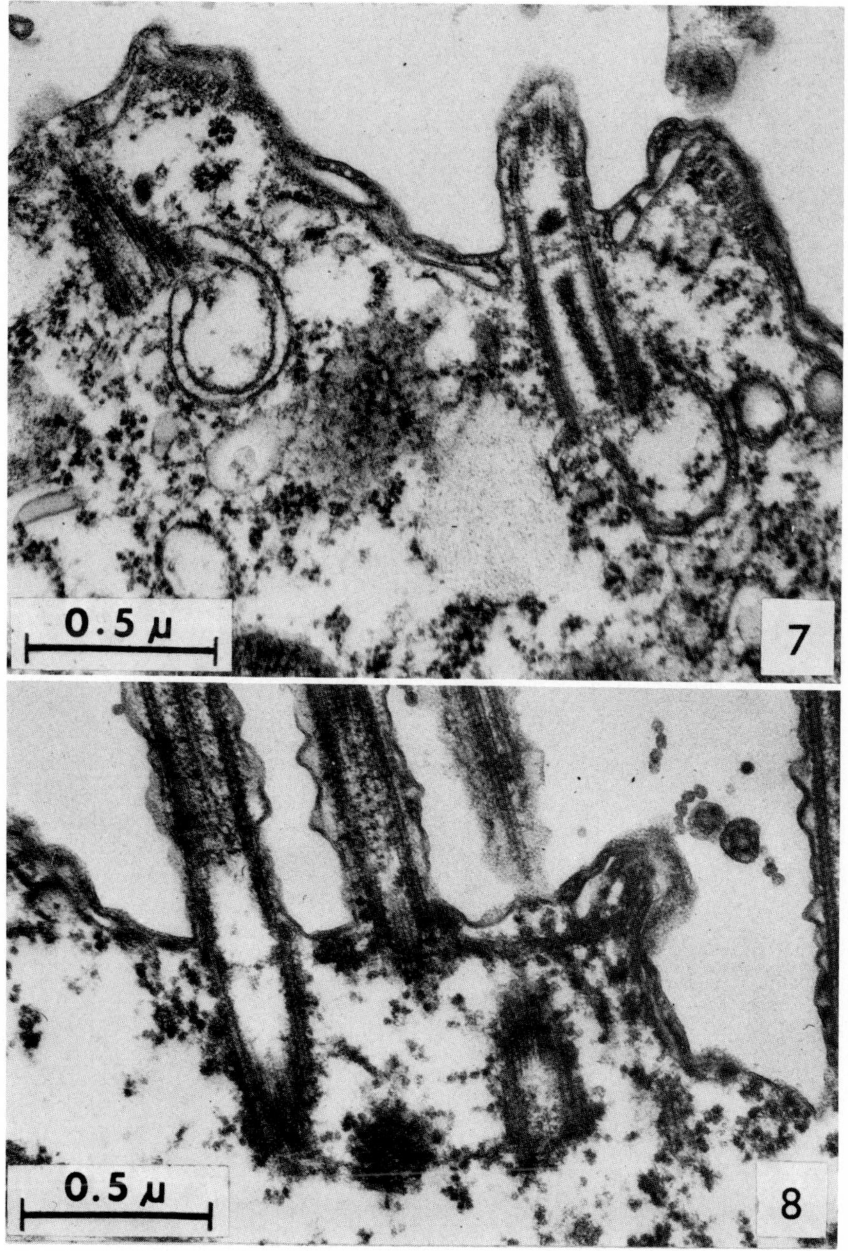
0.5 μ
7
0.5 μ
8

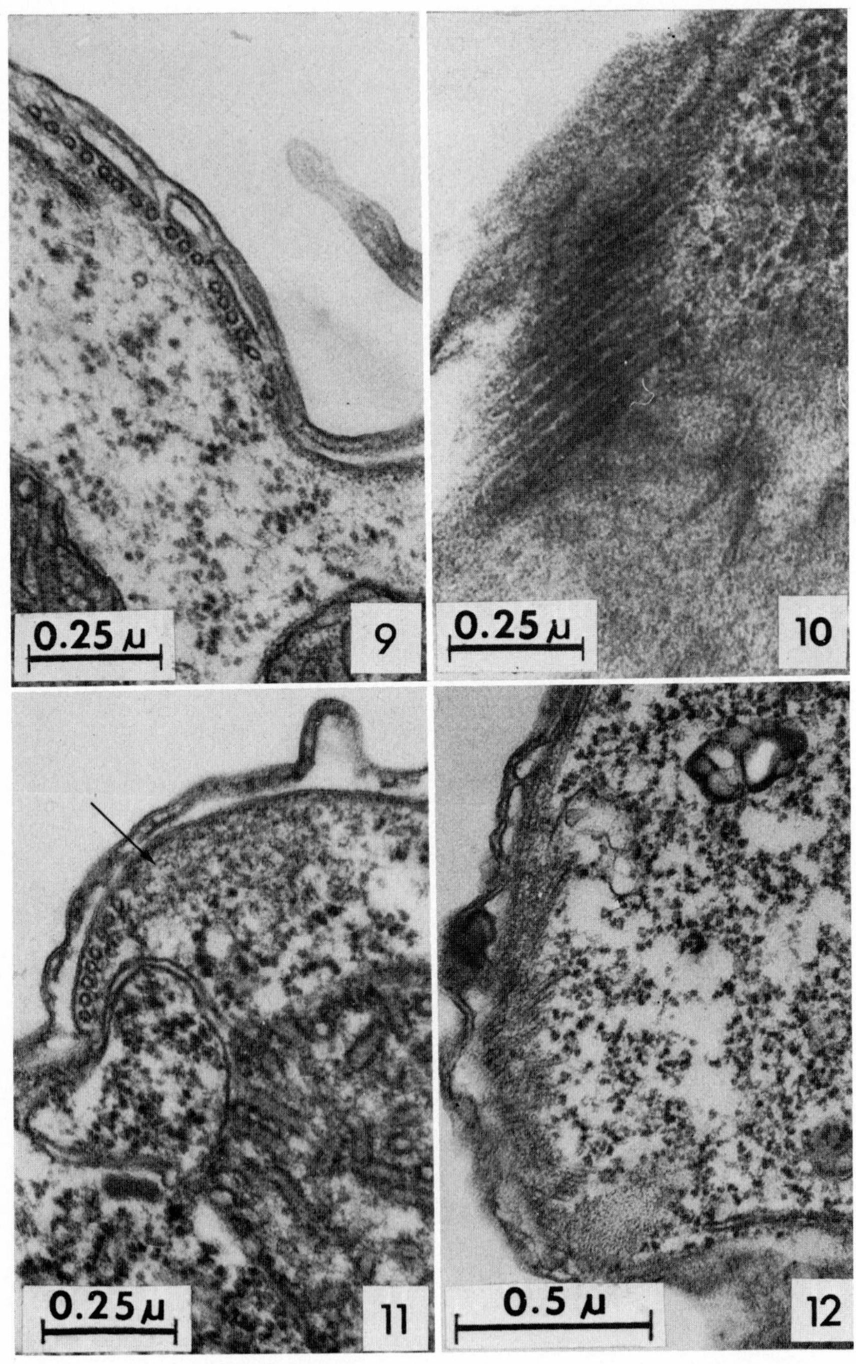
0.25 μ
9
0.25 μ
10
0.25 μ
11
0.5 μ
12

Under hydrostatic pressure, these microtubules became unstable and disintegrated. A section through a cell, fixed under hydrostatic pressure following a 2-minute compression of 7500 psi, shows a progressive disorganization of the longitudinal microtubules; an amorphous dense granular material has replaced the disorganized tubules (Fig. 11). A tangential section through the cell surface of a *Tetrahymena,* subjected to 10,000 psi for 10 minutes, reveals a disorganization of the longitudinal microtubules (Fig. 12).

The endoplasmic reticulum was swollen, distended, and clumped at the highest pressures (10,000 psi for 10 minutes). The lysosomes and plasma membrane appeared unaffected by these pressures. The mitochondria were unchanged except for the possibility that there was an increased number of intramitochondrial masses. The nuclear membrane of the pressurized cells seemed comparable with the controls; however, the chromatin was clumped in the pressurized cells.

III. Effects of Pressure on Division in *Tetrahymena*

Investigations of high hydrostatic pressure on cell division in *Tetrahymena pyriformis* have been made on both synchronized and exponentially growing cells (Lowe-Jinde and Zimmerman, 1969; Macdonald, 1967a,b; Simpson, 1966). In general, these studies showed that pressure can cause delays in division as well as block division of the cells. The manifestation and intensity of these effects depend upon the magnitude of pressure, the length of pressure treatment, and the time at which the pressure is applied during the cell cycle.

In heat-synchronized *Tetrahymena,* pressure pulses (10,000 psi for 2 minutes) applied after the last heat shock resulted in progressive division delays that were directly related to the time after EH that the cells were

---

FIG. 9. Cross section through longitudinal microtubules in *Tetrahymena pyriformis* at atmospheric pressure. Magnification 72,400 ×. (Unpublished electron micrograph courtesy of John Kennedy.)

FIG. 10. Longitudinal section showing longitudinal microtubules at atmospheric pressure. Magnification 72,400 ×. (Unpublished electron micrograph courtesy of John Kennedy.)

FIG. 11. Cross section through *Tetrahymena* subjected to 7500 psi for 2 minutes. Cells were fixed under hydrostatic pressure. Some microtubules are disorganized (arrow) and dense granular material remains. Magnification 72,400 ×. (Kennedy and Zimmerman, 1968.)

FIG. 12. Tangential section through cell surface of *Tetrahymena* subjected to 10,000 psi for 10 minutes. Cells fixed under high pressure. Longitudinal microtubules are disorganized. Magnification 51,600 ×. (Kennedy and Zimmerman, 1968.)

pressurized (Fig. 13A). At this pressure level there was a disruption of cell synchrony when pressure was applied between 43 and 48 minutes after EH. A lower pressure (7500 psi for 2 minutes) initiated at this time also resulted in division delays, but these were shorter than those observed with the higher pressure. However, there were negligible division

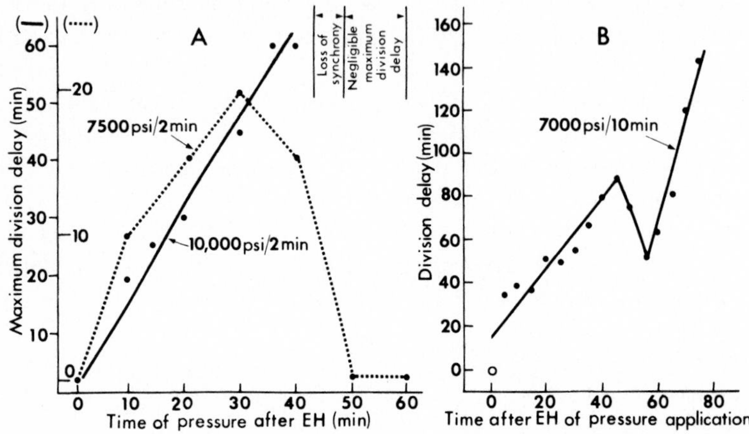

FIG. 13. The effect of pressure on division delay in synchronized *Tetrahymena*. (A). Synchronized cultures of *Tetrahymena pyriformis* GL were subjected for 2 minutes to a pressure of 7500 psi (......) and 10,000 psi (———) at 28°C. The division delays at various times after EH are illustrated. Between 42 and 48 minutes after EH, 10,000 psi pressure treatment causes disruption of cell synchrony; at 50–60 minutes after EH there was negligible delay. (Data of Lowe-Jinde and Zimmerman, 1969.) (B). Graph redrawn from the data of Simpson (1966); the effect of 7000 psi of pressure (for 10 minutes) on synchronized *Tetrahymena pyriformis* GL. The average division delays are plotted as a function of the time after EH at which pressure is applied. The mean division time was recorded to be 94 minutes after EH (open circle). In order to obtain the average division delay, the mean division time was subtracted from the average division time as read from the graph (text Fig. 4) of Simpson (1966).

delays when treatment with either pressure was applied from 50 to 60 minutes after EH (Lowe-Jinde and Zimmerman, 1969). Nevertheless, although the average time of maximum division was not delayed (at EH 50–60 minutes), Lowe (1968) found that furrowing was extended over a period ranging from 60 to 135 minutes following EH.

The work of Simpson (1966) on pressure-induced division delays of heat synchronized *Tetrahymena* concurs with the Lowe and Zimmerman (1969) data, insofar as he found age-dependent division delays during the first 45 minutes following the last heat shock. However, a major discrepancy between the two studies appears to be the absence of a "physiological transition point" in Simpson's investigation. The physiological transition point refers to a time during the division cycle after which the cells are relatively insensitive to chemical and physical agents (Zeuthen, 1964). Simpson (1966) employed pressure pulses of 7000 psi for a duration of 10 minutes which were initiated at EH 5 (see Fig. 13B). The difference between the patterns shown in Figs. 13A and B may be due in part to differences in the duration of the pressure treatment employed in the two experiments. The pressure sensitivity of the synchronized *Tetrahymena* cells increases as the division schedule progresses, and this may be a reflection of the accumulation of pressure-sensitive material necessary to the division process (see discussion, Zimmerman, 1964).

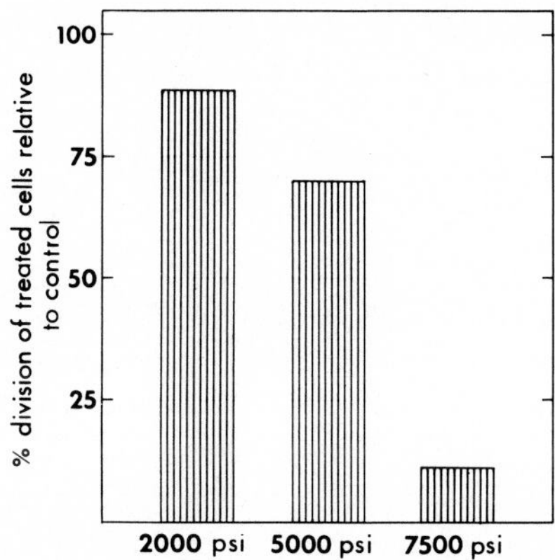

FIG. 14. The effect of pressure on division of synchronized *Tetrahymena*. The cells were subjected to various pressure intensities at 70 minutes after EH for a duration of 20 minutes. Cell counts were determined at the onset of pressure and after decompression. The percent division of treated cells relative to the controls is illustrated at 2000, 5000, and 7500 psi. (From the data of Lowe-Jinde and Zimmerman, 1969.)

Pressurization of heat-treated *Tetrahymena* cells at the time of furrowing can prevent the progress of the furrow reaction as well as cause cells with well-developed furrows to abort. This was shown in the experiments of Lowe-Jinde and Zimmerman (1969) in which cells were subjected to various pressure intensities for 20-minute durations at 65–70 minutes after EH. The percent division of treated cells at three different pressure levels is illustrated in Fig. 14. At a pressure of 7500 psi the furrowing reaction was blocked, i.e., further progression of furrowing was halted, and cleavage furrows that had already formed were abolished. At the lowest pressure, 2000 psi, furrowing was somewhat retarded but not blocked.

After 20 minutes of pressure treatment, the increase in cell count was slightly less than that of nonpressurized controls. The intermediate pressure value of 5000 psi resulted in a partial blockage of furrowing with 30% of the cells being blocked. Recently, Macdonald (1967a) reported that 250 atm of pressure (3700 psi) immediately arrests cell division in logarithmically growing cultures of *Tetrahymena pyriformis* strain W. It would appear from these studies that synchronized *Tetrahymena* are more resistant to pressure inhibiting effects than are logarithmically growing cells. It is interesting to note that Macdonald (1965) found a 14% reduction in oxygen consumption of *Tetrahymena* under 272 atm (4000 psi).

IV. Action of Chemical and Physical Agents on Pressurized Cells

Numerous studies have utilized high pressure as an analytical tool to evaluate the effects of chemical and physical agents on protozoa. In *Amoeba,* high pressures (and low temperatures) produce an exponential weakening in the structural state of the plasmagel layer (Brown and Marsland, 1936; Landau *et al.*, 1954; Landau, 1959). Thus, measurements of pseudopodial stability under pressure may be used to quantitate the effects of chemical and physical agents on gelational reactions. Studies of other protozoa (*Blepharisma, Paramecium, Tetrahymena*) have shown that hydrostatic pressure can affect regeneration and division schedules and cause cell lysis. These factors have also been used to assess the action of chemical and physical agents.

A. Chemical Agents

Analysis of adenosine triphosphate (ATP) treated *Amoeba proteus* (5×10^{-4} *M*) under hydrostatic pressure revealed an increased pseudopodial stability that was presumably related to an increase in the gelational state of the plasmagel (Zimmerman *et al.*, 1958). However, ATP-related compounds (adenosine monophosphate, adenosine pyrophosphate, and Na_2HPO_4) in concentrations comparable with those of ATP did not increase the pseudopodial stability as measured by high pressure. This study was compatible with an earlier investigation of the effects of ATP on marine eggs (Landau *et al.*, 1955), in which small amounts of ATP increased the furrowing potential of the cells. Studies of endogenous ATP in tissue culture cells and marine eggs under pressure were carried out by Landau and co-workers (Landau and Peabody, 1963; Landau, 1966).

In order to elucidate further the role of ATP in ameboid movement, the effects of dinitro-*o*-cresol (DNC) on pseudopodial stability were investigated. This agent was chosen since it acts as an uncoupler of oxidative phosphorylation and, accordingly, would interfere with the action of ATP. At a concentration of 2×10^{-3} *M*, DNC had no effect on the form, movement, or pseudopodial stability of *Amoeba proteus* except at relatively high concentrations and long exposures. The absence of a measurable effect may be a consequence of slow penetration of DNC into the cell, since it was demonstrated that DNC in combination with ATP counteracts the ATP stabilizing action and produces a marked weakening of pseudopodial stability. The potent effect of DNC in combination with ATP is thought to result from ATP providing energy to facilitate transport of DNC across the cell surface (Zimmerman *et al.*, 1958; Zimmerman, 1959). Indeed, the action of ATP on *Amoeba* is complex, since ATP can enter the cell as well as be degraded by enzymes on the surface of the *Amoeba* (Zimmerman, 1962).

Chemical agents, such as mercaptoethanol and heavy water, which alter the stability of structural proteins, affect the reaction of *Amoeba* to hydrostatic pressure. It has been shown that the proper balance of thiol and disulfide groups is essential for the formation and maintenance of gelated structures within the cell (cf. Zimmerman, 1963). Thus when amebae were treated with mercaptoethanol, which readily supplies an excess of thiol groups and reduces protein disulfide bonding, the ability of the pseudopodia to withstand high pressure solation was markedly reduced (Zimmerman, 1964). On the other hand, treatment of amebae with heavy water resulted in a stronger plasmagel system that was more resistant to the solating action of high pressure (Marsland, 1964). It is conceivable that the increased stabilization results from a substitution of the relatively stronger D-bonds for H-bonds in structural proteins.

The action of heavy water has also been shown to counteract the blocking effects of pressure on division in synchronized *Tetrahymena* (Lowe-Jinde and Zimmerman, 1969). When cells were treated with either heavy water (30%) or hydrostatic pressure (3000 psi), cell division was blocked in 20–45% and 31–58% of the cells, respectively. However, the combination of heavy water (30%) and pressure (3000 psi) resulted in an inhibition of only 6–19% of the cells. The data from these studies are expressed as a ratio of treated cells to control cells (Table I). Thus, amelioration of the division-delaying effects of pressure was observed in these cells treated simultaneously with pressure and heavy water. In marine eggs, Marsland and co-workers (cf. Marsland, Chapter 11) have established that heavy water counteracts the solating action of pressure on cytokinesis, meiosis, and mitosis.

However, in the case of *Blepharisma* regeneration, the counteracting effects of heavy water on pressure-induced regeneration delay were not as clearly evident. Giese (1968) found that regeneration of postperistomal portions of *Blepharisma intermedium* was delayed under pressures of 2000 psi as well as by 30% deuterium oxide treatment. When heavy water and pressure treatment were combined, the regeneration delay was comparable with that from heavy water alone. Giese suggests that D_2O and pressure may counteract one another to the extent that the action of these agents were not additive.

TABLE I
PROTECTIVE EFFECTS OF HEAVY WATER ON PRESSURE-TREATED CELLS DURING CYTOKINESIS

Experimental series	Ratios of dividing cells		
	$\frac{3000 \text{ psi}}{\text{control}}$	$\frac{30\% \ D_2O}{\text{control}}$	$\frac{3000 \text{ psi} + 30\% \ D_2O}{\text{control}}$
I	0.69	—	0.94
II	0.54	0.61	0.81
III	0.42	0.80	0.92
IV	0.57	0.55	0.93

In an effort to determine the relationship of the pellicle and the subjacent plasmagel layer to cell form and integrity, Asterita and Marsland (1961) used various enzymes such as chymotrypsin, trypsin, hyaluronidase, amylase, and glucosidase to digest the pellicle in *Blepharisma* and *Paramecium* prior to exposure to high pressure treatment. The cells displayed distinctly different sensitivities to the various enzyme preparations, i.e., *Blepharisma* was more susceptible to the proteases and hyaluronidase, whereas *Paramecium* was affected mainly by β-amylase and hyaluronidase. The authors suggested that these differences might reflect variations in the molecular as well as morphological structure of the pellicular covering and its association with the underlying protoplasm in the two ciliates. Asterita and Marsland (1961) found that after exposure to the enzyme preparations, the resistance of these ciliates to pressure-induced rounding and cytolysis was decreased, indicating that the pellicle plays an important role in maintaining form and integrity.

The solating action of pressure on *Amoeba* has been used to evaluate the action of narcotic agents (Zimmerman, 1967). Morphine concentrations of 2×10^{-3} to 10^{-5} *M* increased the pseudopodial stability of *Amoeba*; treatment of amebae for several days in morphine did not further increase their pseudopodial stability. Cells exposed to morphine for durations of 60 minutes exhibited an increased stability for as long as 3 days. *n*-Allylnormorphine, a specific morphine antagonist, had no effect on pseudopodial stability. However, in combination with morphine, *n*-allylnormorphine reversed the stabilizing action of morphine.

B. Physical Agents

The effects of a physical interference, such as enucleation, on gelation phenomena were investigated by comparing the solational susceptibility of nucleate and anucleate parts of *Amoeba proteus* to hydrostatic pressure (Hirshfield *et al.*, 1958). The results demonstrated that whole amebae and nucleated specimens displayed greater resistance to high pressure solation than did the anucleate ones. These studies suggest that the control of sol–gel transformations is influenced by the nucleus. In view of our present knowledge of macromolecular synthesis, the absence of nuclear RNA would indeed interfere with structural protein synthesis essential to gel formation.

Ultraviolet light affects the sensitivity of both *Amoeba* (Zimmerman *et al.*, 1960) and *Blepharisma* (Hirshfield *et al.*, 1957) to hydrostatic pressure. Amebae were irradiated at wavelengths of 265, 280, 302, and 365 mμ and at dosages of 1500, 3000, and 6000 ergs/mm^2. Dosages of 1500 and 3000 ergs/mm^2 produced no overt effects at any of the wavelengths, and these irradiated specimens showed the same pressure-solation sensitivity as untreated controls. At 6000 ergs/mm^2, similar results were obtained at wavelengths of 365 and 302 mμ; however, at 280, 265, and 230 mμ very definite effects were observed. The 280 and 265-mμ amebae displayed reduced pseudopodia and a vacuolated cytoplasm; however, only the 280-mμ group showed an increased sensitivity to pressure solation. The data suggests that structural proteins are intimately associated with gel maintenance.

Ultraviolet-irradiated *Blepharisma undulans* was studied with reference to sensitivity to lysis by pressures in the range of 10,000 psi (Hirshfield *et al.*, 1957). The nonirradiated controls displayed a minimum sensitivity; the critical lysis pressure was 10,000 psi. Among the irradiated specimens the greatest sensitivity to lysis was found at wavelengths of 230 mμ, slightly less at 280 mμ, and still less at 265 mμ. The critical lysis pressures at 230 mμ, 280 mμ, and 265 mμ (at 6000 ergs/mm^2) were 5000, 6500, and 7000 psi, respectively. The higher wavelengths (302, 313, 334, and 365 mμ) had little or no sensitizing effects. The increased sensitivity to lysis at specific wavelengths (230 and 280 mμ) indicates the importance of structural proteins in maintaining form and structural integrity.

V. Biochemical Studies of *Tetrahymena*

A. Ribosomal Studies

The isolation and characterization of ribosomes in both synchronized and log phase cultures of *Tetrahymena* have been carried out in several laboratories (see Zimmerman, 1969). Microsomal fractions (10,000 *g* supernatant fractions) prepared from synchronized cultures at various times after EH and subjected to sucrose gradient centrifugation displayed distinct optical density profiles (Hermolin, 1967). The ribosomal patterns varied depending upon the time after EH at which the cells were analyzed. Two major peaks were consistently seen in the OD_{260} profiles (Fig. 15). The rapidly sedimenting peak presumed to be polysomal in nature was sensitive to ribonuclease treatment (10 μg/ml for 15 minutes at 4°C) and relatively insensitive to deoxycholate treatment (0.45% for 30 minutes at 4°C). The material in the slowly sedimenting peak (ribosomal peak) was not sensitive to RNase or DOC. Both peaks were sensitive to decreasing concentrations of magnesium. The sedimentation coefficients of these peaks, as determined by analytical ultracentrifuge studies, were found to be 120 S and 84 S.

Analysis of the OD_{260} profiles, following sucrose density gradient centrifugation of the 10,000 *g* fraction, indicating that the relative amount of polysomal material varied depending upon the age of the synchronized culture. A conservative method for evaluating the fluctuations of the ribosomal materials is to measure the area under the polysome peak and compare it with the total amount of ribosomal material present (the polysome and ribosome material). Changes in the ratio of the polysomal area to the total ribosomal area will reflect changes in the amount of polysome material that can be recovered from the cell at any time. There were fluctuations in the ratios during the period preceding the first synchronized division and these are shown in Fig. 16.

The composition of the microsomal fraction (10,000 *g* supernatant) obtained from homogenates of pressurized *Tetrahymena* was distinctly different from that of atmospheric control cells. There is a striking reduction in the amount of polysomes recoverable from pressure treated cells. A typical optical density profile of the 10,000 *g* supernatant fraction prepared from cells (subjected to 5000 psi for 2 minutes) at 60 minutes after EH is shown in Fig. 17. The OD_{260} of the rapidly sedimenting material (presumed polysomal material) collected in the 10–15% fraction from pressure-treated cells is much lower than that found in the controls. The area indicated by the hatching shows the amount of material that could not be recovered as a result of pressure treatment. The greatest loss occurs in the polysomal region.

Evaluation of the ratios (polysomal/total ribosomal) shows that cells are quite sensitive to short pulses of pressure. The ratios were reduced

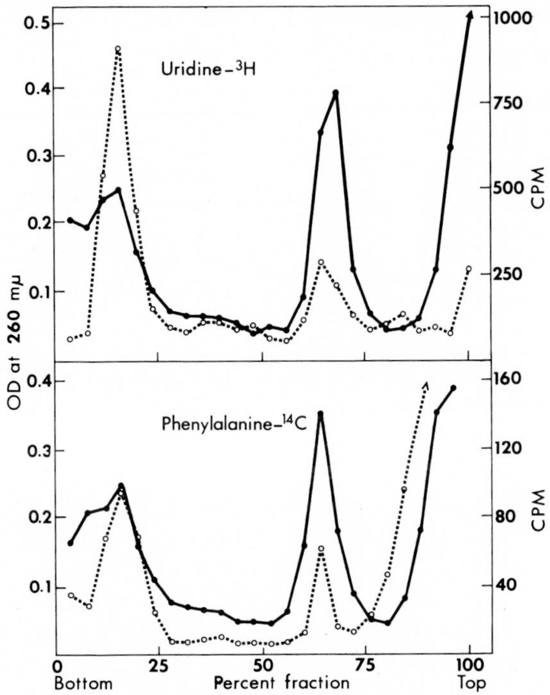

FIG. 15. Sucrose density gradient analysis of microsomal fractions (10,000 *g* supernatant) from heat-synchronized *Tetrahymena*. Cells were labeled with 2 μCi/ml ^{3}H-uridine (specific activity 20 Ci/mmole) for 10 minutes (between 50 and 60 minutes after EH) or in 0.5 μCi/ml ^{14}C-phenylalanine (specific activity 15 mCi/mmole) for 30 minutes (between 30 and 60 minutes after EH). At 60 minutes EH, cells were homogenized and the 10,000 *g* supernatant was analyzed. Two major peaks were observed. A faster sedimenting peak corresponds to polysomal material and the slower sedimenting peak corresponds to ribosomes. Both the OD_{260} (———) and the cpm (○......○) of each fraction are shown. (Data of Hermolin and Zimmerman, 1969.)

at a pressure of 5000 psi; an increase in the magnitude of pressure did not reduce the ratios any further. At 2000 psi (for 2-minute duration) the ratio of the polysome area to the total ribosomal area was only slightly less than the atmospheric controls (Table II).

The previous work clearly demonstrates that immediately following

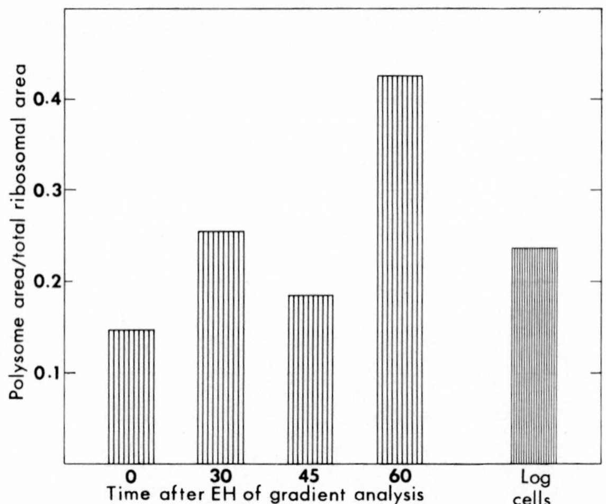

FIG. 16. Fluctuations in the ratio of polysome area to total ribosome area are illustrated at various times after EH. The areas were determined from the sucrose density gradient profiles. (Data of Hermolin and Zimmerman, 1969.)

pressure treatment the quantity of measurable polysomes is reduced. The question of whether or not short durations of pressure affect polysome formation in *Tetrahymena* was investigated by Hermolin and Zimmerman (1969). Heat-synchronized cells were pulsed at 45 minutes after EH (for 2 minutes at a pressure of 5000 psi) and homogenized at 60 minutes after EH. The optical density profiles of the 10,000 *g* supernatant were analyzed following sucrose density gradient centrifugation and the ratios of the polysomal area to total ribosomal area were determined. The relative amount of polysomal material at 13 minutes after decompression (60

TABLE II
CHANGES IN RIBOSOMAL COMPOSITION

Time after EH (min)	Pressure (psi)	Duration (min)	Polysome area / Total ribosomal area
60	atm	—	0.422
60	2,000	2	0.374
60	5,000	2	0.205
60	10,000	2	0.229
60	14,000	5	0.272
45	atm	2	0.186
60	5,000	2 (pressure applied at 45 min EH)	0.368

minutes after EH) was comparable with that of the nonpressurized controls at 60 minutes after EH. The data for these studies are also shown in Table II.

The experiments indicate that the newly formed polysomes are readily

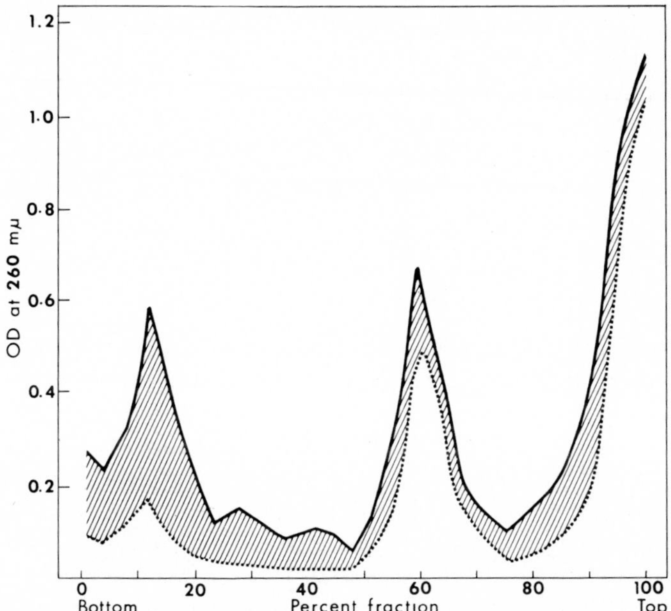

FIG. 17. Sucrose density gradient analysis of 10,000 *g* supernatant fraction derived from pressure-treated *Tetrahymena* (......) is compared with 10,000 *g* supernatant from nonpressurized control cells (———). The pressurized cells were subjected to 5000 psi for 2 minutes at 60 minutes after EH. The control and pressurized cells were homogenized in 0.05 *M* tris, pH 7.4 containing 5 mmoles magnesium. The 10,000 *g* supernatant was centrifuged for 3.5 hours at 23,000 rpm in a sucrose gradient (5–20% sucrose) using a SW 25.2 rotor. The hatched area shows the amount of material that could not be recovered from the pressure-treated cells. The polysome material is recovered in the 10–13% fraction and the ribosomes in the 60–65% fraction. There is a marked decrease in the OD_{260} of the rapidly sedimenting fraction (peak 10–13) from the pressure treated cells. (Data of Hermolin and Zimmerman, 1969.)

disrupted by short pulses of pressure and that synchronized *Tetrahymena* are capable of forming new polyribosomes following a pulse of high pressure. Although pressure does not appear to alter the sedimentation characteristics of monomeric ribosomes, it is quite probable that the ribosomes are also affected by similar pressures and this may be reflected in their ability to control protein synthesis. In order to investigate this possibility, the competence of ribosomes (ability to synthesize protein) from pressure treated *Tetrahymena* was studied (Letts, 1969). Protein synthesis in cell-free preparations containing ribosomes and poly U in an appropriate incubating medium can be recorded as increased incorporation of radioactive phenylalanine. Ribosomes isolated from pressure-treated cells (14,000 psi for 5 minutes) were found to synthesize polyphenylalanine as efficiently as ribosomes isolated from atmospheric control cells. Thus the translational efficiency of ribosomes from pressure-treated cells was equivalent to that from control cells at this level of information output. It is doubtful, therefore, that the decrease in protein synthesis following high pressure (see Section V,B) could be attributed to an alteration of ribosomal competence.

B. RNA and Protein Synthesis

Changes in the relative amount of polysome material recovered from synchronized *Tetrahymena* reflect synthesis essential for division; alterations in the rate of RNA and protein synthesis have been observed in synchronized cultures (see Zimmerman, 1969). Lowe (1968) reported that the rate of incorporation of uridine and phenylalanine into synchronized *Tetrahymena* was reduced following pressure treatment. Moreover, the incorporation rate patterns were dependent upon the age of the cells when the pressure treatment was initiated. Lowe found that cold trichloroacetic acid (TCA) insoluble fractions obtained at various times after EH from synchronized *Tetrahymena* (which were treated for 10 minutes in ^{3}H-uridine) exhibited two peaks of maximum incorporation. Maximum incorporation occurred at 20–30 minutes after EH and just before division. *Tetrahymena* which were pressurized at 15 and 25 minutes after EH (at 10,000 psi for 2 minutes) showed a reduction in the ability of the cells to incorporate the isotope. Furthermore, the displacement of the incorporation curves toward the right indicated that the incorporation was delayed in the pressure-treated cells (Fig. 18).

Comparable studies were conducted using ^{14}C-phenylalanine in which the cells were incubated in the isotope for 10 minutes, lysed, and the acid insoluble fraction was precipitated with hot 5% TCA. The maximum incorporation occurred between 30 and 40 minutes after EH. The lowest rate of incorporation occurred soon after the last heat shock and during the last 20 minutes of the division cycle. Following pressure treatment (10,000 psi for 2 minutes at 15 and 25 minutes after EH), the rates of incorporation of ^{14}C-phenylalanine were reduced and the rate curves were shifted toward the right, similar to the uridine study (Fig. 18).

The previous studies have shown that incorporation of precursors of RNA and protein in synchronized *Tetrahymena* is reduced following pressure treatment. Dr. Shuhei Yuyama, in our laboratory, investigated the kinetics of RNA labeling and the synthesis of various fractions of RNA in synchronized *Tetrahymena*. In selective experiments, synchronized *Tetra-*

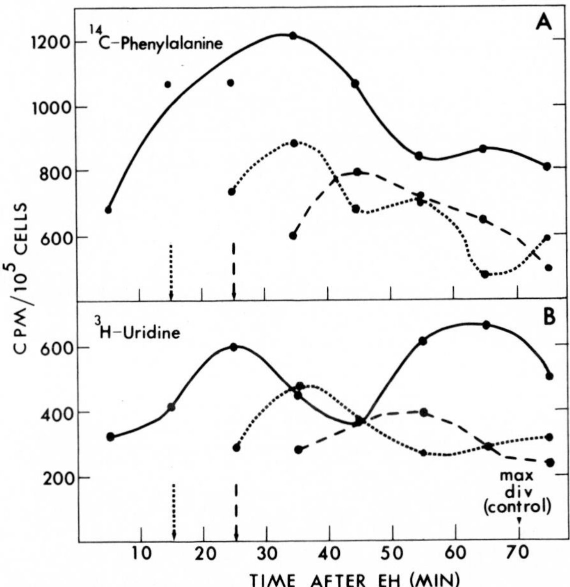

FIG. 18. The effects of pressure on rate of incorporation of ^{14}C-phenylalanine and ^{3}H-uridine. The control cells at atmospheric pressure (———) were pulsed for 10 minutes in the radioisotope, and the radioactivity (cpm/10^5 cells) of the acid-insoluble fraction was plotted as a function of time after EH. In the pressure series, the cells were subjected to 10,000 psi for 2 minutes at 15 minutes (. . . . . .) and at 25 minutes (– – –) after EH. Following pressure treatment the cells were incubated in ^{3}H-uridine (10 μCi/ml) and ^{14}C-phenylalanine (0.25 μCi/ml) for 10 minutes and prepared for counting. (Data of Lowe, 1968, from Zimmerman, 1969.)

hymena were pressurized in the presence of ^{3}H-uridine. RNA was extracted with phenol sodium dodecyl sulfate (Scherrer and Darnell, 1962) and chromatographed on methylated albumin Kieselguhr (MAK) columns (Mandell and Hershey, 1960). In general, pressure caused a reduction in newly synthesized RNA. A representative pattern of cells which were subjected to 5000 psi for 7 minutes beginning 43 minutes after EH and the nonpressurized control cells are shown in Fig. 19. Although there was no appreciable difference in the OD_{260} profile of either the pressurized or control cells, radioactivity of the RNA in various fractions from pressurized cells was about 50% lower than the controls. The results can be interpreted to indicate that the immediate effects of pressure are to reduce synthesis of the RNA in both the low molecular weight (4 S and 5 S)

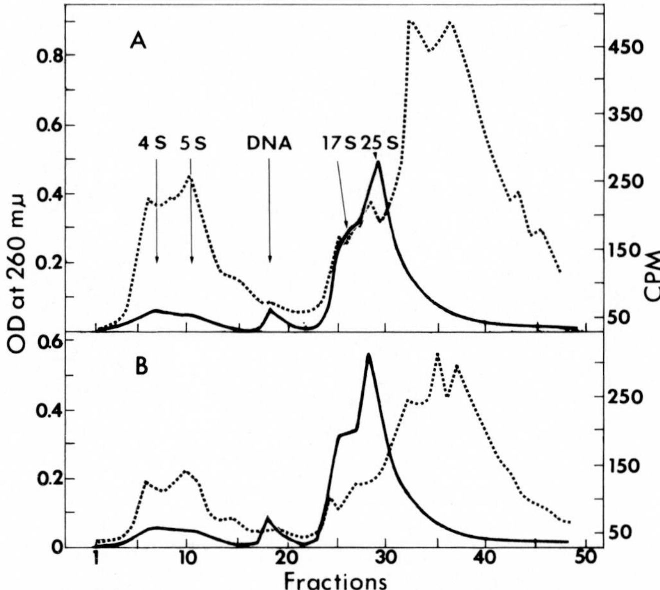

FIG. 19. Influence of pressure on RNA synthesis in synchronized *Tetrahymena*. MAK column chromatography of total cellular RNA from cells labeled for 10 minutes with ^{3}H-uridine (specific activity 20 Ci/mmole). At 40 minutes after EH, cells were placed into ^{3}H-uridine (10 μCi/ml); one sample remained at atmospheric pressure (A) and the other sample was subjected to 5000 psi for 7 minutes (B). At 50 minutes after EH, both samples were chilled in an ice bath and lysed in 1% sodium dodecyl sulfate. Nucleic acids were extracted with cold phenol and fractionated on the MAK column. (———) OD_{260}; (. . . . . .) radioactivity in counts per minute. (Yuyama and Zimmerman, 1969).

and the high molecular weight region (found in fractions 33–48), although there is no effect of pressure on preexisting bulk RNA (Yuyama and Zimmerman, 1969).

Pressure affects not only the rate of incorporation of uridine and phenylalanine into RNA and protein, respectively, but the cumulative incorporation of these labeled precursors as well (Lowe, 1968). Heat-synchronized cultures at EH 0 were placed into ^{14}C-phenylalanine or ^{3}H-uridine. At 15 and 25 minutes after EH the cells were subjected to a pulse of 10,000 psi for 2 minutes; aliquots were taken at 5–10 minute intervals, and incorporation of radioactively labeled material was determined as previously described. These studies demonstrate that there is a marked reduction in RNA and protein synthesis following short pulses of pressure (Fig. 20).

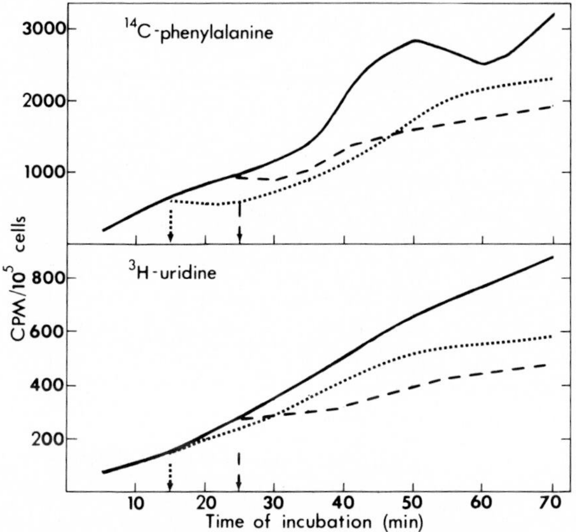

FIG. 20. The effects of pressure on cumulative incorporation of ^{14}C-phenylalanine and ^{3}H-uridine. The synchronized *Tetrahymena* were incubated in ^{14}C-phenylalanine (0.15 μCi/ml) or in ^{3}H-uridine (3.3 μCi/ml) at 0 minutes after EH (———). At 15 (. . . . . .) or 25 (– – –) minutes after EH the labeled cells were subjected to a pulse of 10,000 psi for 2 minutes. Aliquots were removed at various times and the radioactivity (cpm/10^5 cells) of the acid-insoluble fractions was plotted as a function of time of incubation (time after EH). (From the work of Lowe, 1968.)

C. DNA SYNTHESIS

In logarithmically growing cells, morphological changes (as observed in the light microscope) occur in the nuclei at pressures above 6000 psi. Feulgen-stained *Tetrahymena,* which were pressurized at 7500 psi for 10 minutes, displayed swollen nuclei and granulation of Feulgen-positive material. Similar clumping of the chromatin was also observed at the level of the electron microscope (Kennedy and Zimmerman, 1968). At 10,000 psi, there was a prominent pycnosis. These studies suggest that DNA synthesis in *Tetrahymena* is probably affected at this pressure level.

Murakami and Zimmerman (1968) investigated the effects of pressure on DNA synthesis in logarithmic growing and heat-synchronized cultures of *Tetrahymena.* Following a 10-minute exposure to ^{14}C-thymidine at varying pressures (2000–10,000 psi), log phase cells were lysed with sodium dodecyl sulfate (SDS), and the radioactivity in the TCA-insoluble fraction was determined (Fig. 21). At 2000 psi there was a 10% inhibition of incorporation. The Feulgen preparations of these cells showed a negligible pressure effect on the nuclei. At 10,000 psi there was 80% in-

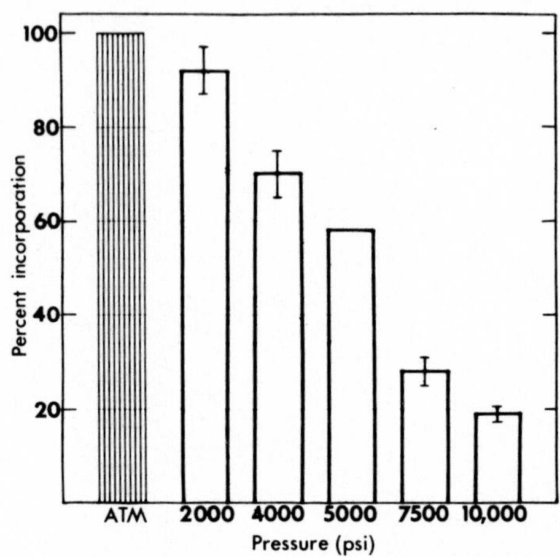

FIG. 21. The effects of pressure on DNA synthesis in log phase cultures of *Tetrahymena*. The cells were incubated in ^{14}C-thymidine (2 μCi/ml) for 10 minutes under varying pressures. The radioactivity in the acid-insoluble fraction was determined as a percentage of the nonpressurized control cells. The standard error at each pressure is shown. (Data of Murakami and Zimmerman, 1968.)

hibition and prominent pycnosis of the nuclei. Similar results were obtained in studies where the cells were pressurized but incubated in the precursor following decompression. In these studies, cells were compressed (2000–10,000 psi for 10 minutes), and 2 minutes after decompression the cells were incubated in the radioactive precursor (for 10 minutes). In other studies it was found that following a 10-minute pulse of pressure at 7500 psi the cells could progressively recover their ability to synthesize DNA; at 65 minutes after decompression the rate of incorporation was comparable with the controls.

The pattern of DNA synthesis in synchronized *Tetrahymena* cultures is more complex than that observed in log phase cells. Control cells (at atmospheric pressure) exhibited maximum incorporation of ^{14}C-thymidine shortly before furrowing (50–60 minutes after EH). In the pressure series, 7500 psi of pressure was applied between 10–20 minutes after EH. The rate of incorporation of ^{14}C-thymidine after decompression was determined at alternate 10-minute intervals. The results from a typical experiment are illustrated in Fig. 22. The decompressed cells exhibited a progressive increase in their rate of DNA synthesis, which at 70–80 minutes after EH exceeded the control values. It is interesting to note that the division of the pressure-treated cells was delayed about 20 minutes.

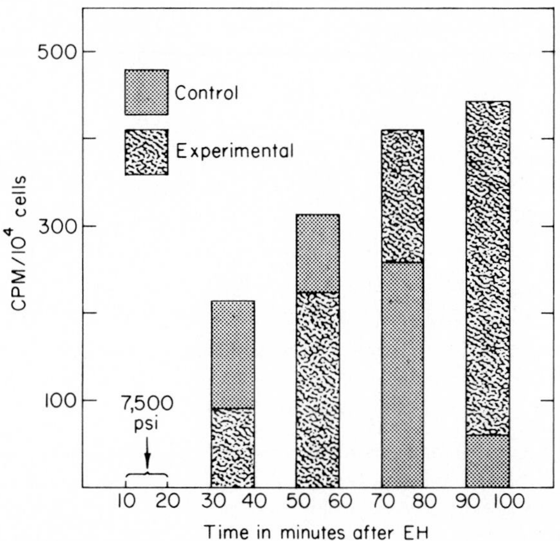

FIG. 22. The effects of pressure on the rate of DNA synthesis in synchronized *Tetrahymena*. Cells were subjected to 7500 psi for 10 minutes between 10 and 20 minutes after EH. Following decompression the cells were incubated in ^{14}C-thymidine (2 μCi/ml) for 10 minutes. The radioactivity in the acid-insoluble fraction was determined. Incorporation in the control and pressure-treated cells are shown. (Data of Murakami and Zimmerman, 1968.)

VI. Concluding Remarks

The action of hydrostatic pressure on protozoa provides a vehicle for study of the structure and function of these organisms and for the analysis of cell synthesis and development. The use of high pressure as an analytical tool with which the effect of various chemical and physical agents on cellular activities may be studied provides another distinct area of investigation. Traditionally, high pressure studies have been concerned with physiological activities such as cell division and movement of cells. More recently, observations of the effects of high pressure have been extended to ultrastructural and molecular levels of the cell. It is now apparent that a more critical analysis of high pressure interference of physiological activities can be made in light of the effects of pressure on the fine structural and biochemical mechanisms of the cell.

The present review has incorporated data from various experiments which show that pulses of pressure affect not only the shape and movement of the organism at the level of the light microscope but also result in biostructural modifications at the level of the electron microscope. Thus the inability of *Tetrahymena* to display normal locomotion can be related to the disorientation of microtubular elements within the cilium. The inhibition and delay in cell division following pressure treatment is similarly presumed to be involved with the disarray of microtubular elements as well as the interference of biochemical synthesis by pressure. The synthesis of protein, DNA, and RNA material has been selectively inhibited by high pressures at various times during the cells' development from cell division to cell division. Such interference with biochemical synthesis as well as information transfer in the cell may also be causally related to an inhibition of cell differentiation.

ACKNOWLEDGMENT

The unpublished experiments from the authors' laboratory have been carried out with the support of research grants from the National Research Council of Canada. Their assistance is gratefully acknowledged.

REFERENCES

Allen, R. D. (1967). Fine structure, reconstruction and possible functions of components of the cortex of *Tetrahymena pyriformis*. *J. Protozool.* **14**, 553–565.

Asterita, H., and Marsland, D. (1961). The pellicle as a factor in the stabilization of cellular form and integrity: Effects of externally applied enzymes on the resistance of *Blepharisma* and *Paramecium* to pressure-induced cytolysis. *J. Cellular Comp. Physiol.* **58**, 49–61.

Auclair, W., and Marsland, D. (1958). Form-stability of ciliates in relation to pressure and temperature. *Biol. Bull.* **115**, 384–396.

Brown, D. E. S., and Marsland, D. (1936). The viscosity of *Amoeba* at high hydrostatic pressure. *J. Cellular Comp. Physiol.* **8**, 159–165.

Byrne, J., and Marsland, D. A. (1965). Pressure–temperature effects on the form-stability and movement of *Euglena gracilis* var. Z. *J. Cellular Comp. Physiol.* **65**, 277–284.

Certes, A. (1884). Note relative à l'action des hautes pressions sur la vitalité des micro-organismes d'eau douce et d'eau de mer. *Compt. Rend. Soc. Biol.* **36**, 220–222.

Chapman-Andresen, C. (1962). Studies on pinocytosis in amoebae. *Compt. Rend. Trav. Lab. Carlsberg* **33**, 73–264.

Giese, A. C. (1968). The effect of hydrostatic pressure and heavy water upon regeneration in *Blepharisma. Exptl. Cell. Res.* **52**, 370–378.

Hermolin, J. (1967). Polysome formation in *Tetrahymena pyriformis.* M.Sc. Thesis, University of Toronto, Toronto, Canada.

Hermolin, J., and Zimmerman, A. M. (1969). The effect of pressure on synchronous cultures of *Tetrahymena:* a ribosomal study. *Cytobios* **1**, 247–256.

Hirshfield, H. I., Zimmerman, A. M., Landau, J. V., and Marsland, D. (1957). Sensitivity of UV-irradiated *Blepharisma undulans* to high pressure lysis. *J. Cellular Comp. Physiol.* **49**, 287–294.

Hirshfield, H. I., Zimmerman, A. M., and Marsland, D. (1958). The nucleus in relation to plasmagel structure in *Amoeba proteus;* a pressure-temperature analysis. *J. Cellular Comp. Physiol.* **52**, 269–274.

Kennedy, J., and Zimmerman, A. M. (1968). Unpublished data.

Landau, J. V. (1959). Sol–gel transformation in *Amoeba. Ann. N.Y. Acad. Sci.* **78**, 487–500.

Landau, J. V. (1965). High hydrostatic pressure effects on *Amoeba proteus.* Changes in shape, volume and surface area. *J. Cell Biol.* **24**, 332–336.

Landau, J. V. (1966). High pressure effects on endogenous adenine nucleotide levels in sea urchin eggs prior to first cleavage. *J. Cell Biol.* **28**, 408–412.

Landau, J. V., and Thibodeau, L. (1962). The micromorphology of *Amoeba proteus* during pressure-induced changes in the sol–gel cycle. *Exptl. Cell Res.* **27**, 591–594.

Landau, J. V., and Peabody, R. A. (1963). Endogenous adenosine triphosphate levels in human amnion cells during application of high hydrostatic pressure. *Exptl. Cell Res.* **29**, 34–60.

Landau, J. V., Zimmerman, A. M., and Marsland, D. A. (1954). Temperature-pressure experiments on *Amoeba proteus:* Plasmagel structure in relation to form and movement. *J. Cellular Comp. Physiol.* **44**, 211–232.

Landau, J. V., Marsland, D., and Zimmerman, A. M. (1955). The energetics of cell division: effects of adenosine triphosphate and related substances on the furrowing capacity of marine eggs. (*Arbacia* and *Chaetopterus*). *J. Cellular Comp. Physiol.* **45**, 309–329.

Letts, P. J. (1969). *In vitro* studies of protein synthesis in *Tetrahymena pyriformis:* A pressure study. M.Sc. Thesis, University of Toronto, Toronto, Canada.

Lowe, L. (1968). The effects of hydrostatic pressure on synchronized *Tetrahymena.* Ph.D. Thesis, University of Toronto, Toronto, Canada.

Lowe-Jinde, L., and Zimmerman, A. M. (1969). Heavy water and hydrostatic pressure effects on *Tetrahymena pyriformis. J. Protozool.* **16**, 226–230.

Macdonald, A. G. (1965). The effect of high hydrostatic pressure on the oxygen consumption of *Tetrahymena pyriformis* W. *Exptl. Cell Res.* **40**, 78–84.

Macdonald, A. G. (1967a). The effect of high hydrostatic pressure on the cell division and growth of *Tetrahymena pyriformis. Exptl. Cell Res.* **47**, 569–580.

Macdonald, A. G. (1967b). Delay in the cleavage of *Tetrahymena pyriformis* exposed to high hydrostatic pressure. *J. Cellular Physiol.* **70**, 127–130.

Mandell, J. D., and Hershey, A. D. (1960). A fractionating column for analysis of nucleic acids. *Anal. Biochem.* **1**, 66–77.

Marsland, D. (1964). Pressure-temperature studies on amoeboid movement and related phenomena. *In* "Primitive Motile Systems in Cell Biology" (R. D. Allen and N. Kamiya, eds.), pp. 173–187. Academic Press, New York.

Marsland, D., and Brown, D. E. S. (1936). Amoeboid movement at high hydrostatic pressure. *J. Cellular Comp. Physiol.* **8**, 167–178.

Murakami, T. H., and Zimmerman, A. M. (1968). Unpublished data.

Pitelka, D. R. (1961). Fine structure of the silverline and fibrillar systems of three tetrahymenid ciliates. *J. Protozool.* **8**, 75–89.

Regnard, P. (1884). Note relative à l'action des hautes pressions sur phénomènes vitaux (movement des cils vibratiles, fermention). *Compt. Rend. Soc. Biol.* **36**, 187–188.

Rustad, R. C. (1964). The physiology of pinocytosis. *Recent Progr. Surface Sci.* **2**, 353–376.

Scherbaum, O., and Zeuthen, E. (1954). Induction of synchronous cell division in mass cultures of *Tetrahymena pyriformis. Exptl. Cell Res.* **6**, 221–227.

Scherrer, K., and Darnell, J. E. (1962). Sedimentation characteristics of rapidly labeled RNA from HeLa cells. *Biochem. Biophys. Res. Commun.* **7**, 486–490.

Simpson, R. E. (1966). The effects of high hydrostatic pressure on cell division and cytodifferentiation in synchronized cultures of *Tetrahymena pyriformis* GL. Ph.D. Thesis, University of Iowa, Iowa City, Iowa.

Tilney, L. G., Hiramoto, Y., and Marsland, D. (1966). Studies on the microtubules in heliozoa. III. A pressure analysis of the role of these structures in the formation and maintenance of the axopodia of *Actinosphaerium nucleofilum* (Barrett). *J. Cell Biol.* **29**, 77–95.

Yuyama, S., and Zimmerman, A. M. (1969). Temperature–pressure effects on RNA synthesis in synchronized *Tetrahymena. Biol. Bull.* **137**, 384.

Zeuthen, E. (1964). The temperature-induced division synchrony in *Tetrahymena. In* "Synchrony in Cell Division and Growth" (E. Zeuthen, ed.), pp. 99–158. Wiley (Interscience), New York.

Zimmerman, A. M. (1959). Effects of selected chemical agents on amoebae. *Ann. N. Y. Acad. Sci.* **78**, 631–646.

Zimmerman, A. M. (1962). Action of ATP on amoeba. *J. Cellular Comp. Physiol.* **60**, 271–280.

Zimmerman, A. M. (1963). Chemical aspects of the isolated mitotic apparatus. *In* "The Cell in Mitosis" (L. Levine, ed.), pp. 159–179. Academic Press, New York.

Zimmerman, A. M. (1964). The effects of mercaptoethanol upon form and movement of *Amoeba proteus. Biol. Bull.* **127**, 538–546.

Zimmerman, A. M. (1967). Sensitivity of *Amoeba proteus* to morphine and N allylnormorphine. A pressure study. *J. Protozool.* **14**, 451–455.

Zimmerman, A. M. (1969). Effects of high pressure on macromolecular synthesis in synchronized *Tetrahymena. In* "The Cell Cycle" (G. M. Padilla, G. L. Whitson, and I. L. Cameron, eds.), pp. 203–225. Academic Press, New York.

Zimmerman, A. M., and Rustad, R. C. (1965). Effects of high pressure on pinocytosis in *Amoeba proteus. J. Cell Biol.* **25**, 397–400.

Zimmerman, A. M., Landau, J. V., and Marsland, D. (1958). The effects of adenosine triphosphate and dinitro-*o*-cresol upon the form and movement of *Amoeba proteus.* A pressure–temperature study. *Exptl. Cell Res.* **15**, 484–495.

Zimmerman, A. M., Hirshfield, H. I., and Marsland, D. (1960). Sensitivity of UV-irradiated *Amoeba proteus* to hydrostatic pressure. *J. Cellular Comp. Physiol.* **55**, 221–225.

CHAPTER 9

THE EFFECTS OF PRESSURE ON MARINE INVERTEBRATES AND FISHES

H. Flügel and C. Schlieper

I. Introduction

All animals living in the oceans of our earth are subject to hydrostatic pressures ranging from less than 1 atm at the surface to about 1000 atm at the bottom of the deep sea trenches. Or, in other words, each square centimeter of body surface is exposed to pressures from slightly more than 1.0332 kg up to values of about 1033.2 kg, 1 atm being equivalent to 14.696 psi. Whereas pelagic, free-swimming animals from the deep scattering layers may endure sudden pressure changes on their daily vertical migrations, benthic animals have to cope with the constant pressure according to the depth of their habitat. Therefore, we have to distinguish between the effects of more or less rapid compression and decompression and the effects of the constant pressure at great depth,

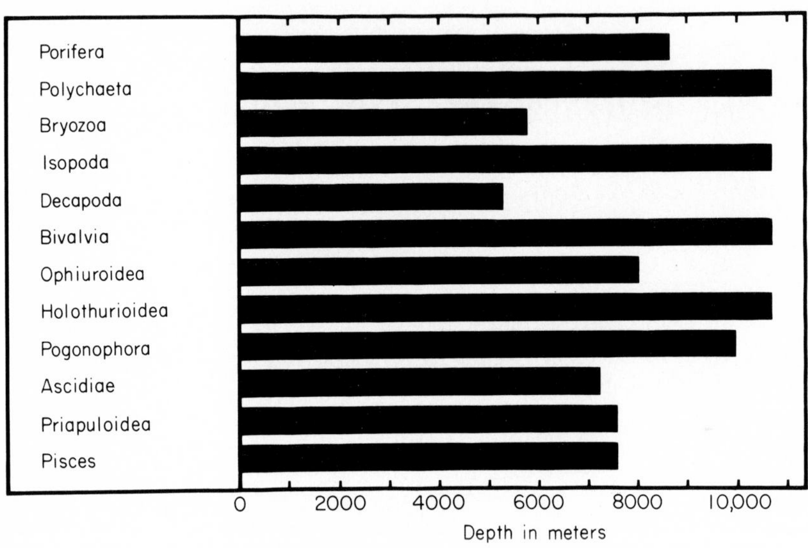

FIG. 1. Greatest depths of distribution of various groups of bottom-living animals (drawn after a table of Zenkevitch, 1963).

whereas, hitherto, most experiments on pressure effects have suffered from the rather rapid changes in hydrostatic pressure. The constant pressure, in addition to the combined effects of low temperature, lack of light, and shortage of food, seems to be responsible for the fact that some invertebrate groups are not represented in the scarce fauna of the depth in excess of 4000 m (Fig. 1). We have to bear in mind, however, that not in the slightest way does the pressure of several hundred atmospheres either damage or injure animals adapted to life at great depths, for there is no pressure gradient between the animal's body and the sea water. In fact, many of the deep sea invertebrates are soft-bodied, delicate organisms like the famous Pogonophora, the ancient mollusk *Neopilina galathea*, or the stilt-legged pycnogonid *Collosendeis collosea*. In the present study we shall review some of the classic works as well as recent investigations of the effects of hydrostatic pressure on whole animals and also on the isolated tissues of marine invertebrates and fish. We shall not, however, review investigations on the responses of marine animals to relatively slight changes in hydrostatic pressure, which have been summarized by Knight-Jones and Morgan (1966).

II. Observations in the Field

During the renowned exploring voyages of the Challenger (1872–1876), the Talisman (1882–1883), the Albatross (1882–1924), the Valdivia (1898–1899), the Discovery (1925 and later), and, more recently, the Galathea (1950–1952) and the Vitiaz (1949 and later), to list just a few, it was well established that animal life exists at all depths. As early as 1877, C. W. Thomson, the scientific director of the Challenger expedition, stated:

> Animal life is present on the bottom of the ocean at all depths. Animal life is not nearly so abundant at extreme, as it is at more moderate depth; but as well developed members of all the marine invertebrates classes occur at all depths, this appears to depend more upon certain causes affecting the composition of the bottom deposits . . . and other materials necessary for their development, than upon any of the conditions immediately connected with depth.

The major task of most of the expeditions in the past was collecting animals rather than analyzing their adaptations to great depths. Experiments on invertebrates and fish taken from the deep sea bottom are still lacking. Aboard the Galathea no precautions were taken to keep bottom animals alive. Lemche and Wingstrand (1959) record the capture of the mollusk *Neopilina galathea* with the following description:

> The sample contained 10 specimens with the soft parts preserved but more or less damaged, and three empty shells. The specimens were washed out of the muddy content of the trawl by means of surface water at a temperature of 28°C, which must have killed the animals immediately, if they did not die already during the transport to the deck of the ship from the abysses with its temperature of only 2°C.

In situ experiments on the effects of hydrostatic pressure on marine invertebrates were performed by Sano (1958) and by Menzies and Wilson (1961). Sano attached several specimens of *Mytilus edulis* and *Bala-*

nus amphitrite to a bathyscaphe. The submarine remained at the maximum depth of 3000 m for 5 hours. The pressure of about 300 atm had apparently no deleterious effects on the animals. Menzies and Wilson selected intertidal invertebrate animals for their field experiments designed to separate pressure from temperature effects. Both the shore crab *Pachygrapsus crassipes* and the common blue mussel *Mytilus edulis diegensis* are normally exposed to a wide range of fast-changing temperatures. In the experiments, five to eight adult crabs were placed in net-covered jars which were attached to the hydrographic wire of the ship and sent to and recovered from the ocean. Experiments were conducted at a range from 469 m to the maximum depth of 3480 m. This depth is equivalent to a pressure of 357 kg/cm^2. Sixty percent of the shore crabs survived compression and decompression equivalent to 92 kg/cm^2 (897 m); beyond that pressure all crabs were found dead upon recovery. An exposure of the crabs to pressures between 76 kg/cm^2 (744 m) and 92 kg/cm^2 (896 m) led to the temporary tetany of the animals. Adjacent specimens firmly clasped each other. They recovered from the stress of pressure gradually, "with the mouth parts showing movements first, followed by the other appendages" (Menzies and Wilson, 1961). Simultaneously, control experiments were also performed keeping crabs for the duration of the experiment at a temperature of 2°C, which was well below the lowest temperature met in the experiments. All crabs survived and were active. The mussels proved to be more resistant to pressure. One hundred percent survived 60 minutes at a depth of 2227 m (227 kg/cm^2). However, after they had been lowered to 3480 m all animals were found dead (357 kg/cm^2). However, Menzies and Wilson emphasized that it is uncertain whether the death of the animals was due to the effects of the compression or of the decompression.

III. The Effects of Pressure on Invertebrates and Fishes Under Laboratory Conditions from Investigations in the Past

A. Whole Animals

To the best of our knowledge, the French physiologist Regnard (1884 a,b,c, 1885, 1891) was the first who succeeded in studying the effects of high hydrostatic pressures on invertebrates and fishes under laboratory conditions. He exposed several freshwater and marine animals to pressures ranging between 100 and 1000 atm, thus successfully demonstrating that sea anemones, echinoderms, mollusks, and annelids are more pressure resistant than decapods and teleosts. In his pioneering experiments he was even able to carry out microscopic examination of small animals such as *Cyclops, Gammarus,* and *Daphnia* through windows set in the sides of the pressure chamber which contained them. He states.

> Au delà de 1000 m (100 atm), ils tombent lentement au fond de l'eau; leurs membres s'agitent avec rapidité, leurs appareils natatoires se raidissent et sont pris d'un tremblement très énergique. Les animaux demeurent à part cela immobiles au fond de l'eau. Ils semblent incapables de se mouvoir, ils sont tétanisés.

The pressure devices, of course, have been much improved and modified since the days of Regnard. According to Ebbecke (1935, 1944), who in many substantial investigations confirmed and completed the early work of Regnard, the capability of various animals to tolerate pressure decreases in the following order: sea anemones, starfish, sea urchins, jellyfish, ctenophores, *Branchiostoma,* gastropods, polychaetes, shrimps, and teleosts. Ebbecke also determined the contraction frequency of *Cyanea* under pressure. These medusae responded to pressure of 100 and 200 atm with increased contraction frequency. Animals subjected to about 300 atm showed no umbrella movements at all. At pressures in excess of 400 atm, the muscles of the specimens remained firmly contracted. Medusae exposed to 600 atm for only a few minutes easily survived the experiments. Ebbecke (1944) has emphasized that both the state of excitation and the state of immobility could be due to nervous reaction, while the third state of muscular tetany seems to be a response to compression of the muscle cells themselves. He has recorded similar responses of shrimps (*Leander, Crangon, Pandalus*) and fishes (*Gobius, Pleuronectes, Spinachia*). Shrimps responded to 50 atm with excitement. At 150 atm the animals seemed to be paralyzed, and at 200 atm and more, only the heartbeat indicated life. Fish were almost instantly killed by the pressure of 200 atm. Ebbecke concluded, therefore, that surface-dwelling fish are not able to cope with the pressure at 2000 m and more. The observations of Regnard and Ebbecke are in accordance with investigations by Fontaine (1928), who measured the oxygen consumption of some marine fish and shrimps. In this experiments, the oxygen consumption of the plaice *Pleuronectes platessa* was increased at 25 atm by 28%, at 50 atm by 39%, and at 100 atm by 58%. At pressures in excess of 100 atm the oxygen consumption decreased.

B. Isolated Tissues

The interpretation of the effects of pressure on whole animals is complicated by the difficulty of separating nervous responses from other cell responses. It is for this reason that the study of the reactions of compressed tissues, organs, and single cells is necessary to an understanding of the influence of hydrostatic pressure on the living system. Many authors in the past have studied the behavior of isolated tissues while under moderate or high pressure (thoroughly reviewed by Cattell, 1936; Ebbecke 1944; see also Murakami, Chapter 5). In this connection, studies on isolated nerves are informative (Grundfest and Cattell, 1935). In the moderately pressurized frog's sciatic nerve (340–544 atm), a single shock led to repetitive responses, normally three and more impulses. At 544–1020 atm the action potentials of the shocked nerve became smaller and the conduction rate slowed, and too high a pressure finally killed the nerve. The responses of nerves resemble those observed in cardiac muscles of the frog, the turtle, and the dogfish (Edwards and Cattell, 1927, 1928). In muscles, three marked effects have been recorded. As the pressure is increased, an augmentation of the twitch tension can be measured. If the pressure is increased even further, smaller and slower twitches, together with a reduction in efficiency, result. Pressures of 800–1000 atm either abolish the action potentials or reduce them to small waves. Eventually, if the pressure is applied for too long a time, the contractibility of the muscles is lost. The observations of Draper and Edwards (1932) have shown how complex the interpretation of the responses to pressure can be. In early investigations the view was supported that an increase in absolute pressure accelerates the rate of the isolated heart. Many later observations showed, however, that the acceleration of the heart is a temporary event only. Draper and Edwards demonstrated clearly that the heart rate of the embryos of *Fundulus* at about 81 atm is slowed down by 9.9%, regardless of whether the hearts are still in place (*in situ*) or carefully dissected from the embryo. Striking parallels between the effects of increasing pressure upon the ciliary frequency of isolated gill filaments of *Mytilus* and upon the cardiac rhythm were found by Pease and Kitching (1939). If the pressure was increased stepwise from 68 atm to about 400 atm, the initial acceleration of the beat frequency, which was measured stroboscopically, was followed by a decrease in the ciliary rate (Fig. 2).

The stimulating and retarding effects of increasing compression on protozoa have already been shown by Regnard (1884a) and, later, by Ebbecke (1935). *Paramecium* swam actively at low pressures; at 50 atm some specimens were slowed down, and at pressures exceeding 400 atm most of the ciliates became motionless. Their slender bodies first changed into short clubs and, finally, into spheres. But decompression gradually revived the ciliates. For a very exact description of the effects of pressure on amebae we are indebted to the important work done by Brown and Marsland (1936) and Marsland and Brown (1936) some 30 years ago. It is well known that amebae move by the formation of pseudopodia. These finger-shaped cell processes consist of a streaming inner portion, called plasmasol or endoplasm, and the more rigid outer plasmagel or ectoplasm. It is by their capability for transformation of the plasmasol into the plasmagel and vice versa that ameboid cells can

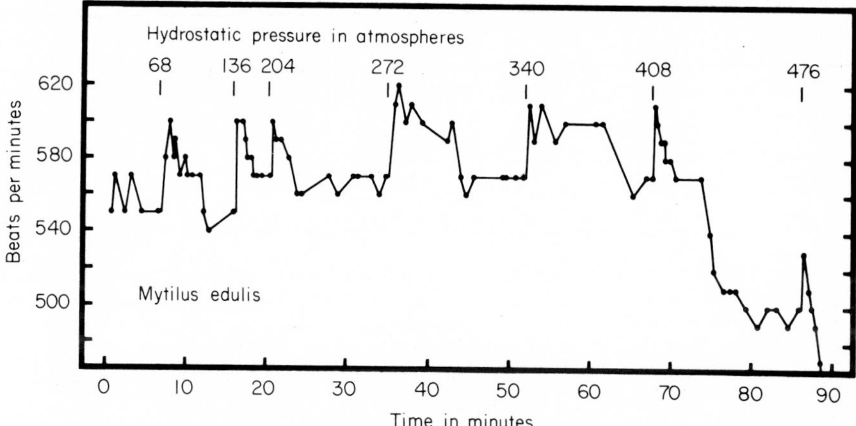

FIG. 2. A single complete record of the ciliary rate of *Mytilus* gills in which the pressure was raised rapidly by successive equal steps of 68 atm (Pease and Kitching, 1939).

move. Below 136 atm the pseudopodia of the pressurized amebae became even longer. At pressures exceeding 136 atm the authors observed the progressive shortening of the pseudopodia. Eventually, around 400 atm, cell extensions could no longer form and the amebae became spherical. The streaming of the plasmasol, necessary for ameboid motion, ceased whenever the cells were subjected to sudden pressures of 200–250 atm. The authors found evidence that the hydrostatic pressure apparently liquefied the rigid plasmagel, thus inhibiting the formation of pseudopodia (see also Zimmerman and Zimmerman, Chapter 8).

Kitching and Pease (1939) subjected the marine suctorian protozoan *Ephelota coronata* to hydrostatic pressures ranging from about 320 atm to about 816 atm. This protozoan is known for its very long and thin tentacles. If the pressure applied reached a certain critical point the tentacles underwent shortening and broke down. The variability proved rather great, but half of the tentacles disintegrated between 450 and 580 atm. In some rare cases, a few of the tentacles withstood liquefaction at a pressure as high as 816 atm. Only the cortical gel of the *Ascaris* egg and apparently the gels of eurybathic marine animals are known to persist at such high pressure (Pease and Marsland, 1939; Pease and Kitching, 1939; Ponat, 1967; see also Kitching, Chapter 7).

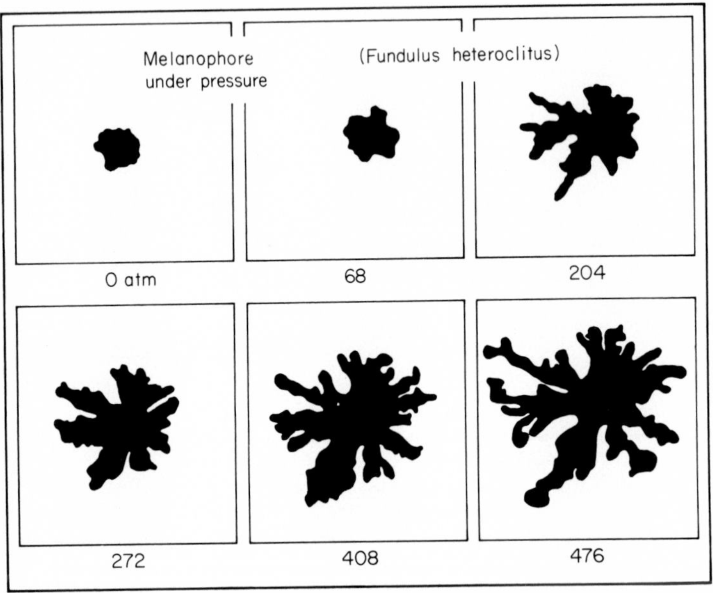

FIG. 3. Pressure effects on the expansion of a melanophore of Fundulus heteroclitus (teleosts) in 0.1*N* KCl solution (drawn after micrographs of Marsland, 1944).

For this same reason, the contracted melanophores of the teleost *Fundulus heteroclitus* displayed an increased expansion of the pigment upon a slow rise in pressure. With pressure release, the melanin contracted again (Fig. 3). Hence, the expansion and contraction of the fish melanophore pigment may result in a reversible gel–sol change. Increased pressure shifts the gel–sol equilibrium toward sol formation (Marsland, 1944).

These hypotheses are in accordance with the behavior of pressurized *Arbacia* eggs spun in the pressure centrifuge. The experiments showed that relatively heavy cell inclusions and organelles moved faster through the cytoplasm of compressed cells than through cells centrifuged at atmospheric pressure (Brown, 1934).

The work reviewed above has shed light upon the general action of hydrostatic pressure on the fundamental processes in animals and tissues. We, however, are primarily involved in ecological physiology, and the responses of marine forms to compression are of great interest to us. The reactions of these forms may elucidate the questions intimately connected with life in the depth of the oceans. One of us (C.S.) has already initiated a series of investigations on the effects of pressure on marine invertebrates and fish (Schlieper, 1963a,b, 1968).

IV. More Recent Investigations on the Effects of Pressure on Marine Invertebrates and Fishes

A. Pressure Effects on the Heart Rate

The investigations of marine animals in the windowed pressure chamber, with its volume of only 75 ml, are handicapped by the fact that toxic metabolic products, as well as lack of oxygen, may interfere with the experiments. These disadvantages are diminished by using either tiny tissue pieces or very small specimens in short-term experiments.

In this way, Naroska (1968) measured the effects of pressure on the heart rate of small, transparent animals like the amphipods *Gammarus duebeni*, *G. oceanicus*, the tunicate *Ciona intestinalis*, and the newborn larvae of the fish *Zoarces viviparus*, which provided excellent material for such studies.

The *in situ* heart of *Gammarus oceanicus* responded to 100 atm with acceleration at first, followed by a gradual reduction in rate. Further stepwise compressions by 100 atm showed the same phenomenon. If the high pressure was not applied for too long a time, rapid decompression was followed by recovery of the heart rate (Fig. 4).

In fish, the heart rate increased temporarily by about 18% at 50 atm. In contrast, at 200 atm the heartbeat of the larvae of *Zoarces* dropped drastically from 38 beats/min to about 12 beats/min.

These experiments revealed a striking agreement with the classic observations on the effects of pressure on the heart rate of *Fundulus* embryos and tadpoles (Draper and Edwards, 1932; J. Landau and Marsland, 1952). They gave further support to the view of the stimulating effects of moderate pressures (up to about 100 atm.) and the retarding influence of high pressure (Edwards and Cattell, 1928; Edwards and Brown, 1934; Ebbecke, 1944).

Measurements at different temperatures provided evidence of the counteracting effects of high temperatures. Amphipods tested at 200 atm in the cold (5°C) responded with depressed heart rate. The initial acceleration proved to be a temporary event only. But at the warmer temperature of 21°C, the heart rate, which dropped slightly at first, recovered to normal after 1 hour of exposure to 200 atm (Naroska, 1968).

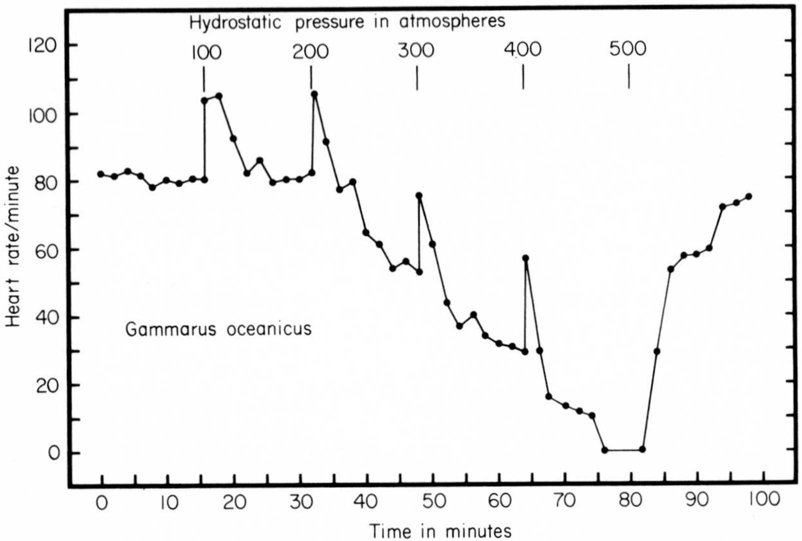

FIG. 4. A single record of the heart rate of *Gammarus oceanicus* in which the pressure was raised rapidly by successive equal steps of 100 atm. The experiment was performed in brackish water of 15‰ salinity at 15°C (Naroska, 1968).

B. PRESSURE EFFECTS ON THE SURVIVAL RATE OF TISSUES AND MARINE ANIMALS

Schlieper (1963a), Schlieper *et al.* (1967), and Ponat (1967), when working with the classic pressure chamber (ZoBell, 1959), used small tissue pieces or small numbers of gill filaments instead of whole animals. The tissue pieces were kept for the duration of the experiments in sea water-filled, lid-covered plastic jars of 25-ml volume. Three of them were placed in the pressure chamber. The entire device was installed in a temperature-controlled room. In addition, experiments were also carried out in the microscope pressure chamber, enabling the experimenter to examine the tissue while compressed. The control tissue was kept at the same temperature in identical jars and simultaneously checked after the experiments. The cellular resistance was judged by the percentage of surviving ciliated cells or the persistence of their beat. In this way, Ponat (1967) determined the lethal limits for the tissues of a series of species from the coastal water of the North Sea and the Baltic. The results proved that the ciliated tissues of the starfish *Asterias rubens* and the bivalve *Mytilus edulis* are more resistant to high hydrostatic pressure than those of the bivalve *Cyprina islandica* and the sea anemone *Metridium senile* from the same habitat (Fig. 5).

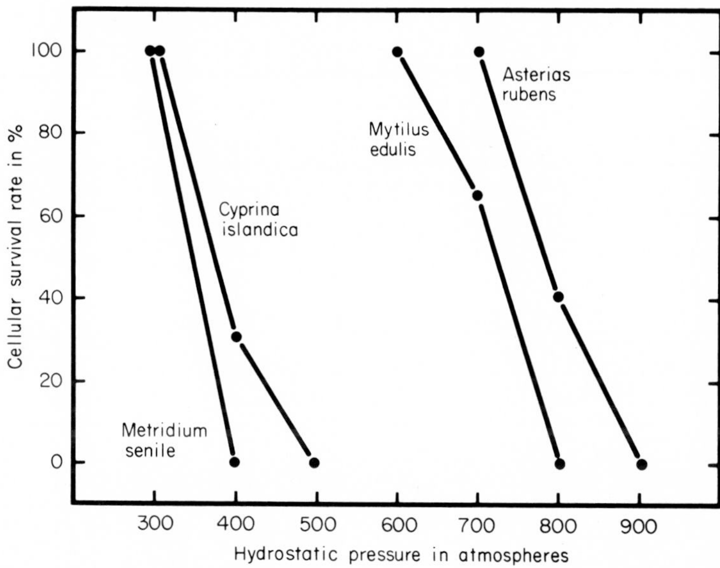

FIG. 5. Pressure effects on the survival rate of ciliated tissues of the sea anemone *Metridium senile* (tentacles), the bivalves *Cyprina islandica* and *Mytilus edulis* (gill filaments), and the starfish *Asterias rubens* (papulae) (Ponat, 1967).

By and large, these experiments were confirmed by Naroska (1968), using a slightly modified method. He subjected many intact marine invertebrates and several fishes to pressures ranging from 100 to 800 atm in 1-hour experiments. After a 24-hour recovery he determined the magnitude of pressure which killed 50% of the specimens (LD_{50} data). In his investigation, 50% of the isopods *Jaera albifrons, Idothea baltica,* the mollusks *Mytilus edulis, Modiolus modiolus,* and *Mya arenaria,* the amphipods *Gammarus duebeni* and *G. oceanicus,* the gastropod *Littorina littorea,* the echinoderm *Asterias rubens,* and the polychaete *Nereis* survived 500 atm and more, thus showing a high resistance to pressures rarely encountered by these specimens in the natural environment (Fig. 6). On the other hand, the decapods, the Mysidacea *Neomysis vulgaris,* the tunicate *Ciona,* and the teleosts are distinctly more sensitive to pressure, as Ebbecke (1944) has already shown. Naroska's laboratory ob-

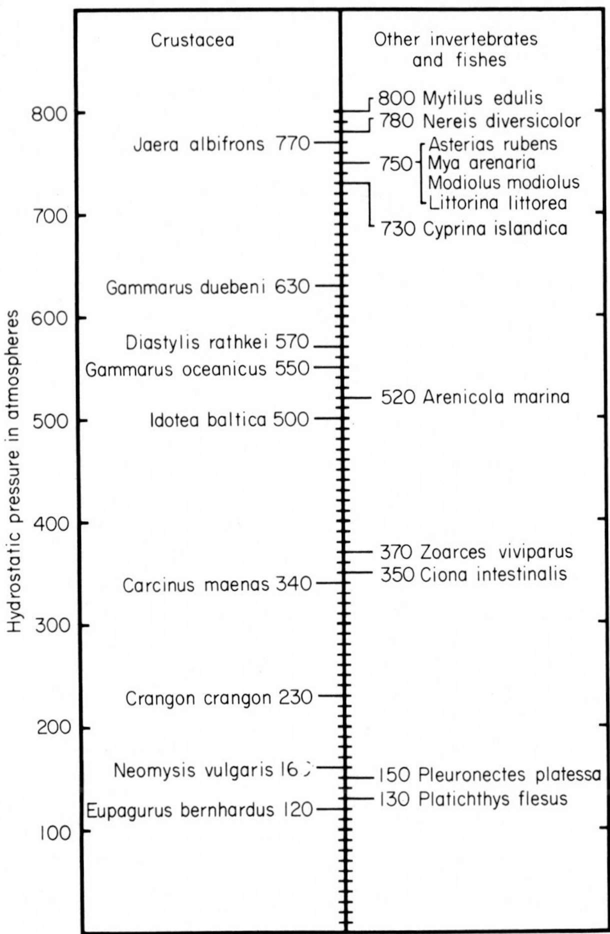

FIG. 6. The lethal pressure in atmospheres which killed 50% of the specimens (LD_{50} data) of various marine invertebrates and fishes from the western Baltic Sea (Naroska, 1968).

servations are in accordance with the fact that the fauna of the deep trenches contains only occasionally decapods (Fig. 1), but relatively many genera of the amphipods and isopods are able to accustom themselves to the conditions prevailing at depths exceeding 5000 m (Dahl, 1954).

C. PRESSURE EFFECTS IN RELATION TO TEMPERATURE

Much attention has been given by Schlieper (1963a, 1968), Ponat (1967), and Naroska (1968) to the additional effects of high or low experimental temperatures on the lethal limits of compressed tissues. Schlieper and Ponat found that the pressure resistance of the tissues of both *Mytilus edulis* from the North Sea and *Mytilus edulis diegensis* from the Pacific was distinctly higher at 15° and 20°C than at 10°C (Fig. 7a). It was for this reason that Schlieper concluded that deep sea animals

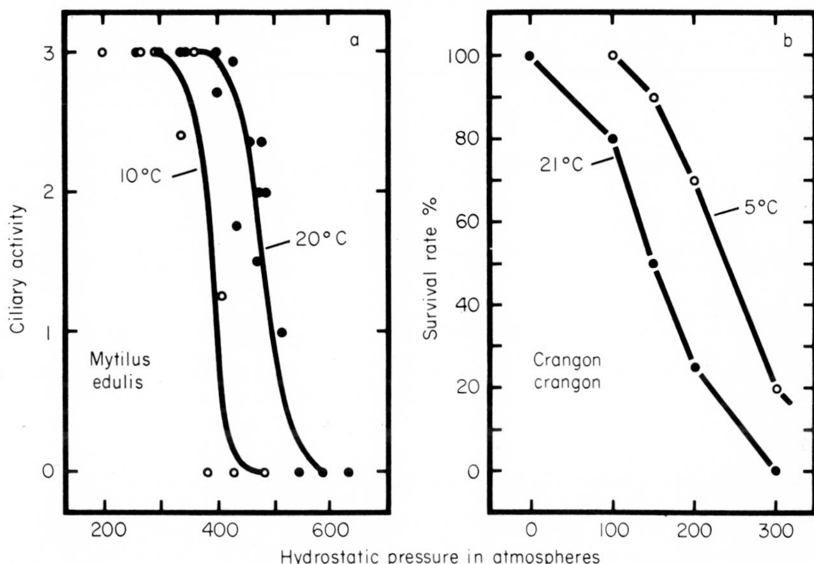

Fig. 7. The effects of temperature on the pressure resistance. (a) Relative ciliary activity of the terminal cilia of *Mytilus* gill filaments (Schlieper, 1963a). (b) Survival rate of the decapod shrimp *Crangon crangon* (Naroska, 1968).

would perhaps cope more successfully with the tremendous pressure at temperatures higher than the normal temperatures of 2°–4°C at the bottom of the deep sea. This view is supported by the fundamental findings of Brown (1934), Marsland and Brown (1942), and their associates that the plasmagel is stabilized at higher temperatures, thus counteracting the deleterious effects of hydrostatic pressure. However, recent investigations demonstrate that the combination of temperatures around 20°C and high pressure by no means increases the pressure resistance in all cases. Naroska, working with well-acclimated amphipods and decapods from the Baltic, found higher pressure resistance at 5° than at 15° and 20°C (Fig. 7b). Although it seems to us insufficient to compare the data obtained from highly organized, osmoregulating crustacea with those obtained from isolated tissues of bivalves, we believe the increase of the pressure resistance of some marine invertebrates in sea water around

15°–20°C is not a phenomenon to be generalized. Obviously, the genetic adaptation and the long-term acclimation to certain temperature ranges play the most important role in the pressure resistance of marine animals.

Ponat (1967) acclimated isolated gill filaments of *Mytilus edulis* from the Baltic Sea to 5°, 10°, and 15°C for 5 days. She then observed the ciliary activity after a 24-hour exposure to 400 atm at 5° and 15°C. The experiment showed that the deleterious combination of a pressure of 400 atm and low temperatures (5°C) was at least partly compensated for by the preacclimation of the tissue to 5°C (Table I).

TABLE I
RELATIVE CILIARY ACTIVITY OF ISOLATED GILL PIECES OF *Mytilus edulis*[a]

Acclimation temperature (°C)	Relative ciliary activity after 24-hour exposure to 400 atm	
	5°C	15°C
5	1.7	2.4
10	0.9	2.5
15	0.6	2.5

[a] Gill filaments of one specimen were investigated at two experimental temperatures after preacclimation to 5°, 10°, and 15°C (Ponat, 1967).

It may well be the case that the pressure resistance of deep sea animals adapted to high pressure as well as low temperatures decreases with rising temperatures. Unfortunately, such experiments on the combined effects of temperature and pressure on marine animals taken from great depth are still lacking.

Recently, the combined effects of pressure and temperature on the activity of alkaline phosphatase of various marine bivalves have been studied in our laboratory. The first preliminary experiments show clearly that the activity of the alkaline phosphatase of *Mytilus* and *Cyprina* could be increased, decreased, and unaffected by pressure (400–800 atm), according to the temperature. After exposure of the enzyme to pressure for 1 hour, the activity decreases at 5°C and increases at 25°C. The enzyme activity was scarcely affected at 15°C.

D. PRESSURE EFFECTS IN RELATION TO SALINITY

When working with animals from the Baltic, the question arose as to the effects of brackish water, i.e., diluted sea water, on the pressure resistance of marine animals. Naroska acclimated the euryhaline amphipod *Gammarus oceanicus* for 9–12 days to salinities ranging from 5‰ to 50‰, and he then determined the survival rate after exposure of the amphipods to pressures between 300 and 700 atm for 1 hour. His experiments showed that 100% of the amphipods survived 500 atm in sea water of full strength. In brackish water of a salinity of only 5‰, the same pressure killed 50% of the animals (Fig. 8). The individual pressure resistance

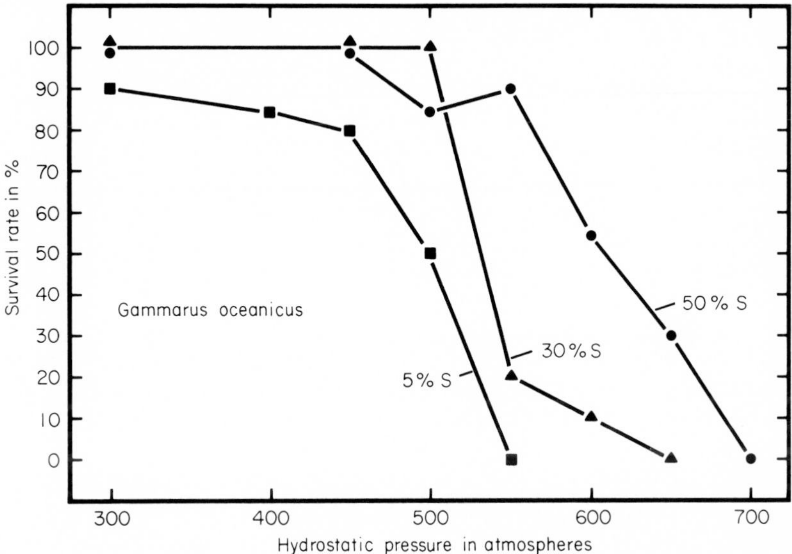

FIG. 8. The effects of various salinities on the pressure resistance of the amphipod *Gammarus oceanicus* from the Baltic Sea. The animals were acclimated to the various concentrations of the external medium for 9–12 days (Naroska, 1968).

was also increased in brackish water of the same strength to which either 0.75 *M* or 1.00 *M* of glycerin were added. Similar responses to pressure were also revealed by the determination of the ciliary rate of isolated gill filaments of *Mytilus* from th Baltic (15‰) and the North Sea (Fig. 9). The ciliary rate of the tissues compressed in sea water of full strength was significantly higher than of those compressed in brackish water (Ponat, 1967).

Various authors hav shown that acclimation of marine invertebrates to brackish water also reduces the cellular resistance to other stress factors such as heat and freezing. It is suggested that this general unspecific decrease of the cellular resistance indicates the destabilization of sensitive protoplasmic components (Schlieper, 1966; Theede, 1965; Theede and Lassig, 1967).

It is interesting to note that not only the osmotic strength of the sea water, but also the addition of certain cations, increases the pressure resistance of isolated tissues and intact invertebrates. Experiments with excess Ca^{++} in the external medium, for instance, were carried out by Ponat (1967) and Schlieper *et al.* (1967). In both investigations the pressure resistance was increased by Ca^{++} excess. Thus Schlieper *et al.* were able to raise the lethal limits of the hermit crab *Eupagurus zebra* from the Red Sea by about 25%. Ponat also demonstrated a considerably in-

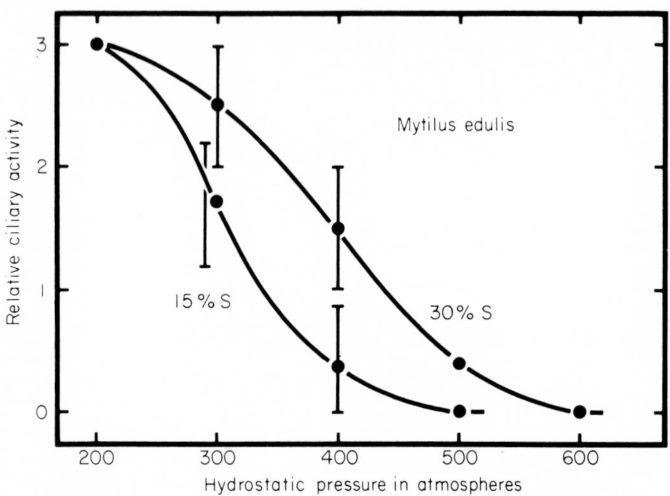

FIG. 9. The effects of a sea water of 15‰ and 30‰ salinity on the pressure resistance of *Mytilus edulis* from the Baltic Sea. Criterion: relative ciliary activity of gill filaments (Ponat, 1967).

creased pressure resistance of the gill filaments of *Mytilus* acclimated to sea water containing twice the normal calcium content. These interesting responses of the compressed organisms and cells to excess Ca^{++} are far from being fully understood, but the authors believe that Ca^{++} is significant in stabilizing the cell membranes and the outer cortex of the cells (see also Heilbrunn, 1952).

Only a very slightly increased pressure resistance of *Arenicola marina* was reported by Naroska, who kept the animals for a couple of days prior to the experiments in brackish water (15‰) in double calcium concentration. In *Gammarus oceanicus* from the same habitat, however, the lethal pressure limits were not altered by excess Ca^{++}. Naroska suggests that in homeoosmotic and ionoregulating animals like *Gammarus*, the internal calcium content is efficiently regulated. In contrast, the more poikilosmotic and stenohaline species do not excrete calcium as quickly from the body. The same holds true for their tissues, which also can be stabilized by excess Ca^{++}.

E. Pressure Effects in Relation to the pH

The pH of sea water changes according to the depth. The pH of about 8 at the surface drops to 7.78 at a hydrostatic pressure of 1000 atm. It is

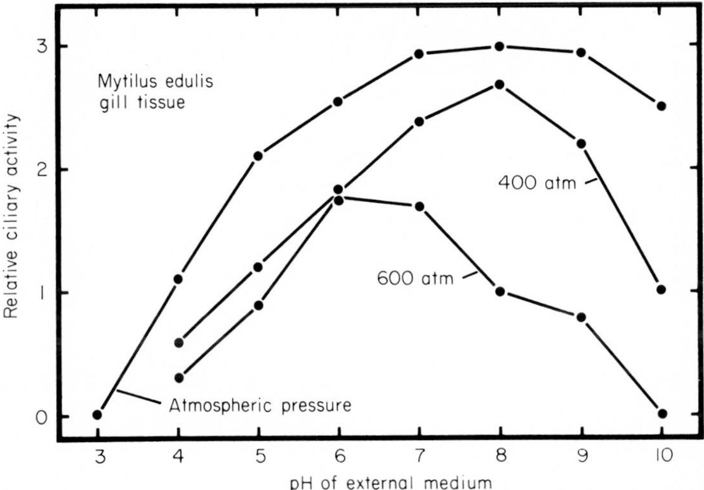

Fig. 10. The influence of various pH values on the cellular pressure resistance of the gill tissue of *Mytilus edulis* (Ponat and Theede, 1967).

for this reason that the responses of cells to compression in a wide range of pH values are of interest to us. Ponat and Theede (1967) have studied the effects of 400 to 600 atm on the ciliated gill tissues of *Mytilus* at pH values ranging from 3 to 10. At atmospheric pressure the optimum of the ciliary rate is not clearly marked. The highest rates were found from about pH 7 to pH 9. After having subjected the tissues to 400 and 600 atm for 3 hours, however, the ciliary rate was clearly reduced and a well-marked optimum had developed. This optimum shifted from pH 8 at 400 atm to pH at 600 atmospheres (Fig. 10). This reaction is not yet understood, but it is suggested that it is evidence of responses at the enzyme level. The authors propose that the relationships between pH values and hydrostatic pressure depend on changes of the chemical bonds. Apparently, these changes do alter the chemical activity as well as the molecular structure of cell enzymes and other macromolecules. Further investigations on the enzyme activity at moderate and high pressure are, therefore, required (see reviews of ZoBell, Chapter 4, and Landau, Chapter 2).

F. Pressure Effects on the Ultrastructure of Cells

In addition, the structural changes of pressurized tissues and epithelia at microscopic and ultrastructural level are of interest and have been subjects for many studies. Ebbecke (1944) has reported on the effects of pressure on the ciliated epithelia and glandular cells of the frog's pharynx. The ciliary movement was slowed after exposure of the epithelia to 1000 atm. A pressure of 1500 atm resulted in a loss of the cilial mobility. The columnar and elongated shape of the cells became more spherical under pressure, indicating structural changes similar to those observed in pressurized *amebas* and other protozoa (Marsland and Brown, 1936).

More recently, electron microscopic investigations on *Amoeba* before, during, and after exposure to high pressures were performed by J. V. Landau and Thibodeau (1962). The plasmalemma and several cell organelles such as mitochondria proved to be relatively stable at pressures of about 533 atm (8000 psi). But the Golgi complexes and the pinocytosis channels, which were readily distinguishable at atmospheric pressure, had disappeared. Tilney *et al.* (1966), who have worked on *Actinosphaerium nucleofilum*, reported that the microtubular elements of the axopodia became unstable under pressure. At 266 atm few tubules remained, but at 400 and 533 atm (4000 and 8000 psi) rapid disintegration took place.

The disorganizational effects of high hydrostatic pressure can be demonstrated even more clearly by ultrastructural studies on the karyoplasm of cells. At atmospheric pressure the chromatin of the macronucleus of Tetrahymena appear to be evenly dispersed. But at high pressure coarse masses of chromatin agglomerate within the karyoplasm (Kennedy and Zimmerman, 1968). Electron microscopic investigations on the pressurized ciliated epithelium of the gills of *Mytilus* (1 hour at 600 atm) have shown that most organelles of these cells are relatively stable, except for the chromatin of the nuclei which seem to form clumps similar to those observed in the nuclei of *Tetrahymena*. Further investigations are needed to elucidate the effects of hydrostatic pressure on the cells of marine invertebrates at ultrastructural level (see also Marsland, Chapter 11; Zimmerman, Chapter 10; and Zimmerman and Zimmerman, Chapter 8).

G. Pressure Effects on the Oxygen Consumption

The necessity of long-term pressure experiments on marine species in running sea water has been repeatedly emphasized by Schlieper (1963 a,b), for such experiments are the only way to eliminate the interfering effects of the rapid compression and the possible lack of oxygen during the experiments. In contrast to long-term field experiments, such laboratory investigations can be performed at well-controlled temperatures. However, these studies were unduly delayed because pressure devices suitable for running sea water were not available. Recently, however, our pressure apparatus, designed by the American Instrument Company, Washington D.C. (AMINCO) was modified and improved and now works quite satisfactorily at 100 to 400 atm. (see also Kitching, Chapter 7).

Fig. 11. Assembled AMINCO pressure apparatus for the measurements of the oxygen consumption at high hydrostatic pressure in steadily running seawater: *a*—piston pump; *b*—manometer; *c*—pressure chamber; *d*—cooling coil; *e*—pressure control unit; and *f*—seawater outlet control valve (Naroska, 1968).

A brief description of this pressure apparatus may be of interest. The sea water is pumped from the reservoir into the high pressure tubes. Simultaneously, the oxygen content of the water can be measured. The sea water is then pumped through ball valves and via the manometer into the stainless steel pressure chamber (Fig. 11). The water leaves the pressure chamber by way of another pressure line which branches at a Y-piece. One of the branches is connected to the pressure control unit, the other tube goes to the sea water outlet control valve. Pneumatic impulses given by the newly installed pressure control unit open or close the sea water outlet control valve, thus maintaining the pressure in the pressure chamber. The water decompresses at the outlet and runs at atmospheric pressure through a cooling coil into the Winkler flask, where the oxygen content can again be measured (for further details, see Naroska, 1968).

The experiments were performed in sea water of 15‰ or 30‰ salinity at a steady speed of 600 to 720 ml/hr. The flounder *Platichthys flesus*, the crab *Carcinus maenas*, and the polychaete *Nereis diversicolor* responded to 100 atm with increased oxygen consumption for the first 3 to 4 hours (Fig. 12). After 7 to 8 hours the metabolic rates had dropped to normal values. On the other hand, the more pressure resistant starfishes did not react distinctly to 100 atm. But at 200 and 300 atm, the metabolic rates of *Asterias rubens* and *Henricia sanguinolenta* dropped immediately after compression. The sea urchin *Psammechinus miliaris* was scarcely affected by either pressure (Fig. 13). If the compression did not exceed the lethal limits of the starfishes, their oxygen consumption rose gradually during the 24-hour acclimation experiments. The tendency for a recovery of the respiration in the 24-hour experiments, together with the fact that the lethal limits of the eurybathic starfishes can be shifted to higher values by a stepwise exposure to increasing hydrostatic pressures, may indicate a nongenetic, individual acclimation (see also Belyaev, 1966).

V. Final Conclusions

Hitherto, experiments on the effects of hydrostatic pressure on marine animals were performed with species from the shallow coastal regions. Fortunately, some of these, such as the starfishes *Asterias rubens*, *Henricia sanguinolenta*, and many bivalves are eurybathic and provide favorable material even for high pressure work. In order to study the significance of the constant pressure as an ecological parameter in the ocean, further long-term experiments in steadily running sea water are required. Such experiments should be extended to several weeks or months, and, if possible, the rearing of marine animals while under pressure also seems to us desirable. We are aware, of course, that this goal would be almost as complicated and costly to achieve as space research.

Furthermore, it also seems necessary to us to improve and put into operation dredges suitable to haul stenobathic deep sea animals aboard the research vessels under controlled temperatures and pressures. Thus, it would be possible to attack many questions concerning the mechanisms developed for life at great depths. How, for example, did some of the invertebrate groups conquer the depths of the oceans whereas others

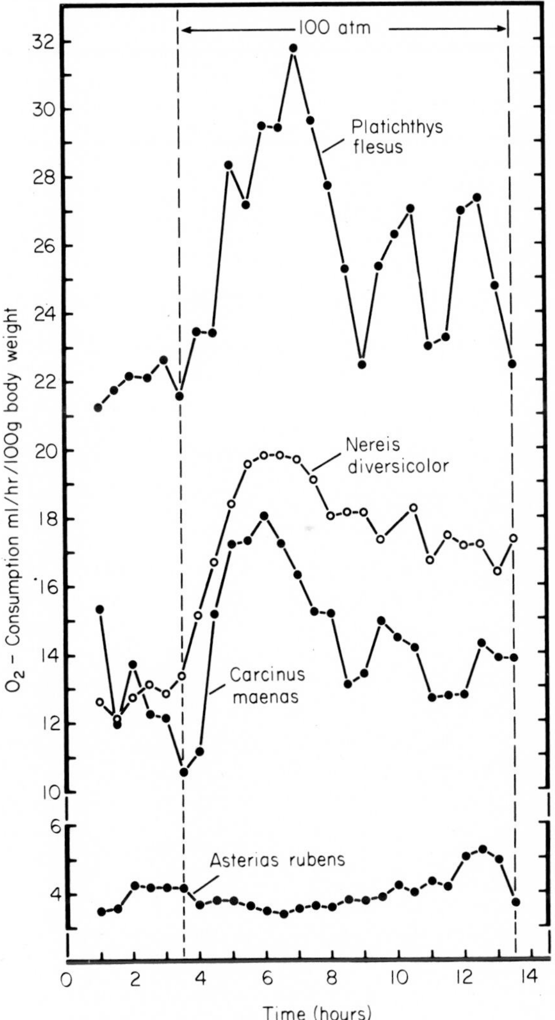

FIG. 12. The oxygen consumption of the starfish *Asterias rubens*, the crab *Carcinus meanas*, the polychaete *Nereis diversicolor*, and the flounder *Platichthys flesus* at 100 atm (Naroska, 1968).

failed to become adapted to the factors governing life of the deep sea? How do stenobathic–barophilic animals respond to temperatures above 2°–4°C? Why are many of the deep sea animals larger than their close relatives from shallow regions? Are shallow water species successfully acclimatable to the conditions prevailing at great depths? Does pressure effect marine forms and their tissues at enzyme level? These are just a few of the many problems still to be solved.

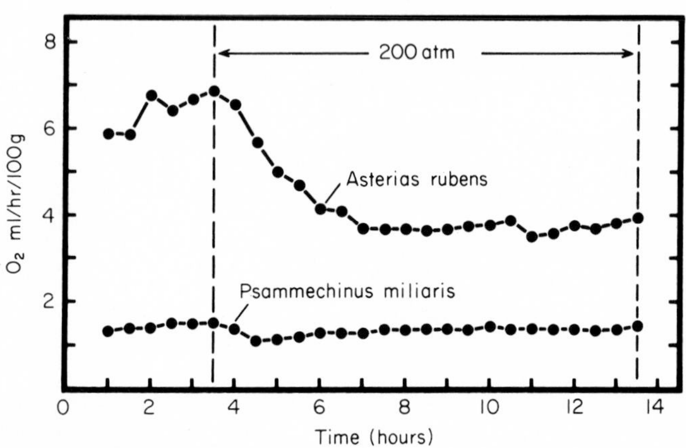

FIG. 13. The oxygen consumption of the sea urchin *Psammechinus miliaris* and the starfish *Asterias rubens* at 200 atm (Naroska, 1968).

Douglas Marsland, one of the leading cell physiologists working on the effects of pressure, finished a review some 10 years ago with the following words:

> Indeed, pressure has begun to take its place with temperature as a fundamental factor governing physiological processes. Physicists and chemists have always recognized the importance of pressure, as indicated by the basic thermodynamic equation: $PV = nRT$. Physiologists are now following suit. (Marsland, 1958).

Since those years, a vast amount of knowledge has been accumulated through the work of the cell physiologists. It is now up to the marine ecologists to follow suit.

REFERENCES

Belyaev, G. M. (1966). "Bottom Fauna of the Ultraabyssal Depths of the World Ocean." Nauka, Moscow.

Bernard, F. (1961). La vie marine en grande profoundeur. *Bull. Soc. Zool. France* **85**, 255–274.

Brown, D. E. S. (1934). The pressure coefficient of "viscosity" in the eggs of *Arbacia punctulata*. *J. Cellular Comp. Physiol.* **5**, 335–346.

Brown, D. E. S., and Marsland, D. (1936). The viscosity of *Amoeba* at high hydrostatic pressure. *J. Cellular Comp. Physiol.* **8**, 159–165.

Cattell, Mck. (1936). The physiological effects of pressure. *Biol. Rev.* **11**, 441–476.

Dahl, E. (1954). The distribution of deep sea Crustaceae. I.U.B.S. (*Intern. Union Biol. Sci.*) *Deep-Sea Colloq., Ser. B* No. 16.

Draper, J. W., and Edwards, D. J. (1932). Some effects of high pressure on developing marine forms. *Biol. Bull.* **63**, 99–107.

Ebbecke, U. (1935). Uber die Wirkungen hoher Drucke auf marine Lebewesen. *Arch. Ges. Physiol.* **236**, 648.

Ebbecke, U. (1944). Lebensvorgänge unter der Einwirkung hoher Drucke. *Ergeb. Physiol. Biol. Chem. Exptl. Pharmakol.* **45**, 34–183.

Edwards, D. J., and Brown, D. E. S. (1934). The action of pressure on the form of the electromyogram of auricle muscle. *J. Cellular Comp. Physiol.* **5**, 1–19.

Edwards, D. J., and Cattell, McK. (1927). Some results of the application of high pressures to the heart. *Proc. Soc. Exptl. Biol. Med.* **25**, 234.

Edwards, D. J., and Cattell, Mck. (1928). The stimulating action of hydrostatic pressure on cardiac function. *Am. J. Physiol.* **84**, 472–484.

Fontaine, M. (1928). Les forte pressions et la consommation d'oxygène de quelques animaux marins. Influences de la taille de l'animal. *Compt. Rend. Soc. Biol.* **99**, 1789–1790.

Grundfest, H., and Cattell, Mck. (1935). Some effects of hydrostatic pressure on nerve action potentials. *Am. J. Physiol.* **113**, 56–57.

Heilbrunn, L. V. (1952). "An Outline of General Physiology." Saunders, Philadelphia, Pennsylvania.

Kennedy, J. R., and Zimmerman, A. M. (1968). Unpublished data.

Kitching, J. A., and Pease, D. C. (1939). The liquefaction of the tentacles of suctorian protozoa at high hydrostatic pressures. *J. Cellular Comp. Physiol.* **14**, 410–412.

Knight-Jones, E. W., and Morgan, E. (1966). Responses of marine animals to changes in hydrostatic pressure. *Oceanog. Marine Biol. Ann. Rev.* **4**, 267–299.

Landau, J., and Marsland, D. (1952). Temperature–pressure studies on the cardiac rate in tissue culture explants from the heart of the tadpole (*Rana pipiens*). *J. Cellular Comp. Physiol.* **40**, 367–382.

Landau, J. V., and Thibodeau, L. (1962). The micromorphology of *Amoeba proteus* during pressure-induced changes in the sol-gel cycle. *Exptl. Cell Res.* **27**, 591–594.

Lemche, H., and Wingstrand, K. G. (1959). The anatomy of *Neopilina galathea* Lemche, 1957 (Mollusca Tryblidiacea). *Galathea Rept.* no. 3.

Marsland, D. A. (1944). Mechanism of pigment displacement in unicellular chromatophores. *Biol. Bull.* **87**, 252–261.

Marsland, D. A. (1958). Cells at high pressure. *Sci. Am.* **199**, 36–43.

Marsland, D. A., and Brown, D. E. S. (1936). Amoeboid movement at high hydrostatic pressure. *J. Cellular Comp. Physiol.* **8**, 167–178.

Marsland, D. A., and Brown, D. E. S. (1942). The effects of pressure on sol-gel equilibria, with special reference to myosin and other protoplasmic gels. *J. Cellular Comp. Physiol.* **20**, 295–305.

Menzies, R. J., and Wilson, J. B. (1961). Preliminary field experiments on the relative importance of pressure and temperature on the penetration of marine invertebrates into the deep sea. *Oikos* **12**, 302–309.

Naroska, V. (1968). Vergleichende Untersuchungen über den Einfluβ des hydrostatischen Druckes auf Uberlebensfähigkeit und Stoffwechselintensität mariner Evertebraten und Teleosteer. *Kiel. Meeresforsch.* **24**, 95–123.
Pease, D. C., and Kitching, J. A. (1939). The influence of hydrostatic pressure upon ciliary frequency. *J. Cellular Comp. Physiol.* **14**, 135–142.
Pease, D. C., and Marsland, D. A. (1939). The cleavage of *Ascaris* eggs under exceptionally high pressure. *J. Cellular Comp. Physiol.* **14**, 407–408.
Ponat, A. (1967). Untersuchungen zur zellulären Druckresistenz verschiedener Evertebraten der Nord- und Ostsee. *Kiel Meeresforsch.* **23,** 21–47.
Ponat, A., and Theede, H. (1967). Die pH-Abhängigkeit der zellurlären Druckresistenz bei *Mytilus edulis. Helgoländer Wiss. Meeresuntersuch.* **16**, 231–237.
Regnard, P. (1884a). Note sur les conditions de la vie dans les profondeurs de la mer. *Compt. Rend. Soc. Biol.* **36**, 164–168.
Regnard, P. (1884b). Effet des hautes pressions sur les animaux marins. *Compt. Rend. Soc. Biol.* **36**, 394–395.
Regnard, P. (1884c). Recherches expérimentales sur l'influence des très hautes pressions sur les organismes vivants. *Compt. Rend.* **98**, 745–747.
Regnard, P. (1885). Phénomènes objectifs que l'on peut observer sur les animaux soumis aux hautes pressions. *Compt. Rend. Soc. Biol.* **37**, 510–515.
Regnard, P. (1891). "Recherches expérimentales sur les conditions physiques de la vie dans les eaux." Masson, Paris.
Sano, K. (1958). Cited after Bernard (1961).
Schlieper, C. (1963a). Biologische Wirkungen hoher Wasserdrucke. Experimentelle Tiefsee-Physiologie. *Veroeffentl. Inst. Meeresforsch. Bremerhaven, Sonderband, Meeresbiol. Symp.* **1**, 31–48.
Schlieper, C. (1963b). Neuere Aspekte der biologischen Tiefseeforschung. *Umschau* **15**, 457–461.
Schlieper, C. (1966). Genetic and nongenetic cellular resistance adaptation in marine invertebrates. *Helgolaender Wiss. Meeresuntersuch.* **14**, 482–502.
Schlieper, C. (1968). High pressure effects on marine invertebrates and fishes. *Marine Biol.* **2**, 5–12.
Schlieper, C., Flügel, H., and Theede, H. (1967). Experimental investigations of the cellular resistance ranges of marine temperate and tropical bivalves: Results of the Indian Ocean Expedition of the German Research Association. *Physiol. Zool.* **40**, 345–360.
Theede, H. (1965). Vergleichende experimentelle Untersuchungen über die zelluläre Gefrierresistenz mariner Muscheln. *Kiel. Meeresforsch.* **21**, 153–166.
Theede, H., and Lassig, J. (1967). Comperative studies on cellular resistance of bivalves from marine and brackish waters. *Helgolaender Wiss. Meeresuntersuch.* **16**, 119–129.
Thomson, Sir C. W. (1877). "The Voyage of 'The Challenger'," The Atlantic, Vols. I and II. Macmillan, New York.
Tilney, L. G., and Hiramoto, Y., and Marsland, D. (1966). Studies on the microtubules in heliozoa. III. A pressure analysis of the role of these structures in the formation and maintenance of the axopodia of *Actinosphaerium nucleofilum (Barrett)*. *J. Cell Biol.* **29**, 77–95.
Zenkevitch, L. A. (1963). "Biology of the Seas of the U.S.S.R." Allen & Unwin, London.
ZoBell, C. E. (1959). Thermal changes accompanying the compression of aqueous solutions to deep-sea conditions. *Limnol. Oceanog.* **4**, 463–471.

CHAPTER 10

HIGH PRESSURE STUDIES ON SYNTHESIS IN MARINE EGGS

Arthur M. Zimmerman

I. Introduction

The earliest investigations concerned with synthesis in marine eggs under the influence of hydrostatic pressure were initiated in the early 1950's when Landau *et al.* (1955) reported that high energy adenine nucleotides increase the strength of cortical cytoplasm surrounding sea urchin eggs; such ATP-fortified cells continue to furrow under hydrostatic pressures which normally block nontreated control cells. Shortly thereafter it was reported that sulfhydryl inhibitors, salyrgan and *p*-chloromercuribenzoate, which effectively inhibit ATPase activity, could weaken the furrowing reaction of marine eggs (Zimmerman *et al.*, 1957). Thus, further evidence was provided that the ATP system plays a vital role in supplying energy for cytokinesis. During the past decade there have been numerous reports on synthesis in various cell systems under hydrostatic pressure (cf. Chapters 2, 11, 8, and 4 by Landau, Marsland, Zimmerman and Zimmerman, and ZoBell, respectively).

In this chapter, the pressure studies discussed will be limited to those on sea urchin cells. Pressure studies concerned with DNA synthesis, the incorporation of radioactively labeled uridine into various cell fractions, and the sedimentation characteristics of microsomal cell fractions will be reviewed. In addition, an analysis of pressure centrifuge-induced furrowing as well as pressure effects on the ultrastructure will be made.

II. Pressure Effects on DNA Synthesis and on the Division Schedule

Cells exhibit differential sensitivity to physical and chemical agents at various stages during the cleavage cycle (cf. Zimmerman, 1968). Prior to discussing the effects of a physical agent such as hydrostatic pressure on DNA synthesis in dividing marine eggs, it is necessary to evaluate the effects of pressure on the cleavage schedule and to review the temporal aspects of DNA synthesis.

It should be emphasized that even short pulses of hydrostatic pressure affect the cleavage schedule as well as structural elements associated with chromosomal movement (cf. Marsland, Chapter 11). Cleavage delays which occur in *Arbacia* eggs following a 1-minute pulse of pressure (7500 psi) applied at various times after insemination are illustrated in Fig. 1 (Zimmerman and Silberman, 1965). Following a 1-minute compression, maximum division delays occurred when the cells were treated at prophase (42 minutes after insemination). Minimum cleavage delays occurred when treatment was initiated at syngamy (union of the pronuclei), in the middle of the cycle just preceding the streak stage (27 minutes after insemination), or just prior to cleavage. Unfertilized eggs were resistant to pressure and they did not display any division delays when inseminated within 10 minutes after pressure treatment. As shown in Fig. 1 the cells were more sensitive to pressure, as reflected in longer division delays, during the organization of the mitotic apparatus, rather than at subsequent stages when the mitotic structure was well formed. Possibly the bonds of the newly polymerized units forming the mitotic apparatus (spindle–aster–chromosome complex) are more labile than the bonds of a well-formed mitotic structure. Thus, disorganization by pressure of the more labile newly formed mitotic structure may be more complete than that of a well-formed structure with equivalent pressure, and hence reorganization may require more time resulting in longer division delays.

The incorporation of radioactive thymidine is a good index for studying

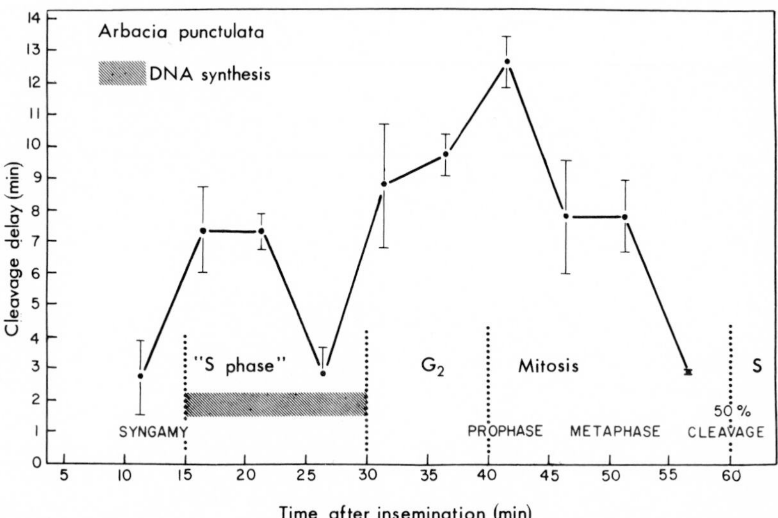

FIG. 1. The sensitivity of fertilized *Arbacia* eggs to short pulses of hydrostatic pressure during the first cleavage cycle. The cleavage delays following a pressure treatment of 7500 psi for 1 minute at various times after insemination (at 20°C) are shown. Each point represents the average cleavage delay for a 5-minute time interval; the standard errors at each time are also shown. The time to 50% cleavage for non-treated control eggs was adjusted to correspond to 60 minutes. The nuclear events and the interval during which the cells incorporate ^{3}H-thymidine during the mitotic cycle are also illustrated. (Data of Zimmerman, 1963; Zimmerman and Silberman, 1965.)

DNA synthesis in marine eggs. Mitotic events and DNA synthetic events (at 20°C) are also shown in Fig. 1. Under atmospheric pressure, fertilized *Arbacia* eggs incorporate ^{3}H-thymidine between 15 and 30 minutes after insemination. During the 15-minute period preceding syngamy (the union of the pronuclei) and during the later half of the first division cycle (30–60 minutes after insemination) the eggs do not actively incorporate ^{3}H-thymidine into nuclear DNA (Zimmerman, 1963). In eggs of the purple sea urchin (*Strongylocentrotus purpuratus*) radioactive thymidine is incorporated into the acid-insoluble fraction (DNA fraction) between 30

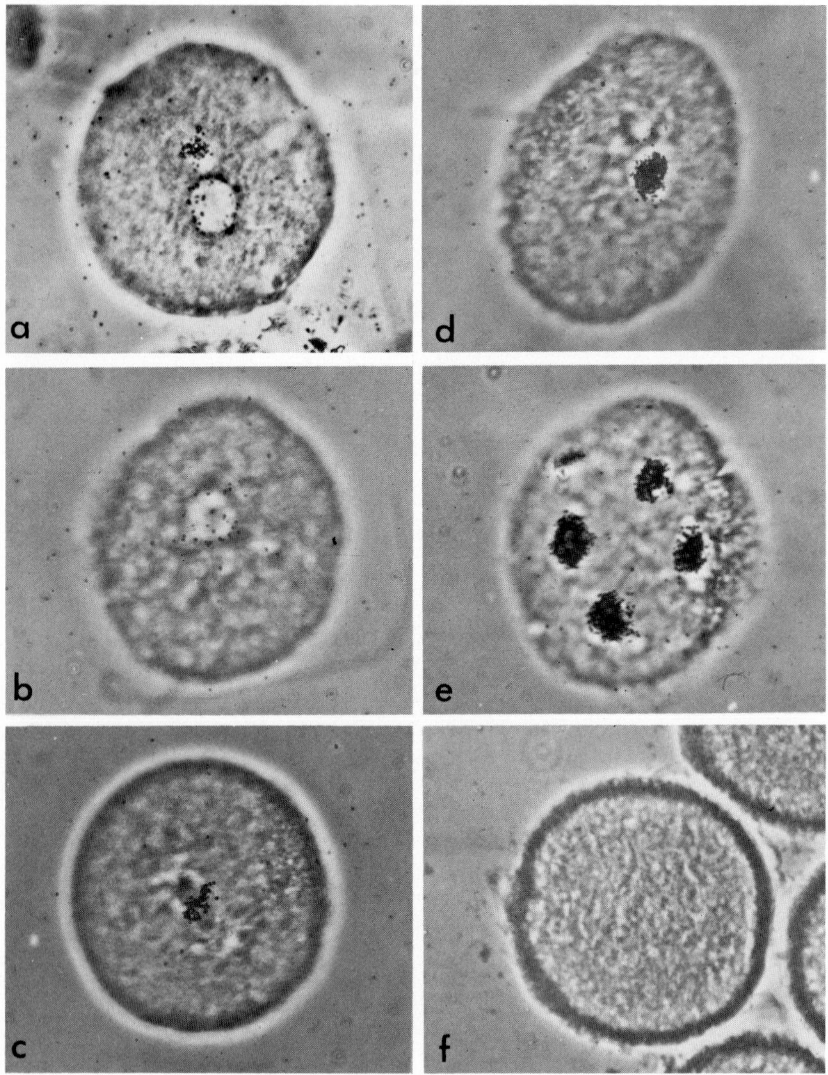

FIG. 2. Autoradiographs showing incorporation of ^{3}H-thymidine into *Arbacia* eggs. (a) eggs immersed in ^{3}H-thymidine 5 minutes after insemination and subjected to 5000 psi for 60 minutes. The autoradiographic silver grains are localized above the male and female pronuclei; the pressure treatment has prevented fusion. In (b), ^{3}H-thymidine–pressure treatment (5000 psi) was initiated at syngamy (15 minutes

and 45 minutes after insemination at 15°C (Hinegardner *et al.*, 1964). It should be noted that the length of the first division schedule for these species is quite different, 60 and 100 minutes, for *Arbacia* and *Strongylocentrotus*, respectively. However, if the duration of the first cleavage cycle is adjusted so that the length of time from insemination to division is considered arbitrarily as 100, then the period during which DNA synthesis occurs, as a proportion of the cleavage cycle, corresponds closely in both species.

As stated earlier, *Arbacia* eggs do not normally incorporate ^{3}H-thymidine into pronuclei prior to syngamy. However, in other marine eggs, such as the sand dollar *Echinarachnius*, incorporation of the precursor occurs in the individual pronuclei prior to fusion (Simmel and Karnofsky, 1961). When fertilized *Arbacia* eggs were subjected to a pressure of 5000 psi, 5 minutes after insemination, pronuclear fusion was blocked. The pronuclei were found separated by varying distances depending upon their specific positions when pressure was applied (Fig. 2a). Under pressure and in the presence of ^{3}H-thymidine, the blocked pronuclei incorporated the radioactive precursor. Autoradiographs of cells compressed for 60 minutes showed silver grains localized above the pronuclei (Zimmerman and Silberman, 1967).

The Colcemid studies by Zimmerman and Zimmerman (1967) shed additional light on the question why pronuclei of pressurized *Arbacia* eggs are able to incorporate ^{3}H-thymidine whereas nonpressurized control cells do not. Does pressure alter the permeability of the cell to ^{3}H-thymidine, or is there a more subtle effect whereby pressure prevents one activity (pronuclear movement) but does not affect certain synthetic events (in this case the sequence of biochemical events essential for DNA synthesis) that are initiated at the time of cell activation? Colcemid, the deacetyl-*N*-methyl derivative of colchicine, reversibly blocks pronuclear fusion in *Arbacia* eggs in concentrations of 2.7×10^{-5} *M* or greater. Fertilized eggs

---

after insemination) for 30 minutes; silver grains are located above the fusion nucleus. In (c), control, the eggs were placed into ^{3}H-thymidine at syngamy for 30 minutes; the cell has progressed to metaphase and the incorporation is localized above the metaphase chromosomes. (d) (e) and (f) illustrate eggs that were immersed in ^{3}H-thymidine at 45 minutes after insemination for a duration of 60 minutes. (d) The eggs were subjected to 5000 psi (for 60 minutes) while being immersed in ^{3}H-thymidine. The silver grains are localized in a mass above the nucleus. The control eggs (e) were incubated at atmospheric pressure in ^{3}H-thymidine for the same duration as the experimentals; these cells exhibited incorporation in all four nuclei. (f) Eggs incubated in ^{3}H-thymidine under a pressure of 7500 psi, there is no incorporation of the radioactive label. (Data of Zimmerman, 1963; Zimmerman and Silberman, 1967.)

treated with radioactive DNA precursor (^{3}H-thymidine) in the presence of blocking concentrations of Colcemid and subsequently prepared for autoradiographic analysis, display silver grains above the separated pronuclei. Since both high pressure and Colcemid have been shown to block pronuclear fusion and cause similar disruptive effects on microtubules (Tilney, 1968; Tilney *et al.*, 1966), it is quite probable that both agents block the movement of pronuclei by altering microtubule structure. Studies on *Arbacia* eggs (see Section III) suggest that pressure does not affect the permeability of unfertilized eggs; therefore, it is improbable that in the fertilized egg pressure favors the penetration of thymidine into the cell. It is more likely, in view of the Colcemid studies, that this magnitude of pressure (5000 psi) merely prevents the movement of the pronuclei, but does not prevent the sequence of biochemical events necessary for DNA synthesis.

It was also found that when pressure (5000 psi) was applied after pronuclear fusion, *Arbacia* eggs could synthesize DNA. When fertilized eggs were treated with ^{3}H-thymidine 15 minutes after insemination (20° C) and compressed for 30 minutes, autoradiographic patterns showed accumulation of silver grains in the area above the fusion nucleus (Fig. 2b). At the time of decompression, nonpressurized control cells displayed well-formed mitotic apparatus with specific localization of ^{3}H-thymidine above the metaphase chromosomes (Fig. 2c). The pressurized cells, however, appeared to have an intact nucleus with silver grains distributed throughout; the chromosomes were not seen. Evidently, pressure prevented the normal progression of nuclear events which would result in the disruption of the nuclear membrane and the formation of the mitotic figure. However, the high pressure did not block the incorporation of ^{3}H-thymidine into nuclear DNA.

In studies where pressure and isotope treatments were initiated at 45 minutes after insemination (a stage just preceding metaphase of Fig. 1) and retained for a duration of 30 minutes, there was no incorporation of labeled thymidine into DNA, although control cells showed uptake and incorporation during this period. However, when the pressure treatment accompanying isotope incubation was extended from 30 to 60 minutes, incorporation of ^{3}H-thymidine was observed (Fig. 2d and e). A possible explanation for the lack of incorporation during the 30-minute pressure treatment is that the pressure (5000 psi) delayed the biochemical events associated with DNA synthesis and the cell could not progress to the next "synthetic phase" (S phase). The incorporation in the control cells during this incubation period occurs because the eggs are permitted to proceed from the nonsynthetic phase to the synthetic phase of the next division cycle. In order for incorporation of ^{3}H-thymidine to occur in the treated eggs they would have to progress to the S phase. Thus, when the duration of the pressure–isotope treatment was extended to 60 minutes, the treated cells progressed to the S phase of the second division cycle and were able to incorporate thymidine. It is clear that the pressure employed does not absolutely block the biochemical processes necessary for the incorporation of ^{3}H-thymidine, but merely introduces a delay. Ultimately, the cells enter the S phase and incorporate thymidine. A major effect of pressure was the modification of the temporal aspects of chromosomal DNA activity. The chromosomes progressed through their DNA synthetic cycle at a reduced rate, despite the fact that normal chromosomal movement was prevented by the disruption of the mitotic apparatus. When cells were pressurized during G_2 or mitosis the chromosomes did not resume DNA synthesis until a time corresponding to the reentry of the chromosomes into their next interphase (See Fig. 1). These studies clearly demonstrate that certain biochemical processes (such as DNA synthesis) continue to function although other basic cellular mechanisms such as formation of the mitotic apparatus, chromosomal movement, and cytokinesis are inhibited by pressure. Moreover, the evidence is compatible with the idea of a "biological clock" (biological time sequence of events) operating at a molecular level which proceeds at a reduced rate in spite of suppression of certain mitotic and cytokinetic events (Zimmerman, 1963). Although DNA synthesis in *Arbacia* can proceed at pressures up to 5000 psi where other activities are inhibited, higher pressures also inhibit DNA synthesis.

Zimmerman and Silberman (1967) found that increasing hydrostatic pressures to 7500 or 10,000 psi had the effect of inhibiting ^{3}H-thymidine incorporation. Cells were incubated in ^{3}H-thymidine at 15 minutes (syngamy) and at 45 minutes after insemination for durations of 30 and 60 minutes. Autoradiographic patterns of incorporation during the first division cycle indicate that pressures in excess of 7500 psi inhibit ^{3}H-thymidine uptake in all but a small percentage of eggs (0–4%), regardless of the stage at which pressure treatment was initiated (Fig. 2f). Most of the eggs subjected to pressures of 7500 to 15,000 psi for a duration of 30 minutes did not show intact nuclei. However, as previously shown, the presence of a structurally intact nuclear membrane is not a prerequisite for ^{3}H-thymidine incorporation. A pressure exposure of 60 minutes at 7500 psi does not completely destroy all cellular integrity even though ^{3}H-thymidine incorporation was blocked, since approximately half the cells go on to cleave after decompression. It should be noted, that these cleavages were abnormal, and subsequent development was grossly aberrant and incomplete.

III. Problems of Permeability under Pressure

Before continuing the discussion of the effects of pressure on synthetic activities of the cell, it might be appropriate at this time to mention briefly some studies which pertain to pressure effects on permeability. For example, if hydrostatic pressure were to effect the cell membrane and alter the rate that precursor was able to enter the cell, either increasing or decreasing a potential substrate, the synthetic process associated with this substrate would indeed be modified. Murakami (1963), studying the inner epidermal cells of growing onion (*Allium cepa*), employed the plasmolysis time (the time for shrinkage of the cell protoplasm away from the cellular wall in hypertonic solutions) and deplasmolysis time (time for the cell protoplasm to return to its normal volume after being placed in a hypotonic solution) as an index of changes in permeability under varying pressures and temperatures. He found that the rate of plasmolysis (shrinkage of protoplasm due to loss of water) and deplasmolysis (the swelling of protoplasm) increased with increasing temperature (2–30°C). Murakami reported that high pressure (7100 psi) delays plasmolysis in hypertonic solutions of electrolytes (KC1, NaCl, $CaC1_2$, and $MgCl_2$) and in nonelectrolytes (sucrose, glycerin, and urea). On the other hand, deplasmolysis is accelerated at 7100 psi in solutions of monovalent cations (NaC1, KC1) and nonelectrolytes (urea, glycerin, and sucrose). However, in solutions of divalent cations ($CaC1_2$, $MgC1_2$) deplasmolysis time is decreased. These studies show that permeability of onion epidermal cells is indeed affected by hydrostatic pressure. However, the apparent delay of plasmolysis and acceleration of deplasmolysis, caused by pressure, is difficult to interpret and remains a puzzling phenomena.

An investigation was undertaken (Murakami and Zimmerman, 1968) to study the effects of hydrostatic pressure on cellular permeability in sea urchin eggs. Unfertilized eggs of *Arbacia punctulata* were placed into various (hypertonic and hypotonic) solutions of electrolytes for 5 minutes. The diameter of the cells were measured with a calibrated ocular micrometer prior to the application of pressure, during compression, and following the return to atmospheric pressure. The investigators concluded that there was no significant difference in the cellular volume between eggs pressurized at 4000 psi and the control eggs at atmospheric pressure in solutions of varying concentrations of monovalent (Na^+, K^+) and divalent (Ca^{++}, Mg^{++}) cations. There was, however, a slight difference in response of the cells to hypotonic solutions of NaCl and KCl at 8000 psi. Thus, at high pressures (8000 psi), cells immersed in 0.3 *M* NaCl (a

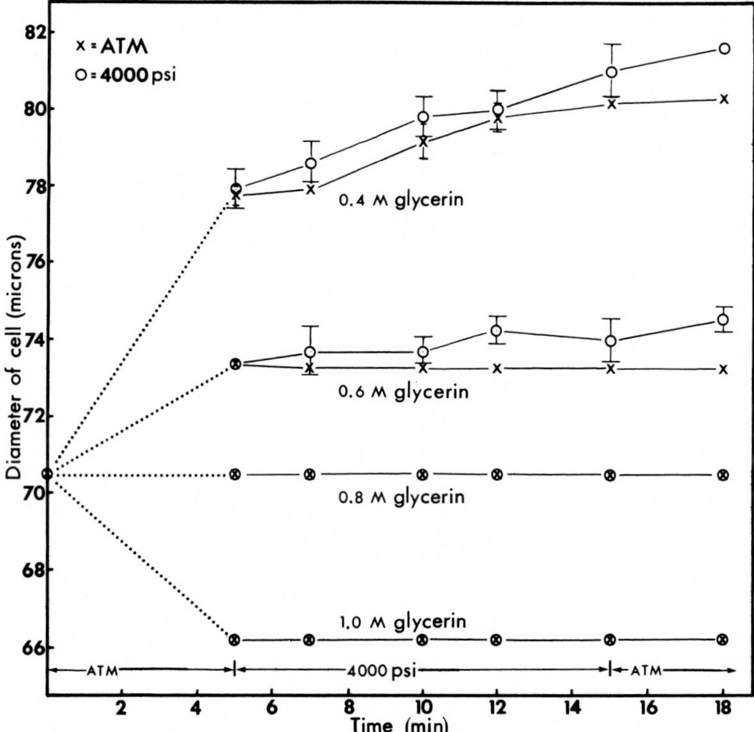

FIG. 3. Effects of pressure on the permeability of *Arbacia punctulata* eggs. Unfertilized *Arbacia* eggs were placed in varying concentrations of glycerin for 5 minutes and then subjected to 4000 psi. The diameter of the cells were measured with an ocular micrometer. There appears to be some difference between the eggs at 4000 psi and control eggs in the 0.4 *M* and 0.6 *M* concentrations; these differences are not statistically significant. The standard errors of each point are shown. (Data of Murakami and Zimmerman, 1968.)

solution hypotonic to *Arbacia* eggs) exhibited swelling, and within 10 minutes almost all the cells lysed, whereas in 0.3 *M* KCl only about 2% of the cells burst under similar pressure conditions.

When hypotonic solutions of nonelectrolytes (sucrose, glycerin, and urea) were employed, there appeared to be slight differences in volume between the pressurized and nonpressurized cells; the volume differences between control and pressurized eggs were not statistically significant. The results of a representative study in which glycerin was used are illustrated in Fig. 3.

In view of the above studies on two different types of cells, it is not possible to make any generalizations concerning the effects of pressure on cell permeability. In onion cells, pressure delays plasmolysis time in hypertonic solutions, whereas, deplasmolysis time is accelerated. On the other hand, hydrostatic pressure does not appear to alter the rate of swelling of unfertilized *Arbacia* eggs in hypotonic solutions. Future investigations employing radioactive cations should prove helpful in clearly resolving the pressure effect on permeability in marine eggs. In spite of the many unanswered questions at this time concerning the effect of pressure on permeability, it is possible to evaluate quantitatively the effect of pressure on various synthetic systems.

IV. Incorporation of Radioactive Uridine

Protein synthesis is initiated soon after insemination in sea urchin eggs (Hultin, 1950; Gross and Cousineau, 1964; Nakano and Monroy, 1958). However, RNA synthesis is negligible prior to fertilization and increases only slightly after fertilization (Nemer, 1963; Wilt, 1963; Brachet *et al.*, 1963). Piatigorsky and Whiteley (1965) reported that fertilized sea urchin eggs (*Strongylocentrotus purpuratus*) accumulate relatively large amounts of radioactively labeled uridine into acid-soluble cell fractions and relatively small amounts into the acid-insoluble fractions. Their studies suggest that penetration of uridine into sea urchin eggs is dependent upon an active transport system and that phosphorylation of uridine may occur at the cell surface and serve as a mechanism by which uridine enters the cell.

In our laboratory, in collaboration with Dr. T. H. Murakami, we investigated the effects of pressure on the uptake of uridine into *Arbacia* eggs (Fig. 4). Fertilized eggs at atmospheric pressure incorporated four to six times as much ^{14}C-uridine as unfertilized eggs. Moreover, the rate of accumulation was lower 30 minutes after insemination than soon after insemination or just prior to cleavage. Following isotope–pressure treatment, in which the ^{14}C-uridine and 14,000 psi of pressure were applied at the same time, the total incorporation of ^{14}C-uridine into the cells was reduced by about one-third to one-half of the nonpressurized controls after insemination (see Fig. 4). As shown earlier in this review (see Section II), unfertilized eggs are relatively resistant to high pressure as compared with fertilized eggs. The present studies confirm this phenomenon by demonstrating that pressures of 2000, 6000, or 14,000 psi for 10-minute durations had no apparent effect on the uptake of ^{14}C-uridine into unfertilized eggs.

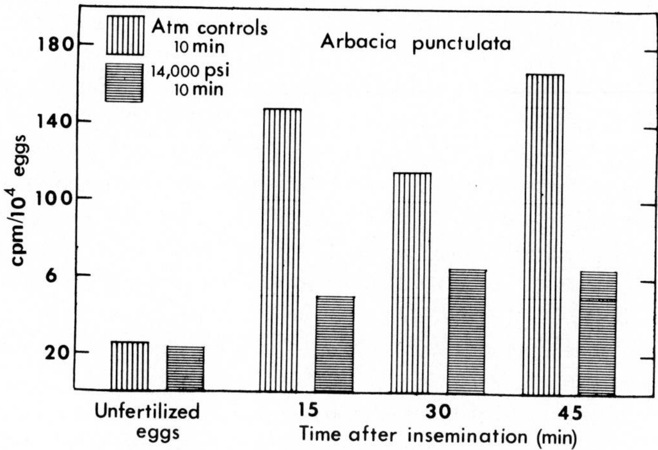

FIG. 4. The effects of pressure (14,000 psi) on the incorporation of ^{14}C-uridine into eggs of *Arbacia punctulata*. *Arbacia* eggs were incubated in ^{14}C-uridine, 0.5 μCi/ml (specific activity 26 mCi/mmole) for 10 minutes at atmospheric pressure and under 14,000 psi (at 20°C). Cells were collected on a Millipore filter and washed with cold sea water containing nonradioactive uridine. The radioactivity of the cells was determined with a Packard Liquid Scintillation Spectrometer. (Data of Zimmerman and Murakami, 1968.)

Piatigorsky and Whiteley (1965) found that most of the radioactive uridine found in *Strongylocentrotus purpuratus* eggs was recovered in the acid-soluble fraction in the form of phosphorylated nucleosides and primary triphosphates. We investigated the nature of acid-soluble material from pressurized cells. Fertilized eggs of *Strongylocentrotus purpuratus* were subjected to pressure of 6000 and 14,000 psi for a duration of 35 minutes in the presence of ^{3}H-uridine. Following decompression, the isotope–pressure-treated cells and the control cells at atmospheric pressure were washed with cold sea water containing nonradioactively labeled uridine and homogenized in trichloroacetic acid (TCA). Radioactivity was determined in the TCA-soluble fraction. Pressures of 6000 psi inhibited the incorporation into the cell of ^{3}H-uridine by only 11% as compared with control eggs at atmospheric pressure; although similar pressures block cytokinesis, alter the division schedule, and disorganize the mitotic spindle material in marine eggs. Pressures of 14,000 psi, on the other hand, inhibited 52% of the uridine incorporation into the acid soluble fraction.

Piatigorsky and Whiteley (1965) reported that inhibition of the phosphorylating capacity of fertilized sea urchin eggs with dinitrophenol or low temperatures prevents labeled uridine from accumulating within the cells. In order to establish whether the cells treated at high pressure recover their phosphorylating capacity, fertilized eggs were pressurized and permitted to recover for a short time prior to incubation in radioactive precursor. Fertilized eggs were subjected to 14,000 psi for 5 minutes and decompressed for 5 minutes at atmospheric pressure. Following decompression they were incubated in ^{3}H-uridine for 35 minutes. The accumulation of radioactive acid-soluble nucleotides was analyzed in the control and pressure-treated cells. The pressurized cells showed a 37% reduction of label as compared with controls. These studies indicate that pressure effects are not readily reversible at these high pressures and may interfere with cell permeability or more probably the mechanism for phosphorylation of uridine.

Although pressure reduces the incorporation of uridine into marine eggs, it may increase the endogenous levels of adenine nucleotides. Landau (1966) found that the total adenine nucleotide level of pressurized *Arbacia* eggs increases about 50% (at 10,000 psi for 15 minutes) if the treatment is initiated at early prophase. Similar results were obtained with eggs of the sea urchin, *Paracentrotus lividus*, where there was a 40% increase of measurable adenine nucleotides. In both organisms the increases were found in ADP and AMP levels. A most important finding was that the increases occurred only at a specific stage in the cell cycle, early prophase, but was not found at other times. Pressure treatment of FL amnion cells also results in an increase of endogenous adenine nucleotides (Landau and Peabody, 1963). It is possible, as Landau proposes, that the pressure treatment may cause alterations in molecular configurations, allowing for reversible disruption of enzyme–substrate complexes.

V. Sedimentation Profiles

Zonal sedimentation analysis of the 10,000 *g* supernatant (microsomal material) obtained from cultures of synchronized *Tetrahymena* indicate that these protozoa are quite sensitive to short pulses of pressure (5000 psi for 2-minute durations). The optical density profiles of the sedimenting microsomal material obtained from pressurized cells show a marked reduction in the ratio of polysome to total ribosomal material, as compared with the ratio found in control cells (Hermolin, 1967; Zimmerman, 1969; Zimmerman and Zimmerman, Chapter 8).

In order to establish whether high pressure exerts a similar action in marine eggs, fertilized and unfertilized eggs of the sea urchin *Strongylocentrotus* were subjected to varying pressures, homogenized, and the sedimentation profiles of the 10,000 *g* supernatants were analyzed (Zimmerman and Murakami, 1968). Analysis by sucrose density gradient centrifugation of 10,000 *g* supernatant fractions (obtained from eggs which were subjected to 14,000 psi for a duration of 5 minutes) yielded OD_{260} profiles in which two components were consistently observed. The rapidly sedimenting material, presumed to be polysomal material, was followed by the ribosomal material which sedimented in the lighter sucrose fractions. There was no difference in the OD_{260} profiles of material obtained from eggs which were pressurized as compared with nonpressurized eggs. The sedimentation profiles were similar.

Analytical ultracentrifugal analysis of the microsomal material (10,000 *g* supernatant) yielded four distinct peaks which could be observed with schlieren optics (Fig. 5). The most rapidly sedimenting material, presumed to be polysomes, is seen as a broad peak (peak number 1). The next fraction which sediments as a sharp peak (peak number 2) contains uniformly sedimenting ribosomal material; the material in this fraction obtained from pressurized eggs has an s_{20} of 74 and that obtained from eggs at atmospheric pressure has an s_{20} of 73. A major slowly sedimenting fraction was found near the meniscus which separated into two fractions (peaks 3 and 4) having s_{20} values of 20 and 4 from the pressurized eggs and s_{20} values of 25 and 4 from control eggs at atmospheric pressure (Fig. 5). Although peak number 3 is seen as a shoulder on Fig. 5, in subsequent frames it is readily discernable as a sharp peak.

Similar analytical ultracentrifugal studies were conducted using the eggs of *Arbacia punctulata*. It was found that there were no apparent effects of pressure on the material obtained from the 10,000 *g* supernatant fraction (Table I). The nature of the microsomal material having sedimentation values of 23–27 S and 6–11 S is not known; however, it is possible that this represents degradation of the ribosome material.

In synchronized *Tetrahymena* the ratio of polysomes to the total ribosomal complement varies depending upon the age of the culture; moreover, there is a marked reduction in the ratio following pressure treatment. However, in naturally synchronized eggs of sea urchins (*Arbacia* and *Strongylocentrotus*), the polysome/total ribosome ratio does not vary during the first division cycle and it is stable to high pressure treatment. The stability of the ribosomal system to high pressure in marine eggs as

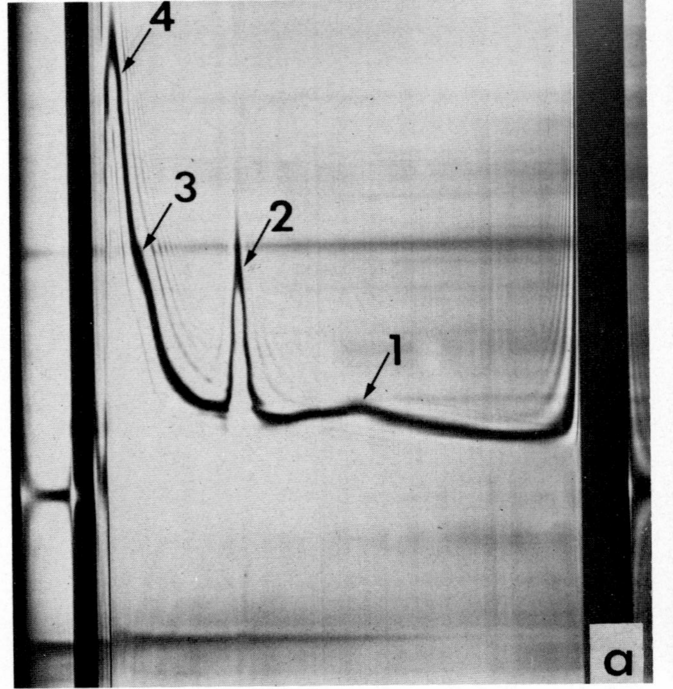

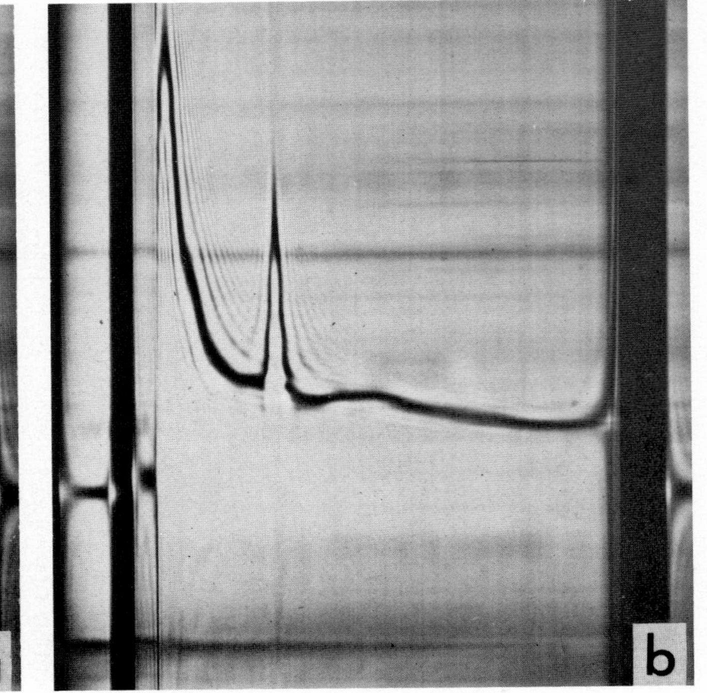

FIG. 5. Analytical ultracentrifuge patterns of the microsomal fraction (10,000 *g* supernatant) obtained from fertilized eggs of *Strongylocentrotus purpuratus* at atmospheric pressure (a) and at a pressure of 14,000 psi (b). The eggs were washed in isotonic NaCl:KCl solution (19:1 v/v), homogenized in 0.01 *M* this buffer, pH 7.6, containing 0.01 *M* $MgCl_2$ and 0.24 *M* KCl. The homogenate was centrifuged at 10,000 *g* for 10 minutes (at 4°C). The sedimentation patterns of the 10,000 *g* supernatant (microsomal fraction) are shown. The centrifugal speed was 34,000 rpm and the phase plate 40°. (a) Ultracentrifuge pattern of material obtained from fertilized control cells at 60 minutes after insemination. The photograph was taken 8 minutes after reaching speed. (See text for explanation of peaks 1, 2, 3, 4.) (b) The pattern obtained from eggs subjected to 14,000 psi for 5 minutes at a stage comparable with the controls. Photograph was taken 9 minutes after reaching speed. The patterns are essentially identical. (Data of

compared with that found in *Tetrahymena* may be related to the organisms ecosystems. The fact that sea urchins are frequently found at ocean depths where high pressures prevail may have been instrumental in evolving a more resistant protein synthesizing system that is better able to cope with high pressures.

TABLE I
SEDIMENTATION ANALYSIS OF 10,000 *g* SUPERNATANT FRACTIONS[a]

			Average s_{20}			
			Sedimenting peaks			
Stage	Organism	Treatment	4	3	2	1
Fertilized	*Stronglyocentrotus*	Atmospheric	4	25	73	133[b]
Fertilized	*Stronglyocentrotus*	14,000 psi/5 min	4	20	74	127[b]
Fertilized	*Arbacia*	Atmospheric	7	27	77	ND[c]
Fertilized	*Arbacia*	14,000 psi/5 min	6	25	75	ND
Unfertilized	*Arbacia*	Atmospheric	11	23	77	ND
Unfertilized	*Arbacia*	14,000 psi/5 min	10	26	75	ND

[a] Obtained from eggs of *Strongylocentrotus purpuratus* and *Arbacia punctulata.* In the fertilized series, the treatment was initiated 30 minutes after insemination for *Arbacia* and 60 minutes after insemination for *Strongylocentrotus.*

[b] Approximate value.

[c] ND represents values not determined.

VI. Action of Inhibitors on Furrow Induction

It was over a decade ago that it was first observed that high hydrostatic pressure combined with high centrifugal forces induce furrowing in *Arbacia* eggs prior to the time for normal cleavage (Zimmerman and Marsland, 1956; Zimmerman *et al.*, 1957). Shortly thereafter, it was reported that furrow induction in *Arbacia* eggs was associated with the disruption of cytoplasmic β granules and rupture of the nuclear membrane (Marsland *et al.*, 1960; Zimmerman and Marsland, 1960). The method for furrow induction consists of applying pressures of 8000 to 14,000 psi accompanied by centrifugal forces in the range of about 33,000–41,000 *g* (for durations of 3 to 5 minutes); induced furrows appear 1–2 minutes after decompression (Fig. 6a).

Recently (S. B. Zimmerman *et al.*, 1968) found that a variety of chemical inhibitors in concentrations sufficient to completely inhibit or markedly retard cell division did not prevent furrow induction (Table II). Metabolic inhibitors of RNA and protein synthesis such as actinomycin D, puromycin and DL-*p*-fluorophenylalanine did not prevent induction (Fig.

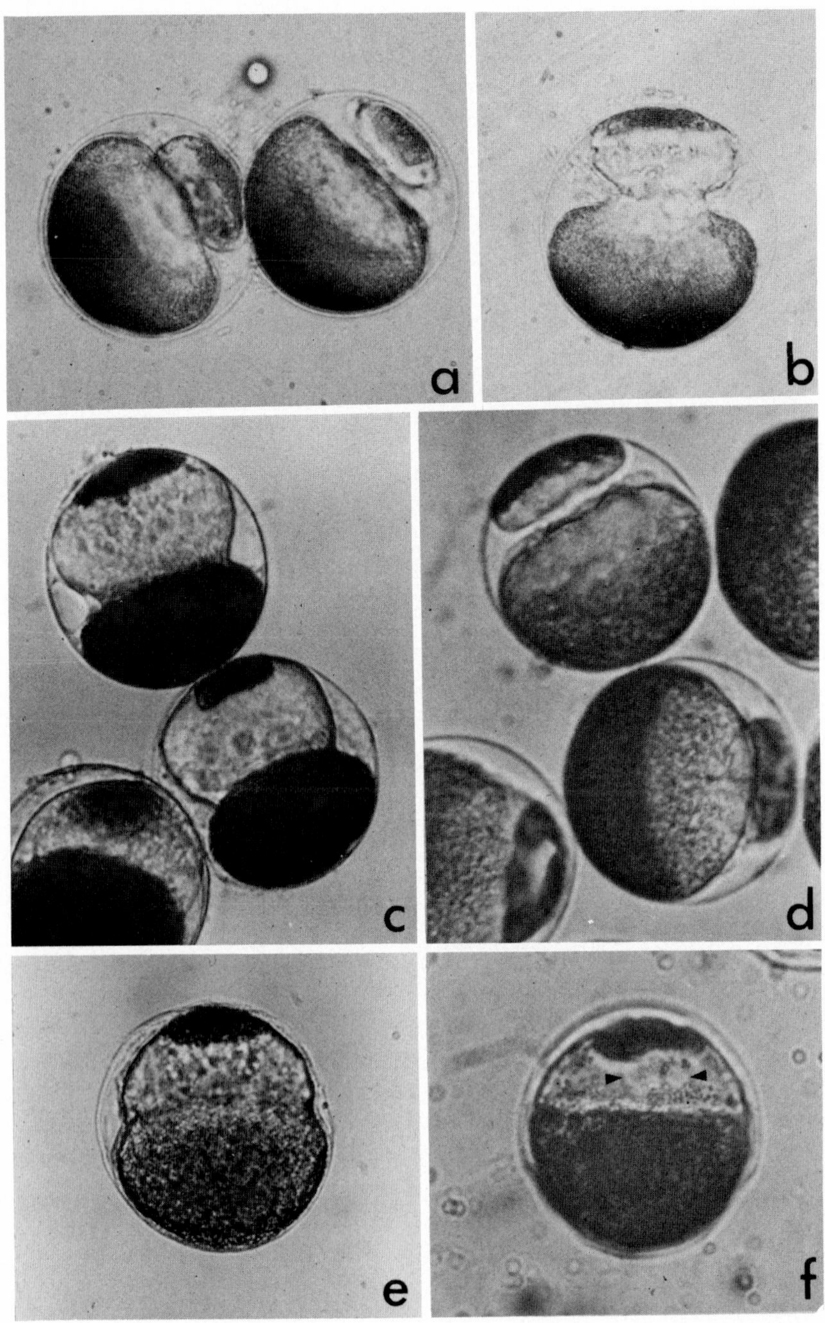

FIG. 6

6b,c). Mitotic inhibitors such as Colcemid and heavy water, which block pronuclear fusion and affect cortical and mitotic gel structures by interference with polymerization of macromolecules (Zimmerman and Zimmerman, 1967; Marsland and Zimmerman, 1965; Marsland and Hecht, 1968), do not prevent furrow induction in *Arbacia* (Fig. 6d,e). When eggs were incubated in metabolic inhibitors such as cornin (a potent inhibitor of oxidative phosphorylation) or 2,4-dinitrophenol and 4,6-dinitro-*o*-cresol (uncouplers of oxidative phosphorylation) in concentrations which prevented development, furrow induction could still be induced.

These studies clearly indicate that neither RNA nor protein synthesis is essential for experimental furrow induction; moreover, interference with the energetics of the egg does not prevent furrow induction. In addition, it was shown (Fig. 6e,f) that fusion of the pronuclei is not essential for furrow induction.

VII. Pressure Effects on Microtubules

Light microscope studies of the mitotic apparatus isolated from pressurized sea urchin eggs show that this highly structured organelle is very susceptible to disorganization by hydrostatic pressure (Zimmerman and Marsland, 1964). Loss of linear organization of the spindle and a loss of radial organization of the asteral region was observed after 1-minute pressure pulses of 7500–10,000 psi (cf. Zimmerman and Marsland, 1964; Zimmerman and Silberman, 1967; Marsland, Chapter 11).

Electron microscope studies of the isolated mitotic apparatus (Kane,

---

FIG. 6. (a) Furrowing reaction induced experimentally 30 minutes after insemination in eggs of *Arbacia punctulata* at 20°C. Induced furrowing treatment consisted of 5 minutes of pressure centrifugation (12,000 psi at 33,000 g). Furrowing was seen 1–3 minutes after treatment. (b) Furrow induction in puromycin-treated cell. Eggs were exposed to 10^{-4} M puromycin 15 minutes after insemination. Pressure centrifugation treatment was initiated 30 minutes after insemination. (c) Cells treated 15 minutes after insemination with 3.3×10^{-3} M *p*-fluorophenylalanine. The eggs were pressure-centrifuged 35 minutes after insemination. The echinochrome pigment is heavily packed in the centrifugal half of the egg. (d) Cells treated with 70% D_20 5 minutes after insemination. Furrow inducing pressure centrifugation treatment was initiated 30 minutes after insemination. (e) Eggs treated with 2.7×10^{-4} M Colcemid 5 minutes after insemination, before pronuclear fusion, and pressure centrifuged 35 minutes after insemination. (f) Colcemid-treated control cell, centrifuged simultaneously as (e), but at atmospheric pressure. Note the presence of two pronuclei shown by arrows; pronuclear fusion was blocked as a consequence of the Colcemid treatment. (From the work of S. B. Zimmerman *et al.*, 1968.)

TABLE II

EFFECTS OF CHEMICAL INHIBITORS ON FURROW INDUCTION

Agent	Concentration	Stage of treatment	Duration of drug treatment prior to induction (min)	Effect on cell division: continuous exposure	Induced furrowing
Actinomycin D	50 μg/ml	15 min before insemination	45	First division delayed	Yes
Actinomycin D	50 μg/ml	15 min after insemination	20	First division delayed	Yes
Puromycin	10^{-4} *M*	15–30 min before insemination	45–60	No division, blocked at streak stage	Yes
Puromycin	10^{-4} *M*	15 min after insemination	20	First division occasionally delayed	Yes
DL-*p*-Flurophenylalanine	3.3×10^{-3} *M*	15 min after insemination	20	Division delayed 10–15 min	Yes
Colcemid	2.7×10^{-4} *M*	5 min after insemination	30	Pronuclear fusion blocked	Yes
D_2O	70–90%	15 min after insemination	20	No division	Yes
Mercaptoethanol	0.09 *M*	15 min after insemination	20	No division	Yes
Cornin	10^{-6} gm/ml	10 min before insemination	25	87% division after 1 hour	Yes
2,4-Dinitrophenol	10^{-5} *M*	5 min after insemination	30	6% division 2–4-cell stage after 20 hours	Yes
4,6-Dinitro-*o*-cresol	10^{-5} *M*	5 min after insemination	30	20% division in 2–4-cell stage after 20 hours	Yes

1962) and the mitotic apparatus *in situ* (Harris, 1961, 1962, 1965) demonstrated highly oriented fibrils (spindle filaments) in sea urchin eggs. Recently, Longo and Anderson (1968) reported on the ultrastructure of the early stages of *Arbacia* development. They showed that as the sperm aster is formed, parallel clusters of microtubules arise from the centriolar fossa of the male pronucleus. Microtubules were also found distributed randomly in the cytoplasm surrounding the male pronucleus.

Recently, Zimmerman and Philpott (1968) studied the effects of pressure on the microtubules in metaphase *Arbacia* eggs. Electron microscope analysis of compressed cells revealed that pressure pulses of 10,000 psi for 1 minute completely disorganize the cytoplasmic microtubules. Microtubules are clearly evident in the cytoplasm of control eggs at atmospheric pressure (Fig. 7).

This investigation complements the study by Tilney *et al.* (1966) who reported that microtubular elements in the axopodia of the heliozoan *Actinosphaerium nucleofilum* are unstable and become disorganized under pressure. There was a retraction of the axopodia accompanying the disappearance of the microtubules. These pressure effects were reversible, and the rate of disintegration of the microtubules was dependent upon the pressure employed (cf. Marsland, Chapter 11).

VIII. Concluding Remarks

Biochemical studies of pressure-treated marine eggs can complement and add new significance to earlier observed physiological effects induced by hydrostatic pressure. Levels of pressure that interfere with cytokinesis and mitosis also affect the synthetic capabilities of the cell. Thus, uridine incorporation and DNA synthesis are markedly reduced by pressure treatment. The sensitivity of the cell to pressure pulses, as measured by division delays, varies depending upon the specific stage of the cell cycle during which treatment is initiated. However, division delays, induced by short pressure pulses, are probably not related to DNA synthesis, since maximum delays occur after the completion of DNA synthesis.

Hydrostatic pressure can inhibit cell division by disorganizing gel systems such as the mitotic apparatus or the cell cortex. Such systems require an expenditure of energy, which may be modified by the action of pressure. Pressure treatment causes a disruption of microtubule structure, which results in a reduction or cessation of chromosomal movement and may play a role in hindering pronuclear movement. Other structural elements that appear to be essential for cytoplasmic movement via gel con-

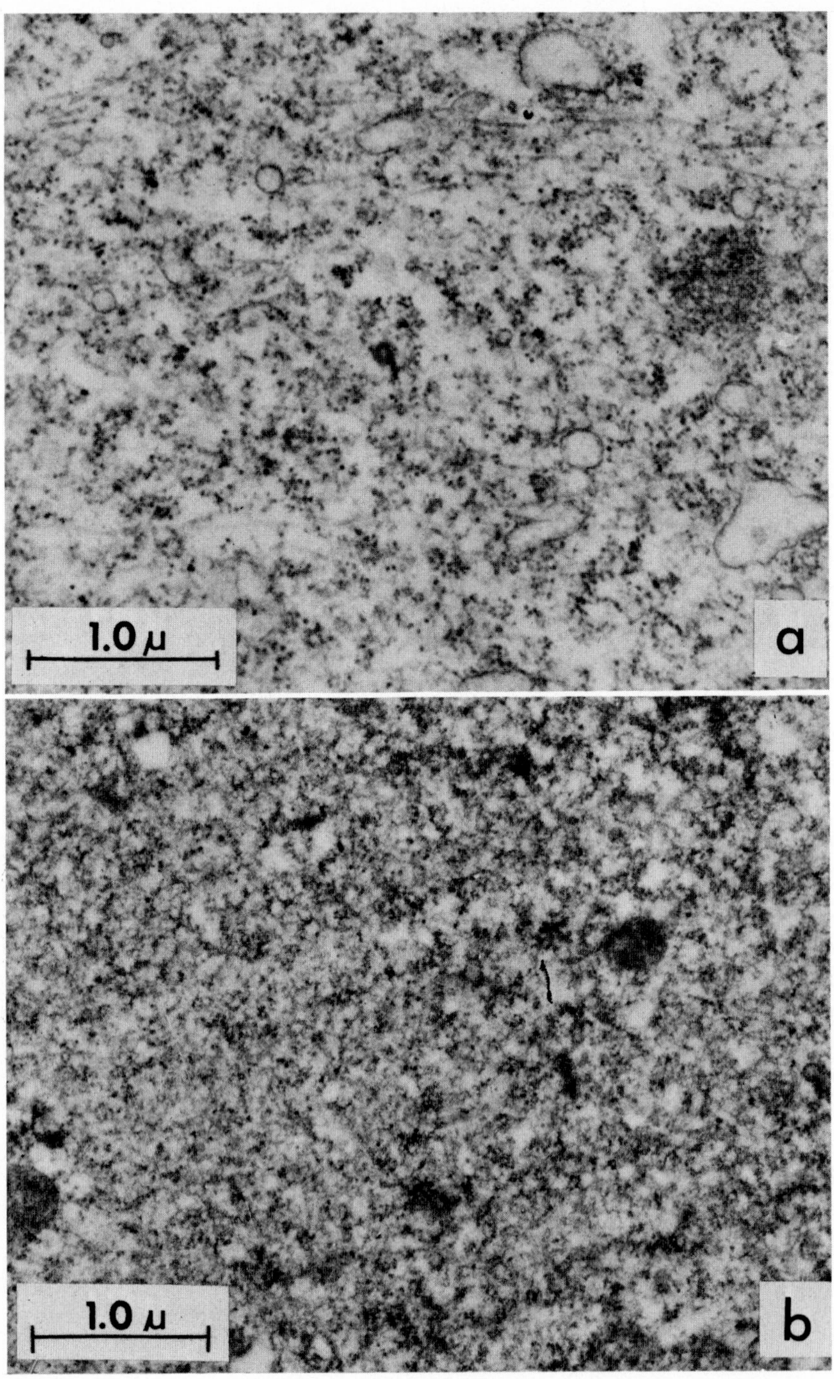

FIG. 7

traction are also disrupted by hydrostatic pressure. The application of pressure at the time of cytokinesis results in a weakening of the structural capacities of the cortical gel, thus preventing cleavage.

Although hydrostatic pressure alters the permeability of onion cells, it does not appear to have any effect on the permeability of sea urchin eggs. In view of the different environmental pressures which marine eggs and onion cells are normally subjected to, further study is needed to elucidate this factor. Future studies should be concerned with marine plants as well as other marine organisms.

Experimentally induced furrowing remains an interesting system for study (cf. Marsland, Chapter 11). Although the absence of effect of various mitotic and cytokinetic inhibitors on furrow induction is puzzling, fine structural studies on cells undergoing experimental furrow induction offers new insight for analyzing cleavage induction.

ACKNOWLEDGMENT

The unpublished experiments from the author's laboratory have been carried out with the support of research grants from the National Research Council of Canada. Their assistance is gratefully acknowledged.

REFERENCES

Brachet, J., Decoroly, M., Ficq, A., and Quertier, J. (1963). Ribonucleic acid metabolism in unfertilized and fertilized sea urchin eggs. *Biochim. Biophys. Acta* **72**, 660–662.

Gross, P., and Cousineau, G. H. (1964). Macromolecule synthesis and the influence of actinomycin on early development. *Exptl. Cell Res.* **33**, 368–395.

Harris, P. (1961). Electron microscope study of mitosis in sea urchin blastomers. *J. Biophys. Biochem. Cytol.* **11**, 419–431.

Harris, P. (1962). Some structural and functional aspects of the mitotic apparatus in sea urchin embryos. *J. Cell Biol.* **14**, 475–488.

Harris, P. (1965). Some observations concerning metakinesis in sea urchin eggs. *J. Cell Biol.* **25**, 73–78.

Hermolin, J. (1967). Polysome formation in *Tetrahymena pyriformis.* M.Sc. Thesis, University of Toronto, Toronto, Canada.

Hinegardner, R. T., Rao, B., and Feldman, D. E. (1964). The DNA synthetic period during early development of the sea urchin egg. *Exptl. Cell Res.* **36**, 53–61.

FIG. 7. Electron micrograph of section through *Arbacia* egg at atmospheric pressure and immediately following pressure treatment. (a) The microtubules in astral region of a metaphase egg at atmospheric pressure are shown. Magnification 26,300 ×. (b) *Arbacia* eggs were subjected to 10,000 psi for a 1-minute duration at metaphase and fixed 30 seconds after decompression. Note the absence of microtubules. Magnification 24,000 ×. The cells were fixed in 5% gluteraldehyde (pH 7.2) in a phosphate–sea water mixture; the cells were postfixed in 1% osmium tetroxide and embedded in Araldite 502. (From the work of Zimmerman and Philpott, 1968.)

Hultin, T. (1950). The protein metabolism of sea urchin eggs studied by means of N^{15}-labeled ammonia. *Exptl. Cell Res.* **1**, 599–602.

Kane, R. E. (1962). The mitotic apparatus: Fine structure of the isolated unit. *J. Cell Biol.* **15**, 279–287.

Landau, J. V. (1966). High pressure effects on endogenous adenine nucleotide levels in sea urchin eggs prior to first cleavage. *J. Cell Biol.* **28**, 408–412.

Landau, J. V., and Peabody, R. A. (1963). Endogenous adenosine triphosphate levels in human amnion cells during application of high hydrostatic pressure. *Exptl. Cell Res.* **29**, 54–60.

Landau, J. V., Marsland, D., and Zimmerman, A. M. (1955). The energetics of cell division: Effects of adenosine triphosphate and related substances on the furrowing capacity of marine eggs (*Arbacia* and *Chaetopterus*). *J. Cellular Comp. Physiol.* **45**, 309–329.

Longo, F. J., and Anderson, E. (1968). The fine structure of pronuclear development and fusion in the sea urchin, *Arbacia punctulata. J. Cell Biol.* **39**, 339–368.

Marsland, D., and Hecht, R. (1968). Cell division: Combined anti-mitotic effects of colchicine and heavy water on first cleavage in eggs of *Arbacia punctulata. Exptl. Cell Res.* **51**, 602–608.

Marsland, D., and Zimmerman, A. M. (1965). Structural stabilization of the mitotic apparatus by heavy water, in the cleaving eggs of *Arbacia punctulata. Exptl. Cell Res.* **38**, 306–313

Marsland, D., Zimmerman, A. M., and Auclair, W. (1960). Cell division: Experimental induction of cleavage furrows in the eggs of *Arbacia punctulata. Exptl. Cell Res.* **21**, 179–196.

Murakami, T. H. (1963). Effect of hydrostatic pressure on the permeability of plasma membrane under the various temperature. *Symp. Cellular Chem.* **13**, 147–156 (in Japanese).

Murakami, T. H., and Zimmerman, A. M. (1968), unpublished data.

Nakano, E., and Monroy, A. (1958). Incorporation of S^{35}-methionine in the cell fractions of sea urchin eggs and embryos. *Exptl. Cell Res.* **14**, 236–244.

Nemer, M. (1963). Old and new RNA in the embryogenesis of the purple sea urchin. *Proc. Natl. Acad. Sci. U.S.* **50**, 230–235.

Piatigorsky, J., and Whiteley, A. H. (1965). A change in permeability and uptake of ^{14}C uridine and response to fertilization in *Strongylocentrotus purpuratus* eggs. *Biochim. Biophys. Acta* **108**, 404–418.

Simmel, E. B., and Karnofsky, D. A. (1961). Observations on the uptake of tritiated thymidine in the pronuclei of fertilized sand dollar embryos. *J. Biophys. Biochem. Cytol.* **10**, 59–65.

Tilney, L. G. (1968). Studies on the microtubules in Heliozoa. IV. The effects of colchicine on the formation and maintenance of the axopodia and the redevelopment of pattern in *Actinosphaerium nucleofilum* (Barrett). *J. Cell Sci.* **3**, 549–562.

Tilney, L. G., and Porter, K. (1965). Studies on the microtubules in Helizoa. I. The fine structure of *Actinosphaerium nucleofilum* (Barrett) with particular reference to the axial rod structure. *Protoplasma* **60**, 317–344.

Tilney, L. G., Hiramoto, Y., and Marsland, D. (1966). Studies on the microtubules in heliozoa. III. A pressure analysis of the role of these structures in the formation and maintenance of the axopodia of *Actinosphaerium nucleofilum* (Barrett). *J. Cell Biol.* **29**, 77–95.

Wilt, F. M. (1963). The synthesis of ribonucleic acid in sea urchin embryos. *Biochem. Biophys. Res. Commun.* **11**, 447–451.

Zimmerman, A. M. (1963). Incorporation of H^3 thymidine in the eggs of *Arbacia punctulata:* A pressure study. *Exptl. Cell Res.* **31**, 39–51.

Zimmerman, A. M. (1968). Die Wirkungen von Druck und Temperatur auf die makromolekularen Systeme. *In* "Uberleben auf See II. Marinemedizinishch-Wissenschafiliches Symposium in Kiel," pp. 165–181.

Zimmerman, A. M. (1969). Effects of high pressure on macromolecular synthesis in synchronized *Tetrahymena. In* "The Cell Cycle" (G. M. Padilla, G. L. Whitson, and I. L. Cameron, eds.), pp. 203–225, Academic Press, New York.

Zimmerman, A. M., and Marsland, D. (1956). Induction of premature cleavage furrows in the eggs of *Arbacia punctulata. Biol. Bull.* **111**, 317.

Zimmerman, A. M., and Marsland, D. (1960). Experimental induction of furrowing reaction in the eggs of *Arbacia punctulata. Ann. N.Y. Acad. Sci.* **90**, 470–485.

Zimmerman, A. M., and Marsland, D. (1964). Cell division: Effects of pressure on the mitotic mechanisms of marine eggs (*Arbacia punctulata*). *Exptl. Cell Res.* **35**, 293–302.

Zimmerman, A. M., and Murakami, T. H. (1968). Unpublished data.

Zimmerman, A. M., and Philpott, D. (1968). Unpublished data.

Zimmerman, A. M., and Silberman, L. (1965). Cell division: The effects of hydrostatic pressure on the cleavage schedule in *Arbacia punctulata. Exptl. Cell Res.* **38**, 454–464.

Zimmerman, A. M., and Silberman, L. (1967). Studies on incorporation of ^{3}H-thymidine in *Arbacia* eggs under hydrostatic pressure. *Exptl. Cell Res.* **46**, 469–476.

Zimmerman, A. M., and Zimmerman, S. (1967). Action of Colcemid in sea urchin eggs. *J. Cell Biol.* **34**, 483–488.

Zimmerman, A. M., Landau, J. V., and Marsland, D. (1957). Cell division: A pressure–temperature analysis of the effects of sulfhydryl reagents on the cortical plasmagel structure and furrowing strength of dividing eggs (*Arbacia* and *Chaetopterus*). *J. Cellular Comp. Physiol.* **49**, 395–435.

Zimmerman, S. B., Murakami, T. H., and Zimmerman, A. M. (1968). The effects of selected chemical agents on furrow induction in the eggs of *Arbacia punctulata. Biol. Bull.* **134**, 356–366.

CHAPTER 11

PRESSURE–TEMPERATURE STUDIES ON THE MECHANISMS OF CELL DIVISION

Douglas Marsland

I. Introduction

A. Historical

Regnard (1884) was one of the earliest to study the physiological effects of pressure per se, i.e., pressure divorced from other factors.[1] Previously, a number of investigators (see Hill, 1912) had studied the effects of pressure transmitted to the organism by way of an intervening gaseous atmosphere, but the results of these studies were greatly influenced by the increased concentrations of dissolved gases in and around the compressed cells. By eliminating the gas phase, Regnard found that much higher pressures are required to produce measurable physiological effects.

Regnard was stimulated by the findings of the Talisman dredging expedition. This group succeeded in recovering a variety of living organisms from oceanic depths up to 12,000 m, corresponding to an environmental pressure of 14,000–15,000 psi, or about 1000 atm. This convinced Regnard that pressure must be regarded as a normal physiological variable and that pressure has a considerable influence upon the lives and destinies of many organisms. Accordingly, Regnard undertook a laboratory study of a variety of unicellular plants and animals, and also of fish and other higher forms. These organisms were subjected to pressures ranging up to about 15,000 psi, and Regnard reported the results of these studies in a series of short papers (1884, 1885, 1886, 1887) and in a book, "La Vie dans les Aux" (1891).

Perhaps the earliest study dealing with the effects of pressure on cell division was that of Chlopin and Tammann (1903), who found that the

[1] Cattell (1936) has provided an excellent review—The Physiological Effects of Pressure—which summarizes much of the work done to the date of publication.

multiplication of various bacteria was markedly inhibited by exposure to pressures of the order of 45,000 psi (3000 atm). Later, however, it was reported (Basset and Macheboeuf, 1932) that certain bacteria retained their reproductive capacity after treatment with somewhat higher pressures (up to about 60,000 psi). It must be said, however, that the minimum pressure required for the inhibition of bacterial cell division has not yet been determined. All the earlier studies were made before the advent of a microscope–pressure chamber and thus they did not allow for a direct observation of cells during exposure to high pressure (Marsland and Brown, 1936; Marsland, 1950, 1958a).

The more complex division processes in the cells of higher organisms appear to be susceptible to inhibition by much lower pressure, usually in the 5000–6000 psi range. This is indicated in Table I, which is taken mainly from Pease and Marsland (1939).

B. Some Thermodynamic Considerations

One can predict the effects of pressure upon isolated physical or chemical reactions on the basis of thermodynamic reasoning, when such reactions are classified according to their energy–volume relationships. Type I reactions may be specified as those that liberate energy ($+\Delta E$), and generally such exergonic reactions are accompanied by a volume decrement ($-\Delta V$) in the system. Accordingly, exergonic reactions tend to occur spontaneously, in the absence of any energy source. The equilibria (or steady states) of such reactions are shifted backwards by increasing temperature, but they are shifted forward by increasing pressure. Moreover, the magnitude of the pressure effect is determined by the magnitude of the volume change—as has been shown for certain gelation reactions by Marsland and Brown (1942).

Type II reactions, in contrast, absorb energy ($-\Delta E$) as they proceed and generally there is a volume increment ($+\Delta V$) in the system. Thus, type II reactions do not proceed spontaneously. They must receive an adequate and appropriate source of energy, which permits them to do work against the environmental pressure. Accordingly, the equilibria or steady states of type II reactions tend to be shifted in the gel direction by increasing temperature and in the sol direction by increasing pressure. This is exemplified in Fig. 1, taken from a study on various isolated gel systems (Marsland and Brown, 1942).

Type III systems appear to be relatively unimportant in relation to the physiological effects of high pressure. Such reactions are virtually isogonic and little if any volume change is involved. Consequently type III reactions are scarcely affected by changes in the environmental tempera-

TABLE I
PRESSURE REQUIRED TO BLOCK DIVISION

Kind of eggs	Pressure (psi)	Observer[a]
ECHINODERMATA		
Arbacia punctulata	5000–6000	Marsland, Pease
Arbacia lixula (*pustulosa*)	5000	Marsland
Echinarachnius parma	5000	Pease
Paracentrotus lividus	5000	Marsland
Psammechinus microtuberculosis	5000	Marsland
Sphaerechinus granularis	6000	Marsland
NEMATODA		
Ascaris megalocephala	Over 12,000	Pease
ANNELIDA		
Chaetopterus pergamentaceus	3500–4000	Marsland, Pease
MOLLUSCA		
Cumingia tellenoides	4000–5000	Pease
Planorbis sp.	4000	Pease
Solen siliqua	3500	Marsland
INSECTA		
Drosophila melanogaster (pole cell division)	5000–6000	Marsland
TUNICATA		
Ciona intestinalis	3000	Marsland
VERTEBRATA		
Fundulus heteroclitus	5000–6000	Marsland
Rana pipiens	5000–6000	Marsland

[a] Data of this table are mainly from Pease and Marsland (1939).

ture or pressure, as is predictable from the work of Freundlich (1937).

Intracellular reactions, of course, are not isolated, and consequently the manner in which they are affected by pressure and temperature is less predictable. Permeability changes may be imposed upon the cellular membrane systems and, of course, the enzyme system catalyzing an intracellular reaction may be altered by the pressure–temperature conditions, as was shown by Brown *et al.* (1942) for bacterial luminescence systems.

C. Importance of Intracellular Gel Structures

The gel structures of the cell are of unique importance, particularly as regards the development of mechanical energy and the performance of various cellular movements. For ameboid movement, contractility developed in the peripheral plamagel system (ectoplasm) is believed to energize the forward flow of plasmasol (endoplasm) during locomotion (Marsland and Brown, 1936; Marsland, 1942, 1964*a*). Also, for stream-

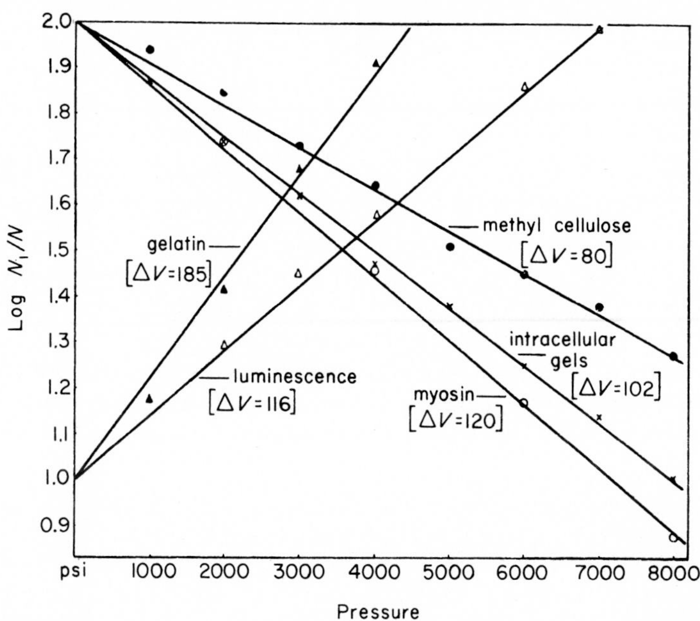

FIG. 1. Effects of pressure on gels (and certain other systems). N_1 = gel strength values at different pressures (psi) and N = values at atmospheric pressure. The slope of each curve = ΔV, the cubic centimeter per mole change in volume which the substance undergoes in passing from the sol to the gel state, being negative for the two curves at the left and positive for the others. Myosin (from rabbit muscle): 2.1% "solution," 36.7°C, pH 6.72. Methyl cellulose: 1.66%, 54°C. Gelatin: *phosphoreum* at 34.0°C. Intracellular gels: averages of the gelational values in the 4%, 22.0°C, pH 6.4. Luminescence: reactivation of light intensity in *Photobacterium* plasmagel systems (cytoplasmic cortices) of *Amoeba* (two species), *Arbacia* eggs (fertilized and unfertilized), *Elodea* leaf cells. Also included are data on the rates of furrowing (*Arbacia*) and protoplasmic streaming (*Elodea*), in which cases N_1/N = the relative rate values. In the luminescence curve, N_1/N_2 = relative light intensity, restored after being dimmed by exposure to high temperature. (From Marsland and Brown, 1942.)

ing in plant cells (Marsland, 1939*b*) and for the movement of pigment granule in the unicellular pigmentary effectors of fish (Marsland, 1944; Marsland and Meisner, 1967), protoplasmic gelations are involved in the development of the mechanical energy. These gelations, moreover, must be classified as type II reactions, since all are weakened progressively by increasing pressure and strengthened by increasing temperature. The cell, apparently, expends metabolic energy in the formation of its gel structures and derives mechanical energy when the gels contract. Sustaining the source of mechanical energy involves a rebuilding of the gel structures, since in the course of contraction they revert toward the sol condition. Thus a continued cycle of gelation and solation is essential to the continued performance of mechanical work.

D. Gel Structures Concerned with Cell Division

The dividing cell, of course, must perform mechanical work in moving the daughter sets of chromosomes apart and (for animal cells) in cleaving the cytoplasm. Two different gel structures appear to be involved: (1) the mitotic apparatus, or spindle–aster complex, for karyokinesis; and (2) a strongly gelated peripheral layer of cytoplasm, the cortex, for cytokinesis. In eggs cells, shortly after fertilization, the cytoplasmic cortex, a layer some 4–5 μ thick (Chambers and Kopac, 1937; Hiramoto, 1957), displays a very great increase in the strength of its gel structure, which attains a maximum at the time when cleavage is to occur (Zimmerman *et al.*, 1957). Both the mitotic apparatus (Zimmerman and Marsland, 1964) and the cytoplasmic cortex (Marsland, 1939a) must be regarded as typical protoplasmic gel structures. Both are progressively weakened, finally to the point of complete solation, by increasing pressure in the range up to 12,000 psi and both tend to be fortified by increasing temperature in the physiological range.

E. Molecular Aspects of Protoplasmic Gelations

All protoplasmic gelations appear to involve processes of polymerization whereby elongate macromolecular complexes are formed by the bonding together of a number of protein subunits, or monomers, initially present in the system prior to its gelation. The evidence indicates that the potential bonding sites of the separate monomeric units are masked by shells of bound water. Thus, bonding cannot occur until a dispersal of the aqueous shells takes place. In the case of the polymerization of the tobacco mosaic virus protein, Lauffer and co-workers (see Lauffer, 1962) have estimated that, on the average, each aqueous shell is comprised of an aggregate of some 300 H_2O molecules. In any event, the dispersal of the protective shells, which is a prelude to polymerization, accounts in part, at least, for the volume increment that the system displays when gelation occurs. The energy requirement for such dispersal must also account for the endergonic nature of protoplasmic gelations and for the fact that such gelations proceed more readily at higher temperatures. Moreover, the evidence indicates that the polymer bonding is partly comprised of hydrogen bonds (Ansevin and Lauffer, 1959) and partly by —S—S—bridges (Mazia and Zimmerman, 1958), although other types of bonding may be involved.

F. Special Techniques and Apparatus

1. *The Microscopic–Pressure Chamber*

A microscopic visualization of cells during exposure to high pressure is essential, especially when cellular movements are being studied. This follows from the fact that most effects, especially in the range of pressure up to about 8000 psi, are rapidly reversed upon decompression. Thus, for example, the very marked effects of pressure upon the rate of protoplasmic streaming in plant cells (Marsland, 1939b) were not revealed in an earlier study (Fontaine, 1930) in which the material was not observed until after it had been removed from the pressure chamber.

The earliest use of a microscope–pressure chamber combination appears to be that of Marsland and Brown (1936), who were able to observe the rather drastic effects of relatively low pressures on the form and movement of free-living amebas. This first microscope–pressure chamber, designed by Dugald Brown, utilized two windows, the lower of which consisted of a simple lens with a fixed focus upon the cells in the chamber. Thus an image of the compressed cells could be picked up and further magnified by a compound microscope. Certain disadvantages developed, however. The first chamber did not allow any flexibility of focus and its total capacity was limited to about 3 ml.

An improved chamber (Fig. 2) was devised by Marsland (1938) and this has been used throughout most of the subsequent work. This is of larger capacity (400 ml) and possesses upper and lower windows made of specially tempered plate glass 7–10 mm thick. Thus it is possible to observe cells directly at pressures up to 20,000 psi and at magnifications up to 600 ×, utilizing a series of long-range objectives in combination with an inverted microscope (Fig. 3).

2. *Pressure Centrifuge Apparatus*

A method of measuring the gelational state of the different cytoplasmic regions and of ascertaining how the gel structure is weakened by increasing hydrostatic pressure was developed by Dugald Brown (1934). This involves centrifuging the cells while they are exposed to high pressures. Thus it is possible to compare the pressure-centrifuged cells with controls centrifuged simultaneously at atmospheric pressure.

Brown's pressure centrifuge device is depicted in Fig. 4. Note that the pressure in the experimental section of the centrifuge head is sealed in by means of a needle valve. This permits a maintenance of the pressure when the head is disconnected from the pressure source and connected

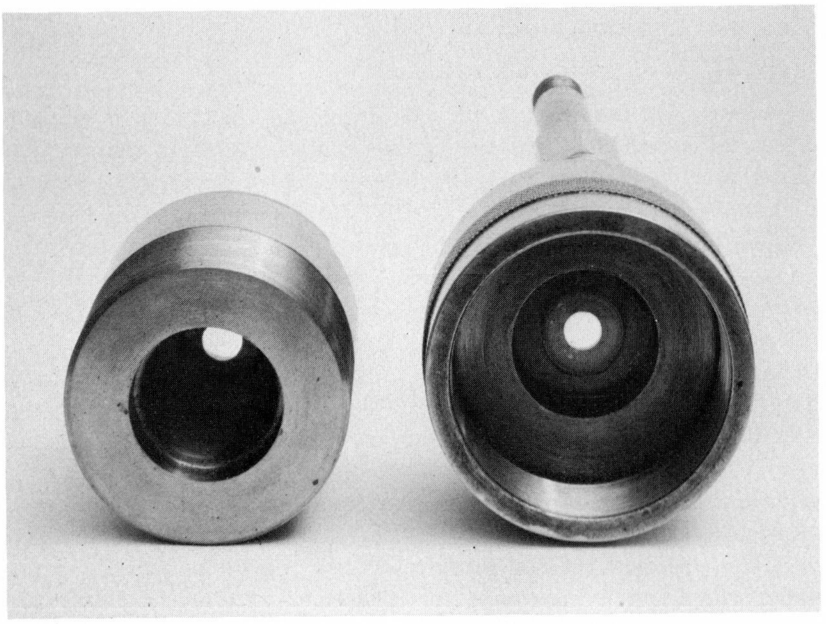

FIG. 2. Upper (right) and lower parts of standard microscope–pressure chamber. Assembled chamber (Fig. 3) measures 6 in. in height. Windows consist of thick (6–8-mm) disks of Herculite plate glass (or sapphire). Sealing between upper and lower chamber parts may be by O-ring or by a specially designed Neoprene washer. Material of the chamber is high tensile phosphobronz initially, stainless steel later. (From Marsland, 1958a.)

to the shaft of a centrifuge motor. The control section, containing the cells at atmospheric pressure, is put in place immediately after the tightening of the needle valve.

The rationale of the foregoing method is, of course, similar to that upon which the centrifugal method for measuring the viscosity of fluid media. In measuring gelational state, however, a certain minimum force must be employed if any displacement of the heavier granules of the cytoplasm is to be obtained. Assuming a constant force above this minimum, the time required to achieve a standardized distance of displacement of

FIG. 3. Assembly for microscopic visualization of cells during a compression period. Light source is provided by a parallel beam passing vertically down through the upper window (A). Image of the cells is picked up by inverted microscope (B), equipped with a 30 X, U.M. 4, long-range objective (Leitz). Exploration of the microscopic field is achieved by means of mechanical platform (C). (From Marsland, 1950.)

cytoplasmic granules is taken to be an index of the relative strength or weakness of the gel structure in which the granules are imbedded. The assumption that the pressure does not induce a significant change in the relative densities of the granules and surrounding gelated medium appears to be justified, since a variety of different granules have yielded almost identical values as to the effects of pressure upon the gelational state of the cytoplasm of various cells. As a general rule, in fact, it may

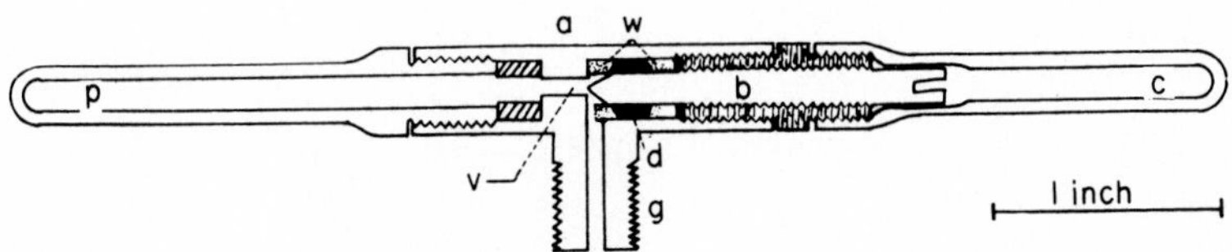

FIG. 4. The pressure centrifuge. The central T-shaped piece (*a*) connects by way of vertical part (*g*), first with the pressure pump, and second with the shaft of a centrifuge motor. The pressure section (*p*) contains the pressurized cells, whereas the control section (*c*) contains the cells which are to be centrifuged simultaneously at atmospheric pressure. Pressure is sealed into compartment (*p*) by means of needle valve (*b*), before putting control section in place. (*w*) = metal washers; (*d*) = rawhide leather packing; (*v*) = valve seat. Construction is of stainless steel. (From Brown, 1934.)

be said that with each increment of 1000 psi (68 atm), the strength of the gel structures in a variety of cells displays a relative decrement of 23.7 ± 1.2% of the value measured at each next lower pressure (Fig. 5; see also Marsland, 1939a). Moreover, a highly similar exponential decrease in gel strength with increasing pressure has been observed *in vitro* with rabbit actomyosin gels (Marsland and Brown, 1942) measured by means of a steel ball falling under gravity through the pressurized gel (Fig. 1). In this latter case, effects of pressure upon the density of the ball, relative to the medium could scarcely be significant.

II. Cytoplasmic Aspects of Cell Division: Cytokinesis

The normal sequence of events in cell division would make it more logical to deal with nuclear division, or karyokinesis, before taking up the mechanisms of cytoplasmic cleavage. However, pressure experiments on cytokinesis preceded those on karyokinesis by some ten years and the results of this earlier work provide a more suitable background for reaching an understanding of those that came later.

A. Pressure-induced Recession of the Cleavage Furrows

This phenomenon was first described more than thirty years ago by Marsland (1938). The original direct observation of a recession of the cleavage furrows when a suitably high pressure is applied during telo-

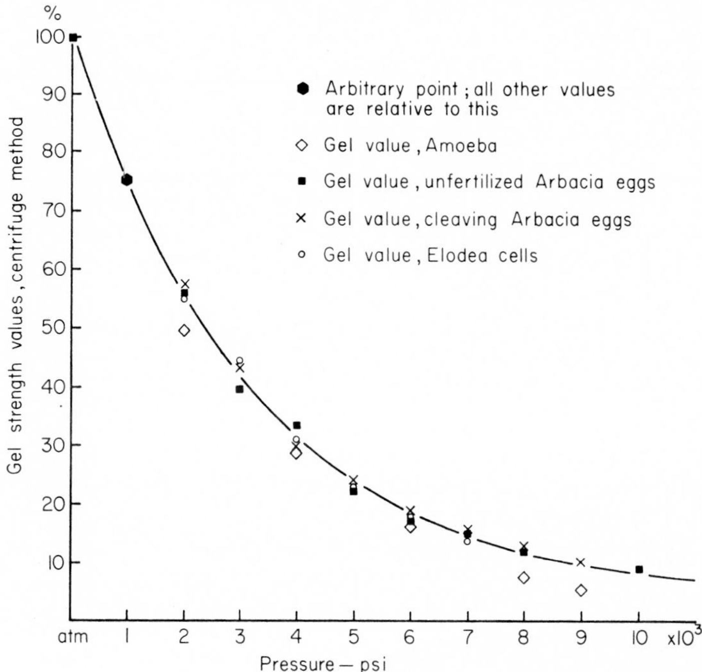

Fig. 5. Direct plot of the gel strength–pressure relationship. Plot of log values of gel strength as a function of pressure yields a strait line such as is shown in Fig. 1. In fact, the intracellular gel plot (Fig. 1) was derived mainly from the above data. All protoplasmic gels studied to date display a similar exponential falling off of strength with increasing hydrostatic pressure. (Data from Marsland and Brown, 1936 (*Amoeba*), Brown, 1934 (*Arbacia* eggs, unfertilized),Marsland, 1939a (*Arbacia* eggs, fertilized), and Marsland, 1939b (*Elodea*).

phase, after the furrow has begun to pinch inwards, was made upon the eggs of a sea urchin (*Arbacia punctulata*). Subsequently, however, the generality of the phenomenon has been established by observations on a wide variety of eggs from animals representing seven different phyla (Table I).

A typical example of pressure-induced furrow recession is provided by a description of events following the compression of the *Arbacia* egg. In these experiments, which were performed at 22°C, a pressure of 6500 psi was applied rapidly (6 seconds to maximum) at the time when 50% of the eggs displayed furrows, some quite shallow, some deep. In order to see the progressing furrows more clearly, the eggs had been denuded of their fertilization membranes by vigorous shaking in a test tube shortly after insemination. The eggs were observed continuously at 450 × magnification throughout the compression period.

Immediately upon establishment of the pressure, there is a very slight and transient deepening of the furrows.[2] Within 1 second, however, all progress ceases, and gradually a slow recession takes place until all the furrows are gone completely. Some eggs become perfectly spherical, although most are slightly ovoid.

The time required for complete recession depends upon the initial depth of a furrow and the intensity of the applied pressure. With minimum blocking intensities (6500 psi at 22°C), the time variation is 1–10 minutes, but at higher pressures recession is much quicker.

If the pressure is released immediately upon completion of recession, furrowing commences again about 1 minute later, and the reinstituted furrows progress to completion. With shallow furrows, indeed, it is possible to block and release cleavage successively four times. However, at the end of 15 minutes (at 22°C) the cleavage tendency is suspended, not to be resumed until the time when second cleavage has started in the nonpressurized control eggs.

In eggs for which the first cleavage is blocked completely, either by the application of sustained or intermittent pressure, two nuclei can be seen in each single cell. At the time of second cleavage, accordingly, a double cleavage is the rule. Sometimes, however, irregularities are observed, resulting in embryos consisting of four cells of unequal size, and sometimes only three cells are formed. However, a large majority of the temporarily pressurized eggs continued development, at least to the extent of forming swimming blastulas of normal appearance.

B. Involvement of the Strongly Gelated Cytoplasmic Cortex in the Cleavage Mechanism

Such an involvement was first proposed by Marsland (1936). Shortly thereafter, however, similar concepts were advanced by Schechtman (1937) and Chambers (1938).

[2] Such a deepening of the furrows is so slight and transient that it escaped notice during the early experiments reported by Marsland (1938). It was first picked up in observations on the furrowing of the eggs of a frog (*Rana pipiens*) reported by Marsland and Jaffe (1949). Subsequently, however, it has been seen (not previously reported) in various other eggs, including those of *Arbacia punctulata* and other sea urchins.

Brown (1934) had demonstrated an exponential weakening of the gel structure of the cytoplasm of unfertilized *Arbacia* eggs with increasing pressure, and Marsland (1936), in some preliminary experiments, noted a marked similarity between Brown's data and the effects of pressure upon the rate at which the cleavage furrow is able to impinge upon the cell. Accordingly, a detailed study (Marsland, 1938) was undertaken, relating the rate of furrow movement to the gelational state of the cytoplasmic cortex.

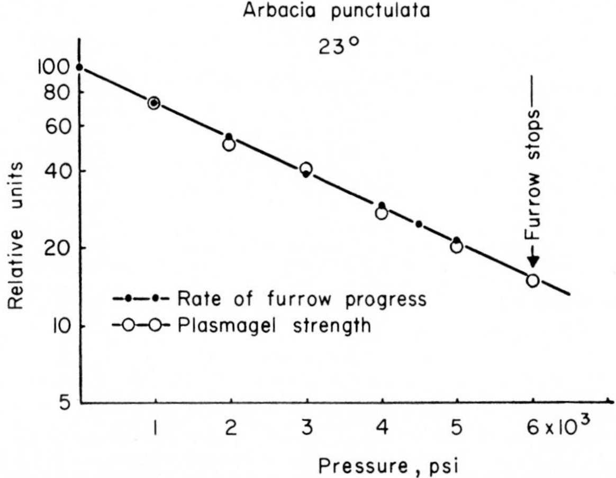

FIG. 6. Relationship between gelational state in the egg cortex and the potency of furrowing. From Marsland, 1956b. (Original data from Marsland, 1938, 1939b.)

Difficulty was experienced in obtaining any displacement of the echinochrome pigment bodies imbedded in the gelated cortex of fertilized *Arbacia* eggs, especially when the centrifuging was done shortly (4–5 minutes) prior to the onset of furrowing. Utilizing the highest force (18,500 *g*) then available at the Stazione Zoologica in Naples, Italy, no displacement was discernible at pressures below 2000 psi. The fact that Brown (1934) had obtained rapid displacement in unfertilized eggs, using a force of only 7200 psi and relatively low pressures, indicated that the gel strength of the fertilized egg was undergone considerable fortification. Subsequent measurements, in fact, showed that the relative gel strength of the egg cortex rises some fiftyfold within 13 minutes after fertilization and that, by the time of furrowing, it has reached a sixtyfold level (Zimmerman *et al.*, 1957).

Although the gel strength measurements were made with an Italian sea urchin, *Arbacia lixula* (formerly *pustulosa*) and the rate of furrow impingement was recorded with the American species, *Arbacia punctulata*, a remarkable coincidence of the curves (Fig. 6) was obtained. These results indicated not only that the effects of pressure upon the structure of protoplasmic gels is a general phenomenon of wide applicability, but also that the gelational state of the cytoplasmic cortex of the dividing cell is intimately related to the magnitude of the force which operates to cleave the cell. The tentative conclusion drawn from these and similar results (Marsland, 1950) was that the cortical gel of the dividing egg represents a contractile system and this conclusion has been sustained in many subsequent studies (Section II, C–G).

C. Effects of Temperature upon the Gelational State of the Cytoplasmic Cortex and the Force Available for Cleavage

The expectation, based on the criteria of Freundlich (1937), that protoplasmic gel structure would be fortified by increasing temperature, at least within the physiological range, has been justified by many experiments. These have dealt not only with the cortical gel systems of dividing eggs (Marsland, 1950; Marsland and Landau, 1954), but also with the plasma gel system of free-living amebas (Landau *et al.*, 1954*a*). Moreover, the capacity of the cells to perform mechanical work (cleavage or ameboid movement) proved to be proportionate to the temperature-induced fortification of the gel structures.

The increasing strength of the cortical gel structure with increasing temperature (Fig. 7) has been amply demonstrated in various dividing eggs (Marsland, 1950; Marsland and Landau, 1954). Also, the relationship between the increase in gel strength and an increased capacity to perform the work of cleavage has been clearly shown (Fig. 8; also see Marsland and Landau, 1954; Marsland, 1956b). Thus, cleavage can proceed only when the temperature–pressure conditions are such that the gel strength of the cortex does not fall below a certain definite minimum (Fig. 8).

D. ATP as a Source of Energy for Fortifying the Cortical Gel Strength and Cleavage Force

1. *Earlier Evidence*

Since protoplasmic gelations are endothermic, one must look for a basic metabolic pattern by which the cell provides energy and determines when and where its essential gel structures shall be formed. Apparently, the metabolic energy which the cell diverts into the building of its gel structures finally appears in the form of mechanical work during the contraction phase of the sol–gel cycle, i.e., when cleavage occurs.

The importance of adenosine triphosphate (ATP) as an energy source in muscular tissues naturally suggested that this ubiquitous metabolite might contribute energy to the sol–gel cycle in cells generally. Indeed, considerable evidence in this regard has begun to come from several directions. Runnström (1949) showed that egg cells immersed in ATP

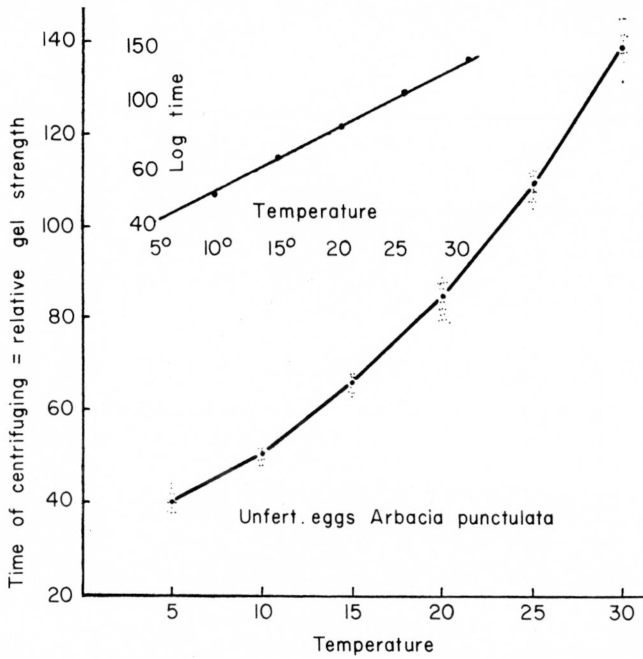

FIG. 7. Temperature effects in *Arbacia* eggs approaching cleavage. Relative changes in the structural strength of the gelated cortical cytoplasm with temperature as measured by the centrifugation method. In the lower curve each of the smaller points represents one determination made at the specified temperature. The time of centrifuging (in seconds) is expressed directly as a function of temperature (centigrade). In the upper curve, the logarithm of the centrifuge time is plotted against temperature. The data show that the cortical protoplasm, in its temperature characteristics, represents a type II gel according to the criteria of Freundlich (1937). (From Marsland, 1950.)

solution became more resistant to hypotonic cytolysis, which seemed to be related to a gelling effect upon the cytoplasm. Kriszat (1949, 1950) found that ATP distinctly modifies the movements of *Amoeba proteus*. Loewy (1952) extracted an actomyosin type of protein from the ameba *Pelomyxa* and showed that this preparation displayed considerable changes in its gelational structure in the presence of ATP and related compounds. H. H. Weber demonstrated that glycerol-extracted fibroblasts underwent a quick and forceful contraction of their elongate pseudopodia when treated with ATP solutions and that this contraction could be stopped quickly and reversibly when the cells were treated with mersalyl acid (salyrgan), a compound that inhibits the hydrolytic splitting of the

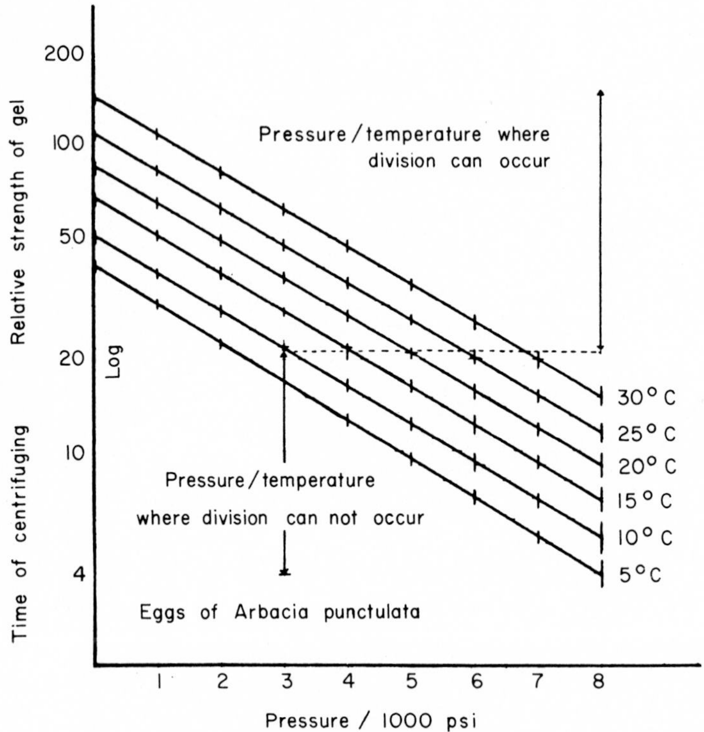

FIG. 8. Family of curves showing the effects of pressure, at different temperatures, on the structural strength of the gelated cortical cytoplasm of the *Arbacia* egg and the relation of these data to the capacity of the egg to perform the work of cleavage. Any combination of temperature and pressure which reduces the cortical gel strength beyond a critical value (about 20% relative to the atmospheric value at 23°C) is just adequate to block division. The centrifuge times are expressed directly in seconds, and the degree of variation is indicated by the length of the markers. (From Marsland, 1950.)

high energy bonds of ATP (cf. Weber and Portzehl, 1954). And finally, Hoffmann-Berling (1954) found that fibroblasts, glycerol-extracted just at the beginning of telophase when shallow cleavage furrows first appeared, showed a remarkable deepening of the furrows, virtually to the point of complete cleavage, when appropriate concentrations of ATP were added to the immersion medium. Accordingly, adenosine triphosphate was chosen as the first metabolite to be investigated in relation to the temperature–pressure parameters of the plasmagel system, and mersalyl acid (salyrgan) was used as an inhibitor of the ATP system.

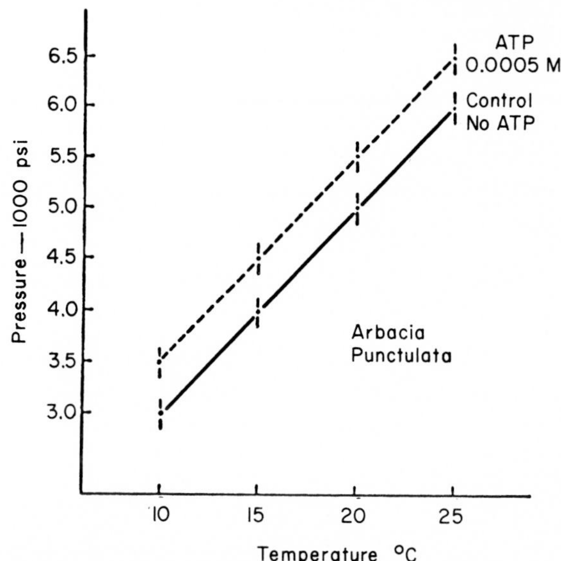

FIG. 9. ATP-induced fortification of cleavage capacity. Each point indicates the pressure level which just suffices to block cleavage at each temperature. From Marsland, 1956b. (Data originally from Landau *et al.*, 1955.)

2. *ATP Effects on Furrowing Strength*

Both *Arbacia* and *Chaetopterus* eggs displayed a distinct increase in the strength of the furrowing reaction as a result of adding ATP (0.0005 *M*) to the sea water, starting approximately 25 minutes prior to the onset of first cleavage. This may be seen in Figs. 9 and 10. These figures also show that, with or without ATP, the minimum pressure required to block the furrows is distinctly higher at each different higher temperature. At atmospheric pressure also, the additional ATP enabled the eggs to complete their furrowing at temperatures that ordinarily are too low to allow for successful cleavage. Specifically for *Arbacia*, more than 90% of the ATP-treated eggs achieved successful furrowing at 9°C (compared with 10% for the untreated specimens), and for *Chaetopterus* at 17°C, 89% of the treated eggs went through (compared with 4% in the untreated specimens). In fact, to achieve an equivalent low temperature inhibition of furrowing in the ATP-treated eggs, it was necessary to reduce the temperature by 2°C, to 7°C in *Arbacia* and to 15°C in *Chaetopterus*. In effect, therefore, it may be said that the ATP treatment permits a 2°C extension of the low temperature range of the species, at least as judged by the achievement of first cleavage.

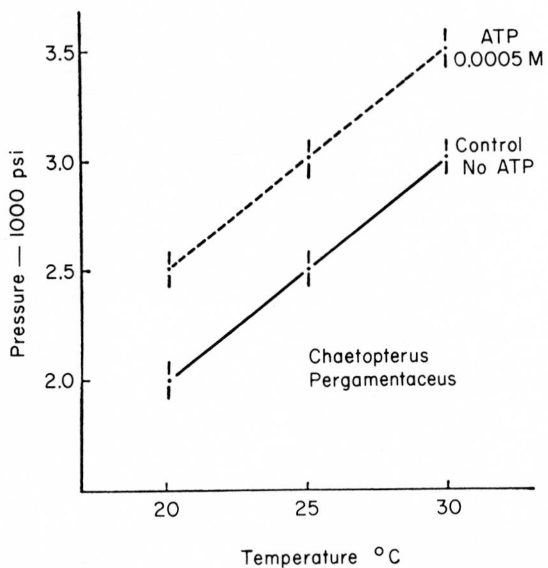

FIG. 10. Another example of how ATP enhances the strength of furrowing. Points indicate minimum pressure required to block cleavage. (From Marsland, 1956b. Original data from Landau *et al.*, 1955.)

3. *ATP Effects on Gel Structure*

The data presented in Figs. 11 and 12 provide further support for the view that the improved furrowing performance of the ATP-treated eggs results from a fortification of the structure of the plasmagel layer.

4. *Experiments with Mersalyl Acid (Salyrgan)*

The experiments of Weber and Portzehl (1954) indicate that salyrgan strongly inhibits the hydrolytic splitting of the high energy phosphate bonds of ATP. Consequently, studies were made on the effects of this compound upon the furrowing strength and gel structure of dividing eggs (*Arbacia* and *Chaetopterus*). These results (Figs. 11 and 12) likewise indicate that energy from the phosphate splitting can be utilized in building up the structure of the plasmagel system and finally serves to fortify the furrowing reaction. Here it may be seen that salyrgan added to the sea water (10 minutes subsequent to fertilization) at a concentration (0.004 *M*) which is not adequate to block the first cleavage does, nevertheless, produce a distinct weakening of the furrowing reaction at each of the temperatures tested. Thus for both *Arbacia* and *Chaetopterus*,

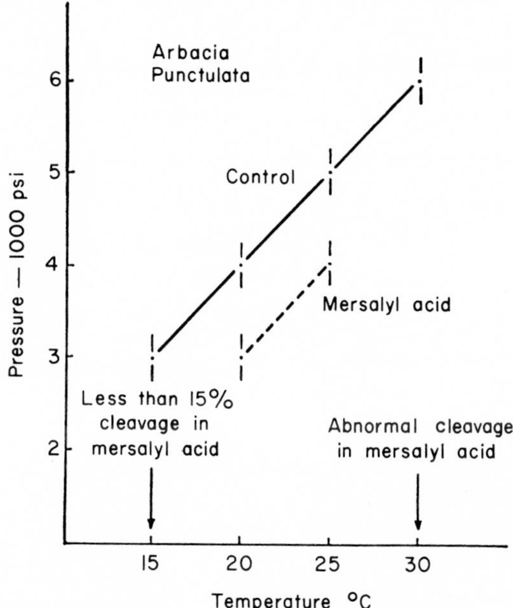

FIG. 11. Reduction of cleavage potency by mersalyl acid (an inhibitor of ATP hydrolysis). Each point indicates the minimum pressure required to block the cleavage furrows. (From Zimmerman *et al.*, 1957).

a lower (by 500 psi) minimum pressure is required to block furrowing at each different temperature (Figs. 11 and 12). This weakening of the furrowing strength is related, moreover, to a concomitant weakening of the cortical gel structure (Fig. 13).

All in all the data indicate that the hydrolysis of ATP constitutes one important metabolic reaction that contributes energy to building up the gel structure of the cortical gel system and that this energy becomes available as mechanical work when the cell executes furrowing. Apparently, the interconnected fibrillar protein components of the cortical gel structure can undergo a forceful shortening, thus converting the protein units into a more globular form, shifting the system toward the solated state (Marsland, 1957).

E. Effects of Heavy Water (D_2O) on the Gelational State of the Egg Cortex

Studies on the antimitotic effects of D_2O were initiated shortly after 1933, when Urey first succeeded in the isolation of deuterium (see

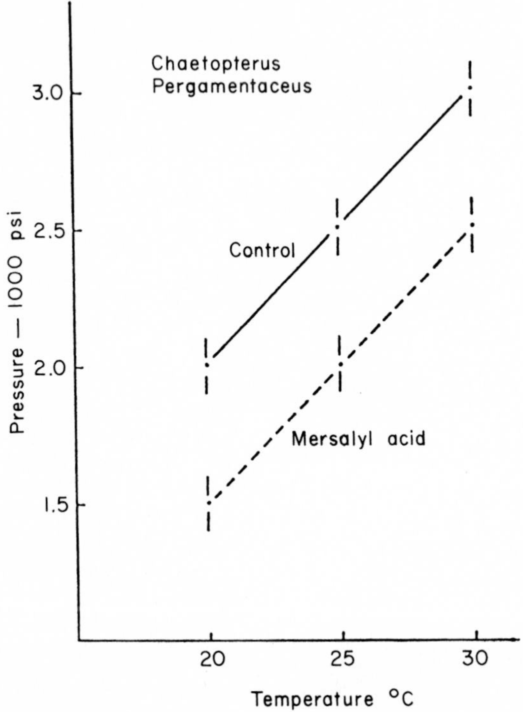

Fig. 12. Another example of the lowering of cleavage potency by an ATP-hydrolysis inhibitor. (From Zimmerman *et al.*, 1957).

Ussing, 1935; Lucké and Harvey, 1935; Gross and Spindel, 1960). Observations on the deuterational effects upon mitotic apparatus, however, have been more extensive than those upon the effects on the egg cortex (Marsland *et al.*, 1962; Marsland and Zimmerman, 1963, 1965).

Generally speaking, a substitution of 70% or more of the H_2O with D_2O, in the surrounding sea water, effects a complete blockage of karyokinesis. Such concentrations, however, do not block cytokinesis, although cleavage does not generally take place unless nuclear division is completed. In fact, relatively low levels of D_2O may even increase the cleavage force (Marsland and Zimmerman, 1963).

With higher concentrations (70% or more) cytoplasmic cleavage can proceed even though the nuclear division is blocked. This is indicated by two facts. First, successful cleavage can occur if the D_2O treatment is deferred until late anaphase,[3] presumably after the reaction that triggers

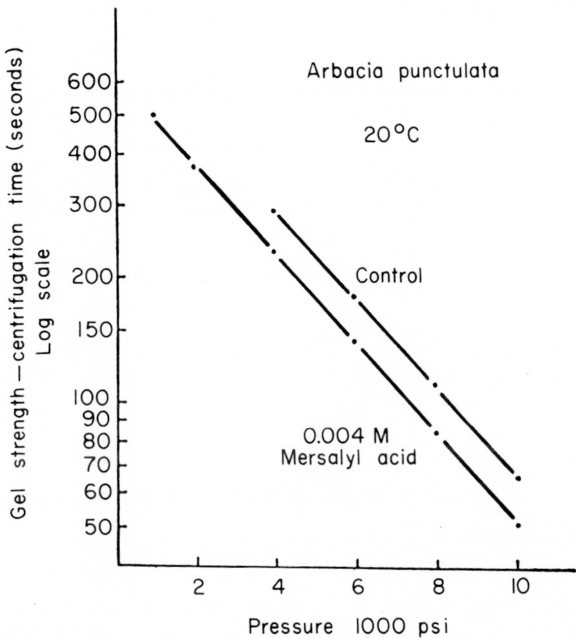

FIG. 13. Weakening of cortical gel by mersalyl acid. Measurements made 5 minutes prior to furrowing. At this time the cortical gel strength has risen to its high maximum. (From Zimmerman *et al.*, 1957).

furrowing has taken place (Section V, B, 3–5). Second, experimentally induced cleavages (Section V, B) can occur even in 90% D_2O, although a more drastic inducing treatment must be employed.

Presumably the effect of D_2O in strengthening the gel structure of the egg cortex involves the substitution of D_2O for H_2O in the aqueous shells which protect the potential polymer bonding sites. Apparently the electrostrictive fields of D_2O molecules are weaker than those of H_2O, thus facilitating a dispersal and fostering the formation of polymer linkages.

[3] This result cannot be attributed to a lag in the penetration of D_2O into the egg cell, since Tucker and Inoué (1963) have shown that the exchange between internal and external D_2O is very rapid, reaching at least 30% completion within 2–3 seconds.

F. Fine Structure of the Egg Cortex in Relation to Furrowing

1. *Differentiation between the Cortical and Medullary Cytoplasm*

It has been difficult to demonstrate a clear fine structural differentiation between the gelated egg cortex and the deeper lying cytoplasm, despite the fact that the reality of the existence of a cortex has been established unequivocally by the many pressure centrifugation experiments of Marsland and co-workers and by conclusive micrurgical experiments (Chambers and Kopac, 1937; Hiramoto, 1957). There is, to be sure, a denser peripheral layer, about 1 μ thick, immediately subjacent to the surface membrane, but the density of this layer grades by imperceptible degrees into that of the deeper cytoplasm (Mercer and Wolpert, 1958; Tilney and Marsland, 1969).

The situation changes, however, when definitive furrows begin to appear or some 2–3 minutes earlier. Then the egg cortex shows a marked increase in electron density (Mercer and Wolpert, 1958) and the differentiation between cortex and deeper cytoplasm becomes quite clear. This is especially true of the cortex along the walls of the advancing furrows (Weinstein and Herbst, 1964; Tilney and Marsland, 1969). In the furrow cortex, moreover, a system of filaments, each some 40–60 Å in diameter, has recently been found embedded in the dense cortical layer (Tilney and Marsland, 1969).

Virtually concurrent with this finding in an echinoderm egg (Tilney and Marsland, 1969) similar furrow cortex filaments have been described in eggs from two other phyla, namely, the Mollusca (Arnold, 1968a and b and the Coelenterata (Szollosi, 1968; Schroeder, 1968).

2. *Contractility of Filaments in Furrow Cortex*

The occurrence of filaments in the furrow cortex of cleaving egg provides strong support for the contractile theory of cytokinesis (Marsland and Landau, 1954; Marsland, 1956b, Wolpert, 1960). Such filaments occur very widely in cells that are distinguished by their contractility. For example, Cloney (1966) reports that filaments appear in the epithelial cells in the ascidian tadpole tail during metamorphosis at the time when these cells shorten rapidly, and that the orientation of the filaments suggests that they may play an active role in the contraction process. Similar filaments have been related to form changes in embryos as during blastopore formation (Baker, 1965), during the formation of the neural tube (Baker and Schroeder, 1967), and during the contraction of the filopodia of secondary mesenchyme cells (Tilney and Gibbins, 1969). Furthermore, in protozoa, filaments of similar dimensions appear to play a contractile role such as in the retraction of the pseudopodia of *Difflugia* (Wohlman and Allen, 1968) and in the contractile stalks of *Stentor* (Randall and Jackson, 1958), *Vorticella* (Favard and Carasso, 1965), and *Spirostomum* (Yaqui and Shigenaka, 1963). Accordingly, it appears very probable that the filaments in the cortex of cleaving eggs play a major role in a contractile process.

Recent evidence indicates that the mobilization of a relative abundance of calcium or other polyvalent cations may be the factor that triggers the deposition of contractile filaments in the egg cortex. Thus Stephens and Kane (1966) have shown that a protein component isolated from the cortices of various sea urchin eggs (the 7 S protein initially described by Kane and Hersch, 1959) can be precipitated in the form of numerous birefringent fibrils upon treatment with divalent cations. These fibrils, moreover, lose birefringence and undergo elongation in the presence of a calcium sequestering agent (disodium ethylenediaminetetracetate, or EDTA), but regain birefringence and become shorter when a critical concentration of divalent cation is restored. Sakai (1968) has described a similar shortening of a KCl-soluble protein isolated from the cortices of eggs when divalent cations or agents which oxidize sulfhydryl groups are added to threads composed of this protein. This shortening was reversed by EDTA. Presumably the 7 S cortical protein of Kane and the 2.3 S protein of Sakai are interrelated, although the relationship remains unclear. In any event, the deposition of filaments in the cell cortex could very well endow this layer with an inherent contractility—as is indicated by the recent work of Gingell and Palmer (1968), which demonstrated a localized contraction of the cortex of the frog's egg upon contact with polyionic solutions micropipetted against a restricted area of the free surface. Moreover, such treatment was not effective in the absence of divalent cations or in the presence of EDTA.

G. The Cortical Gel Contraction Theory of the Cytokinetic Mechanism: A Summary

The simplest, most direct and comprehensive interpretation of the data derived from the various experimental approaches to the problem of cytokinesis is provided by the cortical gel contraction theory (Marsland and Landau, 1954; Wolpert, 1960). An alternative interpretation, namely, the surface expansion theory of Swann and Mitchison (Swann, 1952; Mitchison, 1952) has been advanced but this has not been substantiated by subsequent experiments. In fact, the mechanical difficulties inherent in this latter theory would seem to be insurmountable (Marsland and Landau, 1954; Wolpert, 1960).

A more detailed analysis of the cortical gel contraction theory, presented in schematic form in Fig. 14, can be stated as follows (Sections II, G, 1–3).

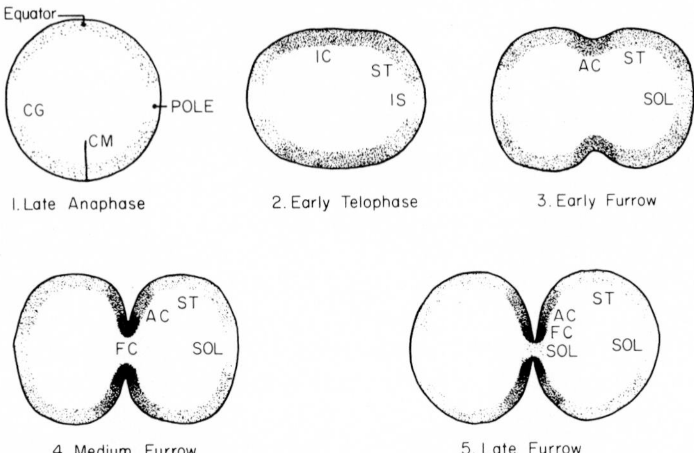

FIG. 14. Schematic representation of the cortical gel contraction theory of cytokinesis, showing progressive changes in the gelational state of cell cortex, and how these changes may be related to the furrowing process. CG, cortical gel; CM, cell membrane; IC, region of incipient contraction; ST, region subjected to stretching; IS, region of incipient solation; AC, actively contracting region; SOL, region of solation; FC, fully contracted (passively pushed) region (further explanation in text). (From Marsland and Landau, 1954.)

1. *Elongation and Early Furrowing (Stages 1–3, Fig. 14)*

The initial deformation of the spherical cell, an elongation in the direction of the spindle axis, appears to result from the initiation of a contraction in a rather broad band of the cortical gel (Fig. 14, Stage 2). Such a contraction is likewise an essential part of the surface expansion theory (Mitchison, 1952). Also it accounts for the observations of Dan *et al.* (1937), Schechtman (1937), and others that the distance between particles adherent to the cell surface and between granules embedded in the protoplasmic cortex, in the region of this broad band, tends to diminish during this early stage of cytokinesis. Also there must be a weakening of the surface layers in the polar areas. Chambers (1938) has demonstrated this by the microdissection technique and by observing the rupture of dividing eggs compressed under a coverslip. Such a polar weakening probably represents an active solation process initiated by relaxing factors equivalent to those concerned with the relaxation phase of muscle activity (Section V, B, 5).

As soon as the actual furrow becomes discernible, the width of the band of contractile cortex diminishes sharply (stage 3). This accounts for the fact that the furrow soon becomes sharp and narrow and also for the numerous observations that the cell surface and subjacent cortex in the peripheral parts of the furrow cease to display evidence of contraction and may display signs of stretching, as judged by the relative displacement of adherent or embedded particles. Such an "expansion" is not to be considered as an active process. It represents a passive stretching of the wall of the furrow in the peripheral region as a result of an active contraction in the deeper part.

2. *Final Furrowing Events (Stages 4–5, Fig. 14)*

The geometrical configuration of the dividing cell is such (stages 4–5) that more and more cortical gel becomes incorporated in the wall as the furrow deepens. Thus, more and more cortical gel possessing an unexpended fund of contractile energy is brought into an operative position as the furrow deepens. Therefore, in the final stages of furrowing, to complete the cleavage, it is only necessary to assume that the region of active contraction shifts from the trough of the furrow to the side walls, first to the region immediately adjacent to the trough, and later to a more peripheral site somewhat removed from the trough (stages 4 and 5). In this way, the gel at the very bottom of the trough, having performed its contractile function, may undergo solation, clearing the way for the approach and final fusion of the cell membrane, which severs the stalk between the daughter cells.

3. *Intracellular Pressure during Cytokinesis*

The foregoing theory provides answers to many questions regarding the mechanics of cytokinesis, except for the one raised by Mitchison (1952), who reported that internal pressure in the cleaving egg cell becomes negative instead of positive as might be demanded by the cortical gel contraction theory. Mitchison's observation is open to serious criticism, however, as has been pointed out by Wolpert (1960). Also, a positive internal pressure is demonstrated by the experiments of Chambers (1938) and of Sichel and Burton (1936), who showed that there is a forceful outpouring of the fluid, deeper cytoplasm through the partially constricted connection between the potential blastomeres of the furrowing egg when one of the blastomeres is destroyed by microdissectional puncture. Moreover, some recent measurements by Wolpert *et al.* (1968) clearly indicate that the cortex of the egg at the time when cleavage is occurring and even some 20 minutes earlier does, indeed, generate a measurable positive pressure within the eggs and that the surface layers of the cytoplasm display a measurable tension. And finally, these latter measurements also show that these mechanical properties of the egg surface are susceptible to drastic reduction, of a reversible nature, upon application of high pressure (10,000–12,000 psi).

III. Nuclear Aspects of Cell Division: Karyokinesis

A. *In situ* Studies on the Effects of Pressure on the Structure and Function of the Mitotic Apparatus

The original observations of Pease (1941, 1946) on the effects of pressure on the structure and function of the spindle–aster complex provide an excellent point of departure. In these early studies, Pease found that the anaphase movement of the sets of daughter chromosomes toward the poles could be retarded and finally stopped by application of high pressure in the 4000–7000 psi range. Such inhibition, moreover, was correlated with structural losses in the linear and radial patterns of the spindle and astral regions, respectively, in cells that were fixed and sectioned immediately after release of the high pressure.

B. Studies on the Isolated Mitotic Apparatus

The excellent method for isolating the mitotic apparatus from egg cells, developed by Mazia and Dan (1952), provided opportunity for following more closely the moment to moment changes in the structural and functional state of the mitotic apparatus, both during the normal course of events and during exposure to high pressure and other agencies. Thus Zimmerman and Marsland (1964) were able to report more fully upon the antimitotic effects of high pressure. In this work, the eggs (*Arbacia punctulata*) were stabilized by immersion in cold (-12°C) 35% ethanol within 15 seconds after release from the pressure treatment. Mitotic isolations were effected by the digotonin (1% solution) method, and observations were made with phase-contrast microscopy.

1. *Effects at Higher Pressures*

Relatively higher pressures (6000–10,000 psi) effected a drastic disorganization of the linear structure in the spindle and astral regions. Also there was a clumping and rounding of the chromosomes, regardless of whether the eggs were in prophase, metaphase, or telophase. These higher pressures produced a complete cessation of chromosomal movement, and eggs compressed at early prophase failed to develop any sign of spindle or aster structure.

In general, the degree of disorganization depended upon the intensity of pressure and the duration of treatment. Complete disorganization was observed after 1 minute at 10,000 psi and 6 minutes at 6000 psi. In some of the metaphase and anaphase preparations, especially in the higher pressure range, the chromosomes became detached from the spindle remnants; at least none could be found associated with these remnants.

2. *Effects at Lower Pressures*

Treatments with relatively lower pressures, namely, 2000 and 4000 psi, even after 5 minutes, yielded isolates which retained a fair degree of normal organization. Chromosomal movement, however, was stopped at 4000 psi and definitely retarded at 2000 psi. At 2000 psi, cells at metaphase reached midanaphase in 10 minutes; but meanwhile, of course, the non-pressurized control cells had finished the first cleavage division.

3. *Postpressure Recovery of Structure and Function*

None of the foregoing treatments imposed any obvious irreversible effects—except when the higher pressure (8000 and 10,000 psi) were maintained for more than 5 minutes. The rate at which structural and functional recovery occurred was slower, however, as the magnitude and duration of the treatments were increased. This was determined by a study of isolates made from pressurized eggs which were not introduced into the chilled ethanol until a suitable recovery period had elapsed following decompression.

Eggs first pressurized at 10,000 psi for 1 minute displayed partial recovery of structural configuration in the mitotic isolates after 5 minutes and what seemed to be full recovery in 10 minutes. This correlates exactly with the observation that the furrowing of a sample of these eggs was delayed just 10 minutes in comparison with the cleavage time of an uncompressed sample. Structural recovery in eggs compressed for 1 minute at 6000 psi, on the other hand, appeared to be complete within 5 minutes and the delay in cleavage was 6 minutes.

4. *Interpretation*

The results of these experiments clearly indicate that the spindle–aster complex must be regarded as a typical protoplasmic gel structure, belonging to type II in the Freundlich classification. The pattern of polymerization is more precisely determined than in the gel structure of the egg cortex, but the disorientation and weakening of the two systems upon exposure to high pressure and low temperature (Inoué, 1964) indicate a fundamental similarity.

C. Fine Structure of the Mitotic Apparatus

Recently (Harris, 1962; Kane, 1962) it has been recognized that an intrinsic component in the structural organization of the spindle–aster complex is constituted by microtubules. Also, microtubules have been found as important components of various other protoplasmic structures, especially those associated with movement. In fact, microtubules have been identified in suctorian tentacles (Rudzinska, 1965); in the cortical cytoplasm of plant cells (Ledbetter and Porter, 1963); in the processes which extend out from melanocyte (Bikle *et al.*, 1966); and in the axopodia of Heliozoa (Kitching, 1964; Tilney *et al.*, 1966). Thus it becomes important to determine how these important fine structural entities may be affected by high pressure, low temperature, and other agencies which are known to influence protoplasmic polymerization and gelation reactions.

D. Pressure Effects on Microtubular Structures

A direct study of the effects of pressure on the microtubular structure of the mitotic apparatus has recently been achieved (see Zimmerman, Chapter 10). Pressure effects on other microtubular structures, namely the axonemic microtubules of the axopodia of Heliozoa, have been extensively studied (Tilney *et al.*, 1966; Marsland *et al.*, 1969), and these results, undoubtedly, have an important bearing on the present problem (see also Zimmerman and Zimmerman, Chapter 8).

The precisely patterned double coil arrangement of microtubules in the axonema of the heliozoan axopod (Fig. 15), first described by Kitching (1964), promised to provide excellent experimental material (see also Tilney and Porter, 1965), and this has proved to be true. In these microtubule experiments (Tilney *et al.*, 1966), specimens of *Actinosphaerium nucleofilum* were observed directly during exposures to pressure in the range up to 10,000 psi. Fixations for the electron microscope preparations were achieved before release from pressure by means of an apparatus designed by Landau (see Landau and Thibodeau, 1962).

Generally speaking, the rapidity and degree of disorganization of the axopodial microtubules, up to the point of complete disintegration, depended on the intensity of pressure treatment. Moreover, each appreciable disorganizational change was accompanied by a beading of the needlelike axopodia and a greater or lesser degree of retraction (Tilney *et al.*, 1966). (For further discussion see Kitching, Chapter 7.)

1. *Effects at Low Pressure*

At a pressure of 4000 psi, the axopodia retracted to about one-tenth their original length within 10 minutes, and at this time no trace of micro-

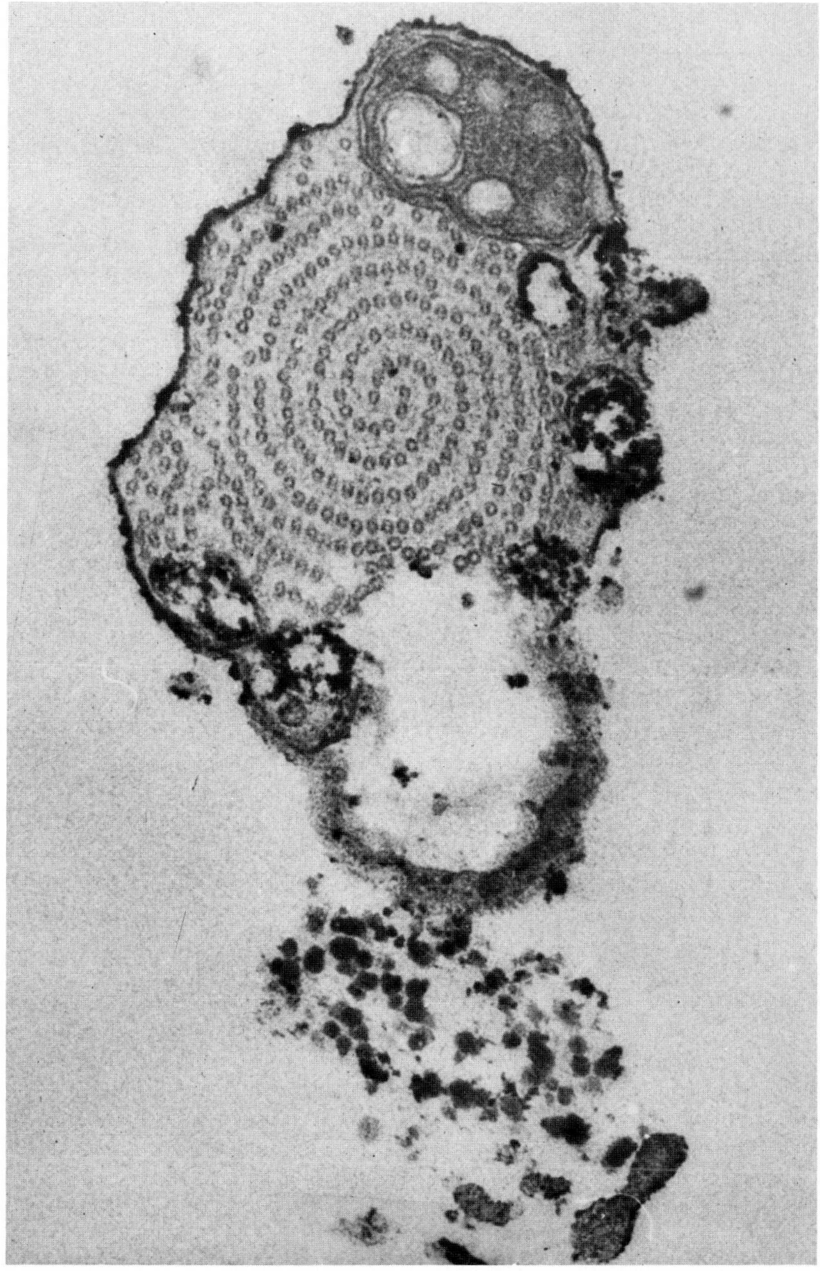

FIG. 15. Normal double coil arrangement of microtubules in the axoneme of an axopodium of *Actinosphaerium nucleofilum* as shown in cross section. Fixation: 1% glutaraldehyde. Magnification: × 75,000. (From Tilney *et al.*, 1966.)

tubular structure remained, either in the beads at the axopodial tips or in the residual length of the drastically retracted axopodia. In the next 10 minutes, however, while the pressure was still maintained, a partial reconstruction of the microtubules occurred and this was accompanied by a small but definite reextension of the axopods.

2. *Effects at Higher Pressure*

At a pressure of 6000 psi, the beading was more rapid, and within 3 minutes it involved the entire length of the axopodia (Fig. 16). Also, there was a complete retraction of the axopodia within 10 minutes, leaving just a few rounded knoblike protrusions at the cell surface. Correlated with the beading and retraction, specimens pressure-fixed at the end of the third minute of exposure to 6000 psi yielded electron micrographs showing a complete absence of microtubules in all parts of the distintegrating axopodia, including the beads (Fig. 17). Also, preparations fixed at the end of the tenth minute showed no sign of microtubular structure in the knoblike protrusions at the cell surface.

A small degree of reconstruction occurred in knoblike protrusions during the next 10 minutes, even when the pressure was maintained. However, the reconstituted microtubules were sparse and the pattern of their arrangement was irregular. No reextension of the axopodia was observed during the 20 minutes of compression.

With higher pressure (8000 and 10,000 psi), the same changes were observed, although they occurred more rapidly. Furthermore, there were no signs of microtubular reconstruction or axopodial reextension during the periods of compression.

All of the disintegrational changes imposed by pressure proved to be reversible, following decompression. Recovery was slower, however, with higher pressures. The first reappearance of short but definitive axopodia occurred in 2, 4, and 6 minutes, following exposures to 6000, 8000, and 10,000 psi, respectively. Attainment of full axopodial length, moreover, required 15, 20, 25, and 50 minutes with pressures of 4000, 6000, 8000, and 10,000 psi, respectively. Examined 24 hours later, all the specimens presented what seemed to be a normal appearance.

3. *Interpretation*

The foregoing experiments indicate not only that certain microtubular structures are subject to the depolymerizational changes imposed by high pressure, but also that these structures play an important role in maintaining the form stability of the axopodia as well as their capacity to perform the work of extending themselves outwards from the cell. Not all microtubular systems are equally susceptible, however, since it has been found (Hinsch and Marsland, 1969) that the microtubules present in the flagella of a variety of sperm cells retain a high degree of structural integrity and normal arrangement despite an exposure to a pressure of 10,000 psi for 10 minutes. With reference to the mitotic apparatus, on the

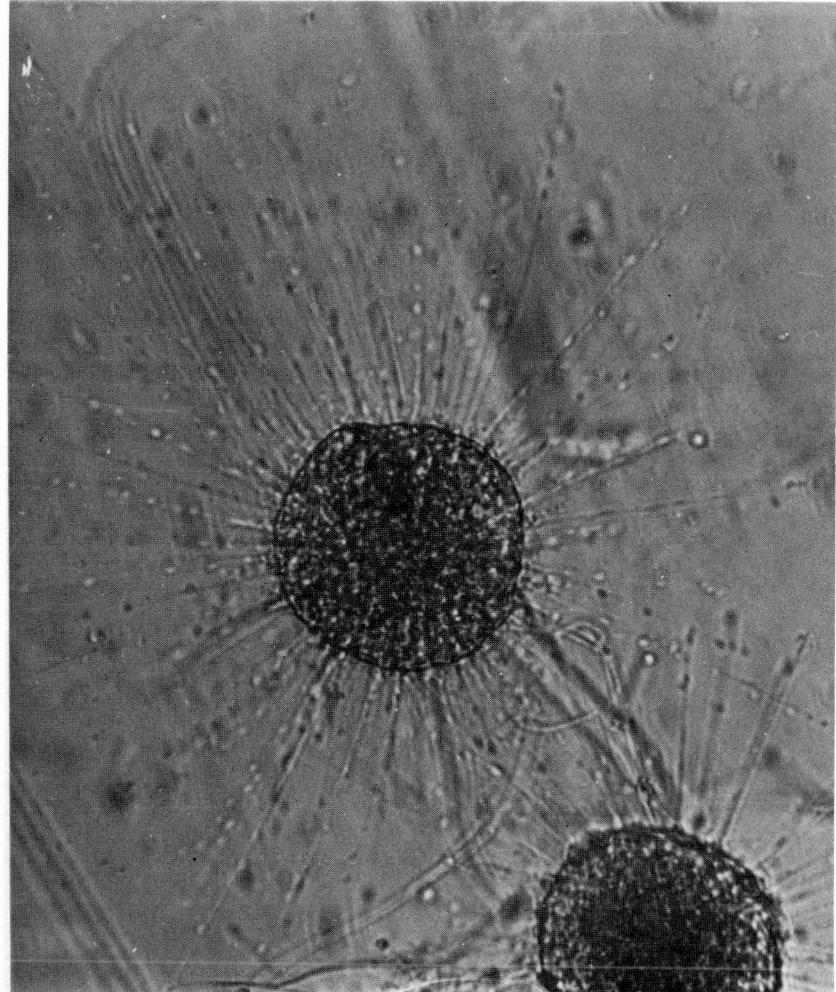

FIG. 16. *Actinosphaerium* as it appears 3 minutes after the imposition of pressure (6000 psi). A drastic solation of the gel structure of the axopodia is indicated by the fact that each becomes beaded, like a slender fluid filament. Concomitantly, there is a disintegration of the microtubular elements (Fig. 17). Magnification: × 100. (From Tilney *et al.*, 1966.)

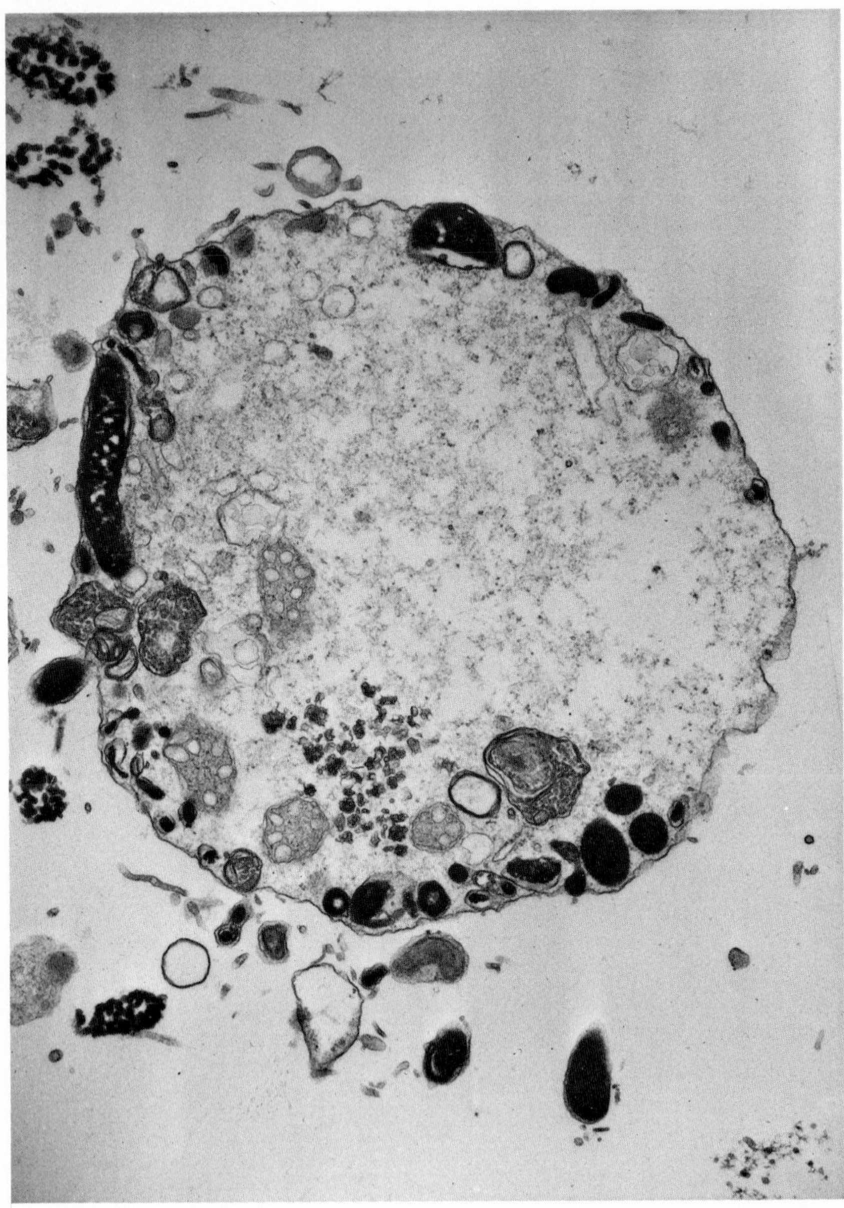

FIG. 17. Section through one of the beads, such as are shown in Fig. 16. Note that no definitive microtubules are present. Pressurized fixation with 1% glutaraldehyde. Magnification: × 50,000. (From Tilney *et al.*, 1966.)

other hand, all the evidence (Pease, 1941, 1946; Zimmerman and Marsland, 1964; Marsland and Zimmerman, 1965; Marsland, 1965) indicates that this system of microtubules is just about as susceptible to high pressure disintegration as is that of *Actinosphaerium.*

E. Stabilization of the Mitotic Apparatus and Other Gel Structures by Heavy Water (D_2O)

1. *Observations*

Evidence that deuteration, i.e., substitution of a significant percentage of H_2O by D_2O in the environing medium of a cell, enhances the polymerization, or gelation, tendencies within the cell has been derived from a significant number of recent investigations. These include studies on: (1) the cortical gel of egg cells (Marsland and Zimmerman, 1965); (2) the plasmagel system of *Amoeba proteus* (Marsland, 1964a); (3) axopodial stabilization in *Actinosphaerium nucleofilum* (Marsland *et al.*, 1969); (4) enhancement of birefringence in the mitotic spindle (Inoué *et al.*, 1963); (5) pigmentary movement in fish melanocytes (Marsland and Cone, 1969); and (6) an *in vitro* analysis of the polymerization tobacco mosaic virus protein (Ansevin and Lauffer, 1959).

With specific reference to the mitotic apparatus, the most pertinent work is that of Marsland and Zimmerman (1965), a study which utilized the method of Mazia and Dan (1952). This permitted microscopic visualization of the structure of the spindle–aster complex in deuterated dividing cells which had been stabilized by immersion in cold ethanol within 15 seconds following exposures to high pressure.

With high concentration (90%) of D_2O, the stabilization effect was very marked. Such deuterated eggs compressed at 12,000 psi for 3 minutes yielded mitotic isolates which were not distinguishable from those derived from the nonpressurized control cells. The linear organization in the spindle and astral regions remained clear, and the arrangement of the chromosomes at metaphase or anaphase appeared to be normal. The pressurized nondeuterated eggs, on the other hand, yielded isolates that were devoid of visible structure, even when the compression period was shortened to 1 minute. Also, the chromosomes were swollen and clumped together.

With lower concentrations of D_2O, the stabilizing effect was less, though easily discernible. With 50% deuteration, for example, quite normal mitotic configurations were obtained after 1 minute exposures to 10,000 psi, whereas this was distinctly not the case with nondeuterated eggs.

2. *Interpretation*

It seems reasonable to conclude that the stabilizing effects of D_2O upon the mitotic apparatus are mediated by effects upon the microtubular components of this structure. This is especially true in light of the recent observations of Marsland *et al.* (1969), which showed approximately the same degree of deuterational stabilization in the microtubular system in the axopodia of *Actinosphaerium*. In fact, it took a pressure of 12,000 psi maintained for 10 minutes to bring about a complete disintegration of the axonemal microtubules when the specimens were immersed in 80% D_2O medium. This is in contrast with the mere 6000 psi for 3 minutes which was required when the animals were kept in H_2O medium (Section II,D,2).

Further support of the foregoing conclusion comes from observations on the effects of temperature upon the stability of microtubular structures. Thus it has been shown (Tilney and Porter, 1965) that a marked decrease in the stability of microtubules (in the axopodia of *Actinosphaerium*) occurs when the temperature is lowered, and Inoué (1952) has demonstrated a fading of the birefringence of the mitotic apparatus with decreasing temperature. Also, the experiments of Marsland *et al.* (1969) show that the susceptibility of the axopodial structure of deuterated *Actinosphaerium* to pressure-induced disintegration tends to be increased at lower and decreased by higher temperatures, within the physiological range. All in all, therefore, the evidence indicates that microtubules must be regarded as typical protoplasmic gel structures in which the process of polymerization, or gelation, requires the absorption of energy and always involves a definitive positive volume change ($+\Delta V$) in the system.

A puzzling question as to the effects of deuteration is why D_2O blocks mitosis when present in the cell at levels of 70% or greater? At present the most probable answer seems to be that such deuteration overstabilizes the gel structure of the mitotic apparatus, preventing the ebb and flow of polymerizations that seem to be essential to the continuance of mitotic activity (also see Inoué, 1964). This viewpoint, indeed, is significantly reinforced by the observations of Marsland and co-workers, reported in the following section. Here it is shown that high pressure is able, in large measure, to counteract the antimitotic effects of heavy deuteration.

F. High Pressure Counteraction of the Antimitotic Effects of D_2O

1. *Rationale*

Since it is known that deuteration tends to fortify the polymer structure of various protoplasmic gel configurations (Sections II,E and III,E) and that high pressure has an opposite effect (Sections II,B and III,B), it seemed likely that pressure might be able, at least to some extent, to antagonize not only the structural effects of D_2O upon the mitotic apparatus, but also its inhibitory effects upon mitotic activity. This prediction has been validated by a considerable number of recent investigations (Marsland, 1964b, 1965; Marsland and Hiramato, 1964, 1966; Carolan *et al.*, 1965, 1966; Marsland and Asterita, 1966).

2. *Observations*

In highly deuterated (70–90%) seawater, none of the eggs (*Stronglocentrotus purpuratus*, immersed during early prophase) succeeded in reaching telophase, i.e., none showed any attempt to cleave. With pressure, however, at levels of 3500–5000 psi, at least 98% developed definitive furrows and many completed first cleavage. The percentage of successful cleavages depended on the intensity of pressure in relation to the percentage of deuteration, being 75 and 70%, respectively, with 70 and 80% deuteration, at 4000 and 4500 psi, respectively. With 90% D_2O, however, a lower percentage of success (15%) was achieved, with an optimum pressure of 5000 psi (Marsland, 1965).

Similar results were also obtained by Marsland and Hiramoto (1966) working on the meiotic divisions of the oocytes of a starfish (*Asterias forbesii*) during their maturation.

In studying their capacity to form and maintain polar bodies, the oocytes were deuterated 60 minutes after shedding and pressurized 6 minutes later. Maximum release from deuterational blockage, as judged by the percentage of persistent polar bodies after 1 hour of treatment, was as follows: 85% at 2000 psi for 50% D_2O; 65% at 2500 psi for 60% D_2O; 77% at 3000 psi for 70% D_2O; 55% at 3500 psi for 80% D_2O, and 65% at 3500 psi for 90% D_2O. Persistence of polar bodies in the nonpressurized deuterated cells did not exceed 15% in any medium containing 60% or more of D_2O, and the average persistence in such media was only 2.5%.

Release from deuterational blockage by application of high pressure is also a temperature sensitive phenomenon, as has been shown in another egg (*Arbacia punctulata*) by Marsland and Asterita (1966). In these experiments the immersions were initiated during early prophase and the deuteration was kept at a constant value of 70%. Without pressure, none of the deuterated eggs achieved successful cleavage.

Maximal release from mitotic blockage tended to decrease with increasing temperature, being 88% at 15°, 78% at 20°, and 71% at 25°C. The pressure required for maximal release was higher for the higher temperatures, being 3500 psi at 15° and 4500 psi at 20° and 25°C. Moreover, in the pressure range below these maxima, each increment of 5°C raised the pressure requirement for equivalent release by 500 to 1000 psi.

3. *Interpretation*

The simplest explanation such a restoration of mitotic activity in deuterated cells would involve the known effects of pressure on the sol–gel equilibrium or steady state. Deuterium, apparently, shifts this equilibrium in the gell direction, carrying it beyond physiological limits and not permitting the shifts which normally occur during mitosis. The action of pressure, accordingly, is to shift the equilibrium in the opposite direction back into the range which is compatible with physiological activity. There must be complicating factors, however, since in no case was the capacity of the eggs to achieve cleavage completely restored.

G. Antimitotic Effects of Colchicine: Intensification by High Pressure; Antagonism by Deuteration

1. *Rationale*

The available evidence indicates that a principal antimitotic action of the colchicine alkaloid results from a weakening of the structure of the mitotic spindle to the extent, with heavy dosage, of complete disintegration (Ludford, 1936). Also Inoué (1964) has shown that colchicine treatment induces a distinct fading of birefringence in the spindle of dividing eggs. Presumably, the effects represent structural changes in the microtubule components of the spindle and an inhibition of the polymerization processes which are essential to their formation (Taylor, 1965). Moreover, Tilney (1968) has found that colchicine treatments produce a disintegration of the microtubules in the Heliozoan, *Actinosphaerium,* and Tilney *et al.* (1966) have found that high pressure, in the range of 6000 to 8000 psi, produces a similar or identical effect.

In short, considerable evidence indicates that colchicine and pressure produce similarly directed changes in the structure of the mitotic apparatus. Consequently, it seemed reasonable to predict that these agencies would have synergistic effects upon mitotic function.

2. *Synergistic Effects of Colchicine and Pressure*

In these combination experiments (Marsland, 1966), the eggs of *Lytechinus variegatus* were utilized. The colchicine immersions and the pressure treatments were administered on a precise time schedule, i.e., the immersions were initiated 45 minutes after insemination and the pressure was applied exactly 10 minutes later. The colchicine concentrations and the pressure intensities were both subliminal with reference to observable antimitotic effects, when these agencies were employed separately.

A very distinct synergism was found between colchicine and pressure, as may be seen in Fig. 18. Pressures well below those required to produce any overt antimitotic effects, in the absence of colchicine, effected a complete blockage of the colchicinized eggs, even though the concentration

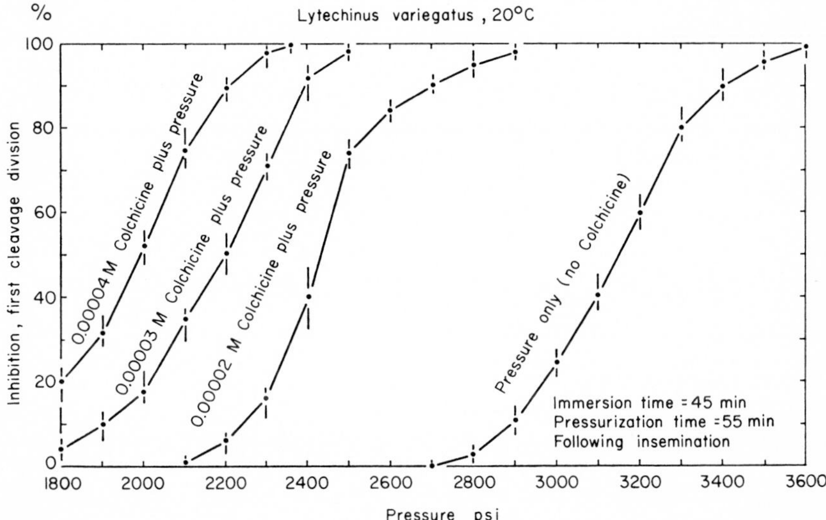

FIG. 18. Inhibition of first cleavage by colchicine and pressure applied concurrently. Each point represents the average of at least three experiments in which about 300 eggs were observed. Vertical lines indicate extremes of variation. (From Marsland, 1966.)

of the alkaloid was definitely subliminal. Moreover, as the concentration of colchicine was increased in the subliminal range, the pressure required to effect a certain degree of mitotic blockage became less and less.

Inspection of the curves presented in Fig. 19 permits one to deduce, rather roughly to be sure, that there is a degree of equivalence in the potencies of these two antimitotic agencies. From the shifting positions of the curves, it would appear that, within the range encompassed by the measurements, each increase of 1.0×10^{-5} in the molal concentration of colchicine was roughly equivalent to a pressure increase of 200–300 psi.

Synergism between colchicine and pressure was also evidenced by the fact that the delay in the appearance of cleavage furrows tended to be greater in the combination experiments than in the experiments dealing separately with colchicine and pressure. The maximum delay, observed when the combined treatments (Fig. 19) resulted in the blockage of 90% or more of the cleavages, averaged at 21 ± 1.5 minutes; whereas in the separate experiments it did not exceed 12 minutes.

The combined effects of pressure and colchicine, when the treatments were not prolonged, displayed a high degree of reversibility. Samples of

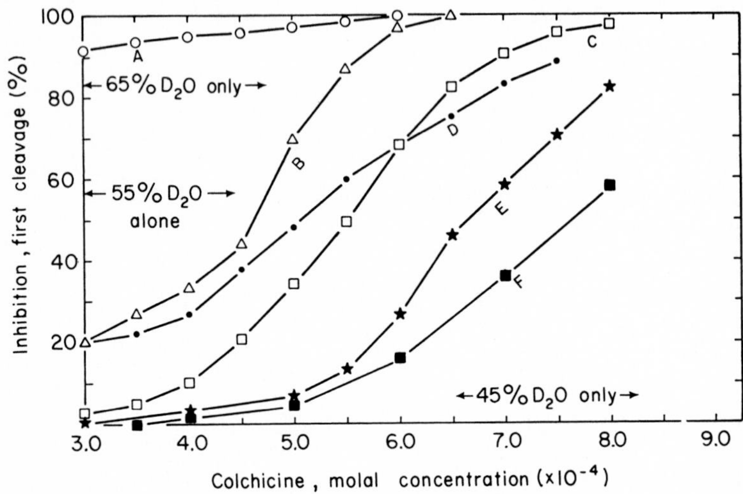

FIG. 19. Combined antimitotic effects of colchicine and D_2O. Treatments were started at early prophase, i.e., 17 minutes before furrowing. Curve *A*—colchicine plus 65% D_2O; *B*—colchicine plus 55% D_2O; *C*—colchicine only; *D*—colchicine plus 45% D_2O; *E* and *F*—colchicine plus 35 and 25% D_2O, respectively. Note antagonism with lower D_2O concentrations, synergism with higher and both effects in the intermediate range (45% D_2O). The points on curve *C* (colchicine only) represent, in each case, an average from at least seven experiments. Other points are averages from a minimum of three experiments. Inhibition, 35% D_2O = 1.0%; 25% D_2O = 0.3%. ○—○ (*A*) colchicine plus 65% D_2O; △—△ (*B*) colchicine plus 55% D_2O; □—□ (*C*) colchicine only; ●—● (*D*) colchicine plus 45% D_2O; *—* (*E*) colchicine plus 35% D_2O; ■—■ (*F*) colchicine plus 25% D_2O. (From Marsland and Hecht, 1968.)

the experimental eggs, examined 22 hours after treatment, were found to have given rise to a high percentage of morphologically normal swimming blastulas, provided that the compression period was limited to 65 minutes and/or the colchicine treatments were limited to 75 minutes. In practice, removal of the colchicine was effected by two successive transfers of one drop of egg suspension through 20 ml volumes of fresh sea water.

3. *Antagonism between Colchicine and Heavy Water*

This relationship is not a very simple one, as has been shown by Marsland and Hecht (1968). With lower (25 to 35%) concentrations of D_2O, however, a distinct antagonism was demonstrated, although some synergism was found at higher concentrations. Moreover, Inoué *et al.* (1965) have reported that D_2O and Colcemid have opposite effects upon the birefringence of the mitotic spindle. Accordingly, it may be said that agencies having similar effects upon mitotic structure tend to have synergistic effects upon mitotic function, whereas those that have opposite effects display antagonism with reference to mitotic function.

4. *Interpretation*

As predicted on the bases of similarity of action upon the structure of the mitotic apparatus, colchicine and high pressure are plainly synergistic in reference to their action upon mitotic function. Treatments which are subliminal when applied separately become definitely effective when applied in combination.

Colchicine, apparently, interferes with polymerization, inhibiting the formation of the microtubular structure by combining with and blocking off the prospective bonding sites by which the protein subunits can be linked together (Taylor, 1965). Pressure, on the other hand, tends to prevent the dispersal of the protective shells of bound water and thus inhibits the formation of such polymer linkages. In any event, the combined action of subliminal treatments with both agencies inhibits polymerization, of fosters depolymerization, to such an extent that the functions of the mitotic spindle can not proceed. Treatments with low concentrations of heavy water, on the other hand, are capable of restoring, to some extent, the polymerization tendencies in the colchicinized spindle.

IV. General Conclusions as to the Mechanisms of Karyokinesis and Cytokinesis

All in all the many experiments reported herewith provide substantial justification for the conclusion that the gel structures formed in animal cells as they are preparing to divide are instrumental in the performance of mechanical work, which leads to the achievement of division. The morphogenesis of both the mitotic apparatus and the strongly gelated cytoplasmic cortex involve the formation of elongate protein filaments, polymerized from subunits present in the cell prior to the time of division. Certainly for cytokinesis and perhaps for karyokinesis, the work process involves a forceful shortening of these filamentous gel structures. In any event, the development of mechanical energy appears to involve some degree of depolymerization. Accordingly, the gel systems tend to transform back toward the sol state. A continued performance of work, therefore, involves a continued reformation of the gel structures.

Metabolic energy, derived at least partly from the hydrolysis of ATP, must be available to support the gelational reactions which, in protoplasmic systems, are always endergonic. Protoplasmic gel structures, therefore, represent mechanisms whereby metabolic energy is transformed into mechanical work.

V. Coordination between Karyokinesis and Cytokinesis

A. A New Approach to the Problem

The precise and orderly sequence of events during cell division presupposes some system of accurate control; but very little was known about this system until recently. In 1956, however, a method was developed which has provided a new approach to the problem. With this method (Zimmerman and Marsland, 1956, 1960; Marsland and Auclair, 1957; Marsland, 1958*b*; Marsland *et al.*, 1960) it has been possible to induce the cleavage reaction prematurely, many minutes before the mitotic apparatus has begun to operate or even before it has been formed. This method involves subjecting egg cells to high force centrifugation while simultaneously exposing them to high pressure. Thus the method provides a means of analyzing some of the factors that initiate cytokinesis immediately following karyokinesis (see also Zimmerman, Chapter 10).

It has been known for some time that the mitotic apparatus exerts considerable control over the position of the prospective furrows in naturally cleaving egg cells. In some manner the spindle–aster complex strongly influences the localization of factors concerned with initiating the furrowing reaction. This conclusion stems from the work of many investigators. However, the centrifugal studies of Harvey (1935, 1956) and the microsurgical experiments of Hiramoto (1956) deserve particular mention. Both of these studies showed that when the mitotic apparatus is displaced prior to anaphase, the furrow is shifted to a new position determined by the changed position of the spindle–aster complex. When, however, the displacement is executed later, the furrow retains the position determined by the original location of the mitotic apparatus. Such results indicate that some sort of triggering device comes into operation during anaphase and that this determines not only the time when furrowing is to occur, but also the place at which the furrows will begin to cut into the egg.

B. Experimental Induction of Premature Furrowing

1. *Experimental Procedure*

This consists of subjecting the eggs (mainly *Arbacia punctulata*) to high speed centrifugation (40,000–50,000 *g*) while simultaneously exposing them to high pressure (8000–12,000 psi) for short periods (3–5 minutes). A small amount of 0.95 molal sucrose solution is placed at the bottom of the centrifuge chambers to a depth of about 1.0 mm, permitting the eggs to remain suspended in the sucrose–sea water boundary and preventing them from being flattened against the bottom of the chamber. The temperature throughout was maintained at 20 ± 0.3°C. Both the samples, the one centrifuged at high pressure and the other simultaneously centrifuged at atmospheric pressure, continue development, at least through several further division cycles. Full speed centrifugal acceleration was achieved within about 15 seconds and complete deceleration required only 8 seconds. Decompression was instantaneous and an examination of the eggs was initiated 1 minute later. Timing of the treatment varied from 15 to 55 minutes subsequent to insemination, while the eggs were maintained at the experimental temperature. The centrifugal force, in most of the experiments, was kept at 41,000 *g*, the speed being checked periodically by means of a stroboscope. None of the nonpressurized centrifuged eggs ever showed any sign of cleavage induction.

2. *Observations on Experimentally Induced Premature Cleavage*

Examined 1-minute after pressure centrifugation, the eggs are virtually spherical, although occasionally a slight elongation along the centrifugal axis can be observed. The stratification of the pressurized eggs, as compared with those that were centrifuged at atmospheric pressure, is more pronounced, most notably in regard to the displacement of the pigment bodies from the egg cortex into a definitive pigment zone at the centrifugal pole.

Some 3–5 minutes later, the pressurized centrifuge eggs begin to display incipient cleavage furrows, generally displaced slightly, sometimes drastically, in the direction toward the lighter hemisphere (Fig. 20). In most cases the percentage of furrowing eggs was 90–100%, and usually the furrows continued to advance until successful cleavage was achieved. In some cases such cleavage was induced as much as 45 minutes prematurely, at the time when syngamy was occurring, and generally it occurred well before prophase, i.e., before any normally occurring disintegration of the nuclear membrane and before the appearance of the mitotic apparatus.

FIG. 20. Incipient stage of an experimentally induced cleavage. Eggs photographed 3.5 minutes after inducing treatment which consisted of centrifuging the eggs at 41,000 *g* for 4 minutes, while compressed at 12,000 psi. Note that no nucleus can be found in the hyaline zone, that pigment is also absent from this zone, and that a distinct pigment zone is present at the centrifugal pole. In this experiment, 100% of the eggs showed induction and 98% of the furrows progressed to completion of cleavage. Eggs were maintained at 20°C and treatment was initiated 35 minutes after insemination. (From Tilney and Marsland, 1969.)

The duration of the pressure centrifugation was important in determining the percentage of experimentally induced cleavages, as is shown in Fig. 21.

3. *Involvement of Both Nuclear and Cytoplasmic Factors in the Induction Process*

The fact that a disintegration of the nuclear membrane in normally dividing cells is generally followed, with greater or lesser lag, by the onset of furrowing gave rise many years ago to the idea that substances liberated into the cytoplasm at this time play a role in initiating the cleavage reaction. Prior to the development of the technique of artificial cleavage induction, however, no experimental method for testing this hypothesis was available.

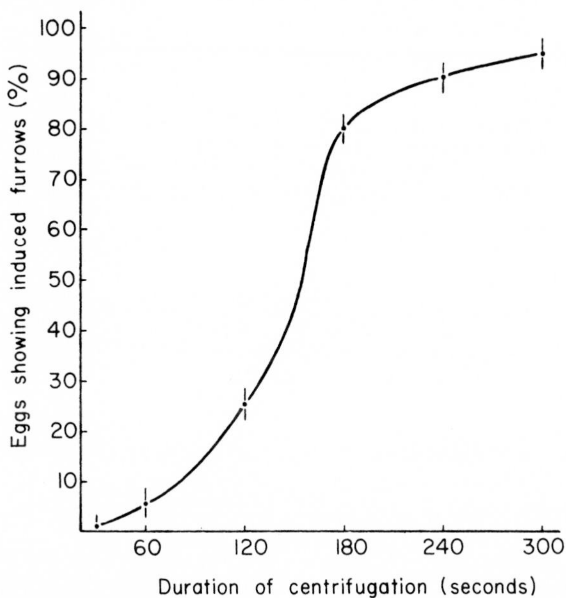

FIG. 21. Effects of duration of pressure centrifugation on the frequency (percentage) of cleavage induction in the eggs of *Arbacia punctulata*. Pressure, 12,000 psi; force, 41,000 g; treatments started 35 minutes after insemination; temperature, 20°C. Each point represents an average of five or more experiments. Vertical lines indicate extremes of variation. (From Marsland *et al.*, 1960.)

The new evidence unequivocally indicates that substances liberated upon the breakdown of the nuclear membrane do indeed play a role in cleavage induction. Also there is good evidence that certain cytoplasmic factors are also involved (Marsland, 1958b).

4. *Involvement of a Nuclear Factor*

The evidence for nuclear involvement is derived mainly from the experiments on premature cleavage induction. Throughout hundreds of experiments involving observation of thousands of eggs, experimentally induced furrowing has never occurred unless the nucleus was disrupted by the pressure centrifugation treatments. In eggs subjected to high-force centrifugation without pressurization, an intact nucleus can always be found situated (in the egg of *Arbacia punctulata*) in the hyaline zone, in close proximity to the oil cap. On the other hand, intact nuclei are never found in pressure centrifuged eggs, if the treatment has been sufficient to elicit the induced premature cleavage phenomenon. Instead, one finds small clumps of Fuelgen-positive material which vary somewhat as to their position along the axis of centrifugation (Marsland *et al.*, 1960). Moreover, Marsland has found (unpublished data) that among five kinds of echinoderm eggs studied only one kind (the eggs of *Sphaerechinus esculentis*) proved to be recalcitrant with reference to experimental cleavage induction, and this was the only one in which it was not possible to cause a breakdown of the nuclei using maximum available intensities of pressure and centrifugal force.[4]

5. *Possible Role of Calcium in Cleavage Induction*

The precise nature of the intranuclear furrow inducing factor remains problematical. However, there is some evidence (Tilney and Marsland, 1969) indicating that it may be ionic calcium, since calcium appears to be essential for the deposition of filaments in the wall of the incipient and advancing furrow (Section II,F).

The nuclear content has been found to be rich in calcium and other sources of polyvalent ions (Steffensen and Bergeron, 1959; also see Mirsky and Osawa, 1961). Accordingly, it may not be purely coincidental that nuclear breakdown almost universally constitutes a prelude to cytokinesis, not only in naturally occurring cell divisions, but also in experimentally induced cleavages.

6. *Identity and Role of a Cytoplasmic Factor*

In addition to a contraction of the cytoplasmic cortex in the equatorial zone of the dividing egg, a relaxation, or solation, of the gel structure in the polar regions would seem to be a necessary factor in the achievement of successful cleavage (see Marsland and Landau, 1954; Wolpert, 1960). Such a weakening of the gelated cytoplasmic cortex shortly prior to the appearance of definitive cleavage furrows has been demonstrated experimentally by Just (1922), Chambers and Chambers (1961), and by

[4] Exceptions to the general rule that cytokinesis follows in due time after nuclear breakdown may depend on special circumstances. For example, Harvey's observations (1936) of cleavage without nuclei in the parthenogenetic merogones of *Arbacia punctulata* may have been dependent on the release of materials from the germinal vesicle during maturation of the eggs.

Marsland (1939a). Accordingly, it is important to identify the mechanism which causes polar relaxation, as well as that which causes equatorial contraction.

It would be difficult to escape the conclusion that a relaxing factor analogous to the relaxing grana, first detected by Marsh (1952), must be involved. This is especially true in light of the recent experiments of Kinoshita and Yazaki (1967). These workers succeeded in obtaining preparations of relaxing grana from sea urchin embryos and have shown that the relaxing material in the egg becomes more concentrated at the poles, as compared with the equatorial region, when the furrow is developing. Undoubtedly, the grana effect a relaxation of the polar cortex by serving as calcium sequestering agents. In fact, Kinoshita and Yazaki have measured the calcium-binding capacity of their preparation, finding it to be very considerable, although not as great as grana derived from muscle.

Efforts to identify the cytoplasmic relaxing factor among the known structural elements present in the cell have proved difficult. The work of Kojima (1959) with echinoderm eggs suggests a relationship with certain vitally stainable cytoplasmic granules, especially those that stain with neutral red. Also, the experiments of the Marsland group (Marsland and Auclair, 1957; Zimmerman and Marsland, 1956; Marsland, 1958b; Marsland *et al.*, 1960) seem to indict certain vacuole-contained granules, identified by their susceptibility to metachromatic staining with toluidine blue.

Perhaps the most persuasive evidence as to the identity of the polar relaxation factor is provided by the recent work of Tilney and Marsland (1969). In his study the various fine structures identifiable in the *Arbacia* egg were followed during the course of naturally occurring cleavage and during experimentally induced cleavage.

The results of this work show that among the known fine structural elements present in the egg, only the annulate lamellae (Anderson, 1968) behave in a manner which is consistant with the assumption that they are closely related to, or identical with, the relaxing grana of Kinoshita and Yazaki (1967). In naturally occurring cleavage, the annulate lamellae disintegrate—in the sense that they no longer can be found—shortly before the appearance of the cleavage furrows, only to reappear shortly after telophase has been completed. Likewise, it was found that successful cleavage induction does not occur unless the pressure centrifugation treatment is adequate to cause a disintegration of the annulate lamellae. A similar involvement of remnants of the disintegrated nuclear membrane can not be ruled out, however. But since the annulate lamellae are known to be derived from nuclear envelope (Kessel, 1963), this distinction does not appear to be inportant. Tentatively it is postulated, accordingly, that the annulate lamellae and probably other derivatives of the nuclear envelope play a significant role in polar relaxation.

During normal cleavage, presumably, his relaxing material is carried to both poles by the mitotic apparatus (see below); but for experimentally induced cleavage the transporting agency must be centrifugal force. With subliminal treatments, in which the centrifugal force or the hydrostatic pressure or both are insufficient to trigger the cleavage reaction, intact nuclei and annulate lamellae are always found in the hyaline zone, near the centripetal pole and, undoubtedly, the breakdown products of these elements, at least initially, would occupy the same position. Accordingly, it almost invariably happens that the cleavage furrows of induced cleavages are displaced in greater or lesser degree from the equator, in the direction of the centripetal pole.

In later cleavages, when cytoplasmic annulate lamellae appear to be scarce or absent, it seems reasonable to assume that the much smaller cells of the embryo have a more modest requirement of polar relaxation factor, which could be provided by the disintegration of the nuclear envelope or from intranuclear annulate lamellae (Everingham, 1968).

In a broad sense the annulate lamellae and related elements may be regarded as analogous to the central vesicles of the membranous triads present in the Z-line of striated muscle. These elements of the complex sarcoplasmic reticulum (Porter and Franzini-Armstrong, 1964) likewise seem to serve as calcium sequestering loci and thus to play a dominant role in muscle relaxation. Moreover, the foregoing analogy is strengthened by the observation (Tilney and Marsland, 1969) that the annulate lamellae in egg cells often display definitive connections with the endoplasmic reticulum.

The assumption that a translocation of relaxing factors to the polar regions as cleavage approaches is achieved by virtue of activity centered in and around the mitotic apparatus is based upon a number of early observations as to cytoplasmic currents and the movement of microscopically visible elements in the cytoplasm of dividing eggs. More recently, moreover, Kojima (1959) has observed a polar displacement of neutral red-stained granules accumulated at the periphery of the asters and Auclair and Marsland (1965) have made confirmatory observations in other echinoderm eggs.

VI. Summary of Tentative Conclusions as to the Mechanisms of Cell Division

In summary, several general hypotheses, in support of which is greater or lesser fund of evidence has been cited in this chapter, are submitted, as follows.

1. The mitotic apparatus, including its component complex of microtubules, must be regarded as a typical protoplasmic gel structure, susceptible to solational disintegration and malfunction upon treatment with high pressure or low temperature. These agencies, as well as others (e.g., colchicine and heavy water), interfere with the processes of polymerization and depolymeration (sol–gel reactions) which are essential to the formation and functioning of the mitotic structure (Section III).

2. Cytokinesis is achieved in many, if not all, animal cells by virtue of contractile processes localized in the equatorial zone of the strongly gelated cytoplasmic cortex (Section II).

3. Contractility in the cortex of the dividing animal cell is associated with the deposition of numerous fine filaments (40–60-Å diameter) polymerized from protein subunits present in this part of the cytoplasm (Section II,F).

4. The deposition of contractile filaments in the cytoplasmic cortex is triggered by a mobilization calcium and/or other polyvalent cations derived from material liberated by the disintegration of the nuclear envelope (Section V,B,4).

5. A relaxation of the cytoplasmic cortex in the polar regions of the dividing cell is achieved by relaxing factors which are carried to the mitotic poles by virtue of mechanical activity centered in and around the mitotic apparatus (Section V,B,5). The polar localization of relaxing factors also accounts for the fact that contractile filaments are not formed in the polar cortex.

6. The relaxing factors are derivatives of the annulate lamellae, both cytoplasmic and intranuclear, and also of the nuclear envelope, from which the annulate lamellae originate (Section V,B,5).

Finally, a very broad conclusion, which was reached relatively early in these pressure–temperature studies, is that sol–gel transformations in cells generally and the underlying processes of polymerization and depolymerization serve primarily as mechanisms for the development of mechanical energy, which is derived initially from the energy of metabolism.

ACKNOWLEDGMENT

Work supported by grant series CA-00807 from the National Cancer Institute and GM-15893 from the National Institute of General Medical Sciences.

REFERENCES

Anderson, E. (1968). Oocyte differentiation in the sea urchin, *Arbacia punctulata,* with particular reference to the origin of cortical granules and their participation in the cortical reaction. *J. Cell Biol.* **37**, 514–529.

Arnold, J. M. (1968a). Formation of the first cleavage furrow in a telolecithal egg (*Loligo peallii*). *Biol. Bull.* **135**, 408–409.

Arnold, J. M. (1968b). An analysis of cleavage furrow formation in the egg of *Loligo peallii. Biol. Bull.* **135**, 413.

Ansevin, A. T., and Lauffer, M. A. (1959). Native tabacco mosaic virus protein of molecular weight 18,000. *Nature* **183**, 1601–1602.

Auclair, W., and Marsland, D. (1965). Unpublished observations made upon the eggs of *Arbacia lixula,* at the Stazione Zoologica, Naples, Italy.

Baker, P. C. (1965). Fine structure and morphogenetic movements in the gastrula of the treefrog, *Hyla regilla. J. Cell Biol.* **24**, 95–116.

Baker, P. C., and Schroeder, T. E. (1967). Cytoplasmic filaments and morphogenetic movement in the amphibian neural tube. *Develop. Biol.* **15**, 432–450.

Basset, J., and Macheboeuf, M. A. (1932). Etude sur les effets biologiques des ultrapressions: Résistance des bactéries, des diastases et des toxins aux pressions très élevées. *Compt. Rend.* **195**, 1431–1433.

Bikle, D., Tilney, L. G., and Porter, K. R. (1966). Microtubules and pigment migration in the melanophores of *Fundulus heteroclitus* L. *Protoplasma* **61**, 322–345.

Brown, D. E. S. (1934). The pressure coefficient of "viscosity" in the eggs of *Arbacia punctulata. J. Cellular Comp. Phsyiol.* **5**, 335–346.

Brown, D. E. S., Johnson, F. H., and Marsland, D. A. (1942). The pressure–temperature relationships of bacterial luminescence. *J. Cellular Comp. Physiol.* **20**, 151–168.

Carolan, R. M., Sato, H., and Inoué, S. (1965). A thermodynamic analysis of the effect of D_2O and H_2O on the mitotic spindle. *Biol. Bull.* **129**, 402.

Carolan, R. M., Sato, H., and Inoué, S. (1966). Further observations on the thermodynamics of the living mitotic spindle. *Biol. Bull.* **131**, 385.

Cattell, McK. (1936). The physiological effects of pressure. *Biol. Rev.* **11**, 441–476.

Chambers, R. (1938). Structural and kinetic aspects of cell division. *J. Cellular Comp. Physiol.* **12**, 149–165.

Chambers, R., and Chambers, E. L. (1961). "Explorations into the Nature of the Living Cell," Chapter 19, p. 236. Harvard Univ. Press, Cambridge, Massachusetts.

Chambers, R., and Kopac, M. J. (1937). The coalescence of sea-urchin eggs with oil drops. *Ann. Rept. Tortugas Lab., Carnegie Inst. Wash. Yearbook* **36**, 88–89.

Chlopin, G. W., and Tammann, G. (1903). Über den Einfluss höher Drucke auf Mikro-organismen. *Z. Hyg. Infectionskrankh.* **45**, 171–204.

Cloney, R. A. (1966). Cytoloplasmic filaments and cell movements: Epidermal cells during ascidian metamorphosis. *J. Ultrastruct. Res.* **14**, 300–328.

Dan, K., Yanagita, T., and Sugiyama, M. (1937). Behavior of the cell surface during cleavage I. *Protoplasma* **28**, 68–81.

Everingham, J. W. (1968). Attachment of intranuclear annulate lamellae to the nuclear envelope. *J. Cell Biol.* **37**, 540–550.

Favard, P., and Carasso, N. (1965). Mise en évidence d'un réticulum endoplasmique dans le spasmonème de ciliés péritriches. *J. Microscopie* **4**, 567–572.

Fontaine, M. (1930). Recherches expérimentales sur les réactions des êtres vivants aux fortes pressions. *Ann. Inst. Oceanogr. Monaco* **8**, 1–99.

Freundlich, H. (1937). Some recent work on gels. *J. Phys. Chem.* **41**, 901–310.

Gingell, D., and Palmer, J. F. (1968). Changes in membrane impedance associated with a cortical contraction in the egg of *Xenopus laevis*. *Nature* **217**, 98–102.

Gross, P. R., and Spindel, W. (1960). Heavy water inhibition of cell division: An approach to mechanism. *Ann. N. Y. Acad. Sci.* **90**, 500–522.

Harris, P. (1962). Some structural and functional aspects of the mitotic apparatus in sea urchin embryos. *J. Cell Biol.* **14**, 472–488.

Harvey, E. B. (1935). The mitotic figure and cleavage planes in the egg of *Parechinum microtuberculatus,* as influenced by centrifugal force. *Biol. Bull.* **69**, 287–299.

Harvey, E. B. (1936). Parthenogenetic merogony or cleavage without nuclei in *Arbacia punctulata*. *Biol. Bull.* **71**, 101–121.

Harvey, E. B. (1956). "The American *Arbacia* and Other Sea Urchins." Princeton Univ. Press, Princton, New Jersey.

Hill, L. E. (1912). "Caisson Sickness, and the Physiology of Work in Compressed Air." Arnold, London.

Hinsch, G., and Marsland, D. (1969). Manuscript in preparation.

Hiramoto, Y. (1956). Cell division without mitotic apparatus in sea urchin eggs. *Exp. Cell Res.* **11**, 630–638.

Hiramoto, Y. (1957). The thickness of the cortex and the refractive index of the protoplasm in sea urchin eggs. *Embryologia* (*Nagoya*) **3**, 361–374.

Hoffman-Berling, H. (1954). Die Glycerin-Wasserextrahierte Telophasenzelle als modell der Zytokinese. *Biochim. Biophys. Acta* **15**, 332–339.

Inoué, S. (1952). Effect of temperature on the birefringence of the mitotic spindle. *Biol. Bull.* **103**, 316.

Inoué, S. (1964). Organization and function of the mitotic spindle. *In* "Primitive Motile Systems in Cell Biology" (R. D. Allen and N. Kamiya, eds.), pp. 549–598. Academic Press, New York.

Inoué, S., Sato, H., and Tucker, R. W. (1963). Heavy water enhancement of mitotic spindle birefringence. *Biol. Bull.* **125**, 380–381.

Inoué, S., Sato, H., and Ascher, M. (1965). Counteraction of Colcemid and heavy water on the organization of the mitotic spindle. *Biol. Bull.* **129**, 409–410.

Just, E. E. (1922). Studies on cell division. I. The effect of dilute sea-water on the fertilized egg of *Echinarachnius parma* during the cleavage cycle. *Am. J. Physiol.* **61**, 505–515.

Kane, R. E. (1962). The mitotic apparatus. Fine structure of the isolated unit. *J. Cell Biol.* **15**, 279–287.

Kane, R. E., and Hersch, R. T. (1959). The isolation and preliminary characterization of a major soluble protein of the sea urchin egg. *Exptl. Cell Res.* **16**, 59–69.

Kessel, R. G. (1963). Electron microscope studies on the origin of annulate lamellae in öocytes of *Necturus*. *J. Cell Biol.* **19**, 391–414.

Kinoshita, S., and Yazaki, I. (1967). The behaviour and localization of intracellular relaxing system during cleavage in the sea urchin egg. *Exptl. Cell Res.* **47**, 449–458.

Kitching, J. A. (1964). The axopods of the sun animalcule *Actinophrys sol* (Heliozoa). *In* "Primitive Motile Systems in Cell Biology" (R. D. Allen and N. Kamiya, eds.), pp. 445–455, Academic Press, New York.

Kojima, M. K. (1959). Relation between the vitally stained granules and cleavage activity in the sea urchin. *Embryologia* **4**, 191–199.

Kriszat, G. (1949). Die Wirkung von Adenosinetriphosphat auf Amöben (*Chaos chaos*). *Arkiv Zool.* **1**, 81–86.

Kriszat, G. (1950). Die Wirkung von Adenosinetriphosphat und Calcium auf Amöben (*Chaos chaos*). *Arkiv Zool.* **2**, 477–490.

Landau, J. V., and Thibodeau, L. (1962). The micromorphology of *Amoeba proteus* during pressure-induced changes in the sol–gel cycle. *Exptl. Cell Res.* **27**, 591–594.

Landau, J. V., Marsland, D., and Zimmerman, A. M. (1954a). Kinetics of cell division: Action of mersalyl acid (MA) on the pressure–temperature cleavage block in the eggs of *Arbacia* and *Chaetopterus*. *Anat. Record* **120**, 789–790.

Landau, J. V., Zimmerman, A. M., and Marsland, D. (1954b). Temperature–pressure experiments on *Amoeba proteus*: *Plasmagel* structure in relation to form and movement. *J. Cellular Comp. Physiol.* **44**, 211–232.

Landau, J. V., Marsland, D. A., and Zimmerman, A. M. (1955). The energetics of cell division: Effects of adenosine triphosphate and related substances on the furrowing capacity of marine eggs (*Arbacia* and *Chaetopterus*). *J. Cellular Comp. Physiol.* **45**, 309–329.

Lauffer, M. A. (1962). Polymerization-depolymerization of tobacco mosaic virus protein. *In* "The Molecular Basis of Neoplasia," pp. 180–206. Univ. of Texas Press, Austin, Texas.

Ledbetter, M. C., and Porter, K. R. (1963). A "microtubule" in plant cell fine structure. *J. Cell Biol.* **19**, 239–250.

Loewy, A. G. (1952). An actomyosin-like substance from the plasmodium of a myxomycete. *J. Cellular Comp. Physiol.* **40**, 127–156.

Lucké, B., and Harvey, E. N. (1935). The permeability of living cells to heavy water (deuterium oxide). *J. Cellular Comp. Physiol.* **5**, 473–482.

Ludford, R. J. (1936). The action of toxic substances upon the division of normal and malignant cells *in vitro* and *in vivo*. *Arch. Exptl. Zellforsch.* **18**, 411–441.

Marsh, B. B. (1952). The effects of adenosine triphosphate on the fibre volume of a muscle homogenate. *Biochim. Biophys. Acta* **9**, 247–260.

Marsland, D. (1936). The cleavage of *Arbacia* eggs under hydrostatic compression. *Anat. Record* **67**, 38.

Marsland, D. (1938). The effects of high hydrostatic pressure upon cell division in *Arbacia* eggs. *J. Cellular Comp. Physiol.* **12**, 57–70.

Marsland, D. (1939a). The mechanism of cell division. Hydrostatic pressure effects upon dividing egg cells. *J. Cellular Comp. Physiol.* **13**, 15–22.

Marsland, D. (1939b). The mechanism of protoplasmic streaming. The effects of high hydrostatic pressure upon cyclosis in *Elodea canadensis*. *J. Cellular Comp. Physiol.* **13**, 23–30.

Marsland, D. (1942). Protoplasmic streaming in relation to gel structure in the cytoplasm. *In* "The Structure of Protoplasm" (W. Seifriz, ed.), pp. 127–161. Iowa State Coll. Press, Ames, Iowa.

Marsland, D. (1944). Mechanism of pigment displacement in unicellular chromatophores. *Biol. Bull.* **87**, 252–261.

Marsland, D. (1950). The mechanism of cell division: temperature-pressure experiments on the cleaving eggs of *Arbacia punctulata*. *J. Cellular Comp. Physiol.* **36**, 205–227.

Marsland, D. (1956a). Protoplasmic contractility in relation to cytokinesis. *Pubbl. Staz. Zool. Napoli* **28**, 182–203.

Marsland, D. (1956b). Protoplasmic contractility in relation to gel structure: Temperature-pressure experiments on cytokinesis and amoeboid movement. *Intern. Rev. Cytol.* **5**, 199–227.

Marsland, D. (1957). Temperature–pressure studies on the role of sol–gel reactions in cell division. *In* "Influence of Temperature on Biological Systems" (F. H. Johnson, ed.), pp. 111–126. Waverly Press, Baltimore, Maryland.

Marsland, D. (1958a). Cells at high pressure. *Sci. Am.* **199**, 36–43.

Marsland, D. (1958b). Nuclear and cytoplasmic factors in cleavage induction. *Anat. Record* **132**, 473.

Marsland, D. (1964a). Pressure–temperature studies on ameboid movement and related phenomena: an analysis of the effects of heavy water (D_2O) on the form, movement, and gel structure of *Amoeba proteus*. *In* "Primitive Motile Systems in Cell Biology" (R. D. Allen and N. Kamiya, eds.), pp. 173–187, Academic Press, New York.

Marsland, D. (1964b). High pressure reversal of the antimitotic effects of heavy water (D_2O). *Biol. Bull.* **127**, 356.

Marsland, D. (1965). Partial reversal of the anti-mitotic effects of heavy water by high hydrostatic pressure. *Exptl. Cell Res.* **38**, 592–603.

Marsland, D. (1966). Anti-mitotic effects of colchicine and hydrostatic pressure: Synergistic action on the cleaving eggs of *Lytechinus variegatus*. *J. Cell Physiol.* **67**, 333–338.

Marsland, D., and Asterita, H. (1966). Counteraction of the antimitotic effects of D_2O in the dividing eggs of *Arbacia punctulata*: A temperature-pressure analysis. *Exptl. Cell Res.* **42**, 316–327.

Marsland, D., and Auclair, W. (1957). A further study on the induced furrowing reaction in *Arbacia punctulata*. *Biol. Bull.* **113**, 348–349.

Marsland, D. A., and Brown, D. E. S. (1936). Amoeboid movement at high hydrostatic pressure. *J. Cellular Comp. Physiol.* **8**, 167–178.

Marsland, D. A., and Brown, D. E. S. (1942). The effects of pressure on sol–gel equilibria, with special reference to myosin and other protoplasmic gels. *J. Cellular Comp. Physiol.* **20**, 295–305.

Marsland, D., and Cone, V. (1969). Effects of D_2O on pigment movement and fine structure in the protoplasmic processes of fish melanocytes. Current observations, to be reported shortly.

Marsland, D., and Hecht, R. (1968). Cell division: Combined antimitotic effects of colchicine and heavy water on first cleavage in the eggs of *Arbacia punctulata*. *Exptl. Cell Res.* **51**, 602–608.

Marsland, D., and Hiramoto, Y. (1964). Partial reversal by high pressure of the anti-meiotic effects of heavy water in the öocytes of the starfish, *Asterias forbesi*. *Biol. Bull.* **127**, 380.

Marsland, D., and Hiramoto, Y. (1966). Cell division: Pressure-induced reversal of the anti-meiotic effects of heavy water in the öocytes of the starfish, *Asterias forbesi*. *J. Cellular Physiol.* **67**, 13–21.

Marsland, D. A., and Jaffe, O. (1949). Effects of pressure on the cleaving eggs of the frog (*Rana pipiens*). *J. Cellular Comp. Physiol.* **34**, 439–450.

Marsland, D. A., and Landau, J. V. (1954). The mechanisms of cytokinesis: temperature-pressure studies on the cortical gel system in various marine eggs. *J. Exptl. Zool.* **125**, 507–539.

Marsland, D., and Meisner, D. (1967). Effects of D_2O on the mechanism of pigment dispersal in the melanocytes of *Fundulus heteroclitus*: A pressure-temperature analysis. *J. Cell Physiol.* **70**, 209–216.

Marsland, D., Tilney, L. G., and Hirshfield, M. (1969). Stabilization of microtubular structure in the axopodia of a heliozoan (*Actinosphaerium nucleofilum*) by D_2O: A pressure analysis. To be submitted shortly.

Marsland, D., and Zimmerman, A. M. (1963). Cell division: Differential effects of heavy water upon the mechanisms of cytokinesis and karyokinesis in the eggs of *Arbacia punctulata. Exptl. Cell Res.* **30**, 23–35.

Marsland, D., and Zimmerman, A. M. (1965). Structural stabilization of the mitotic apparatus by heavy water, in the cleaving eggs of *Arbacia punctulata. Exptl. Cell Res.* **38**, 306–313.

Marsland, D., Zimmerman, A. M., and Auclair, W. (1960). Cell division: Experimental induction of cleavage furrows in the eggs of *Arbacia punctulata. Exptl. Cell Res.* **21**, 179–196.

Marsland, D., Zimmerman, A. M., and Asterita, H. (1962). Effects of D_2O on the cortical gel structure and cleavage capacity of *Arbacia* eggs. *Biol. Bull.* **123**, 484.

Mazia, D., and Dan, K. (1952). The isolation and biochemical characterization of the mitotic apparatus of dividing cells. *Proc. Natl. Acad. Sci. U. S.* **38**, 826–838.

Mazia, D., and Zimmerman, A. M. (1958). SH compounds in mitosis. II. The effect of mercaptoethanol on the structure of the mitotic apparatus in sea urchin eggs. *Exptl. Cell Res.* **15**, 138–153.

Mercer, E. H., and Wolpert, L. (1958). Electron microscopy of cleaving sea urchin eggs. *Exptl. Cell Res.* **14**, 692–732.

Mirsky, A. E., and Osawa, S. (1961). The interphase nucleus. *In* "The Cell' (J. Brachet and A. E. Mirsky, eds.), Vol. 2, pp. 677–770. Academic Press, New York.

Mitchison, J. M. (1952). Cell membranes and cell division. *Symp. Soc. Exptl. Biol.* **6**, 105–127.

Pease, D. C. (1941). Hydrostatic pressure effects upon the spindle figure and chromosome movement. I. Experiments on the first mitotic division of *Urechis* eggs. *J. Morphol.* **69**, 405–441.

Pease, D. C. (1946). Hydrostatic pressure effects upon the spindle figure and chromosome movement. II. Experiments on the meiotic divisions of *Tradescantia* pollen mother cells. *Biol. Bull.* **91**, 145–169.

Pease, D. C., and Marsland, D. A. (1939). The cleavage of *Ascaris* eggs under exceptionally high pressure. *J. Cellular Comp. Physiol.* **14**, 1–2.

Porter, K. R., and Franzini-Armstrong, C. (1964). Sarcolemmal invaginations and the T-system in fish skeletal muscle. *Nature* **202**, 355–357.

Randall, J. T., and Jackson, S. F. (1958). Fine structure and function in *Stentor polymorphus. J. Biophys. Biochem. Cytol.* **4**, 807–830.

Regnard, P. (1884). Recherches expérimentales sur l'influence des très hautes pressions sur les organismes vivants. *Compt. Rend.* **98**, 745–747.

Regnard, P. (1885). Influence des hautes pressions sur l'éclosion des oeufs de poisson. *Compt. Rend. Soc. Biol.* **37**, 48.

Regnard, P. (1886). Actions des hautes pressions sur les tissues animaux. *Compt. Rend.* **102**, 173–176.

Regnard, P. (1887). Influence des hautes pressions sur la rapidité du courant nerveux. *Compt. Rend. Soc. Biol.* **39**, 406–408.

Regnard, P. (1891). "Recherches expérimentales sur les conditions physiques de la vie dans les eaux." Masson, Paris.

Rudzinska, M. A. (1965). The fine structure and function of the tentacle in *Tokophrya infusionum*. *J. Cell Biol.* **25**, 459–477.

Runnström, J. (1949). The mechanism of fertilization in metazoa. *Advan. Enzymol.* **9**, 241–327.

Sakai, H. (1968). Contractile properties of protein threads from sea urchin eggs in relation to cell division. *Intern. Rev. Cytol.* **23**, 89–112.

Schechtman, A. M. (1937). Localized cortical growth as the immediate cause of cell division. *Science* **85**, 222–223.

Schroeder, T. E. (1968). Cytokinesis: Filaments in the cleavage furrow. *Exptl. Cell Res.* **53**, 272–318.

Sichel, F. J. M., and Burton, A. C. (1936). A kinetic method of studying surface forces in the egg of *Arbacia*. *Biol. Bull.* **71**, 397–398.

Steffensen, D., and Bergeron, J. A. (1959). Autoradiographs of pollen tube nuclei with calcium-45 *J. Biophys. Biochem. Cytol.* **6**, 339–342.

Stephens, R. E., and Kane, R. E. (1966). Studies on a major protein from isolated sea urchin egg cortex. *Biol. Bull.* **131**, 382–383.

Swann, M. M. (1952). The nucleus in fertilization, mitosis and cell division. *Symp. Soc. Exptl. Biol.* **6**, 89–104.

Szollosi, D. (1968). The contractile ring and changes of the cell surface during cleavage. *J. Cell Biol.* **39**, 133a.

Taylor, E. W. (1965). The mechanism of colchicine inhibition of mitosis. I. Kinetics of inhibition and the binding of H^3-colchicine. *J. Cell Biol.* **25**, 145–160.

Tilney, L. G. (1968). Studies on the micortubules in Heliozoa. IV. The effect of colchicine on the formation and maintenance of the axopodia and the redevelopment of pattern in *Actinosphaerium nucleofilum* (Barrett). *J. Cell Sci.* **3**, 549–562.

Tilney, L. G., and Gibbins, J. R. (1969). Microtubules in the formation and development of the primary mesenchyme in *Arbacia punctulata*. II. An experimental analysis of their rule in development and maintenance of cell shape. *J. Cell Biol.* **41**, 227–250.

Tilney, L. G., and Marsland, D. (1969). A fine structural analysis of cleavage induction and the furrowing mechanism in the eggs of *Arbacia punctulata*. *J. Cell Biol.* **42**, 170–184.

Tilney, L. G., and Porter, K. R. (1965). Studies on the microtubules in Heliozoa. II. The effect of low temperature on these structures in the formation and maintenance of the axopodia. *J. Cell Biol.* **34**, 327–343.

Tilney, L. G., Hiramoto, Y., and Marsland, D. (1966). Studies on the microtubules in Heliozoa. III. A pressure analysis of the role of these structures in the formation and maintenance of the axopodia of *Actinosphaerium nucleofilum* (Barrett). *J. Cell Biol.* **29**, 77–95.

Tucker, R. W., and Inoué, S. (1963). Rapid exchange of D_2O and H_2O in sea urchin eggs. *Biol. Bull.* **125**, 395.

Urey, H. C. (1933). The separation and properties of the isotopes of hydrogen. *Science* **78**, 566–571.

Ussing, H. H. (1935). The influence of heavy water on the development of amphibian eggs. *Skand. Arch. Physiol.* **72**, 192–198.

Weber, H. H., and Portzehl, H. (1954). The transference of the muscle energy in the contraction cycle. *Progr. Biophys. Biochem.* **4**, 60–111.

Weinstein, R. S., and Herbst, R. R. (1964). Electron microscopy of cleavage furrows in sea urchin blastomeres. *J. Cell Biol.* **23**, 101A.

Wohlman, A., and Allen, R. D. (1968). Structural organization associated with pseudopod extension and contraction during cell locomotion in *Difflugia*. *J. Cell Sci.* **3**, 105–114.

Wolpert, L. (1960). The mechanics and mechanism of cleavage. *Intern. Rev. Cytol.* **10**, 164–216.

Wolpert, L., Marsland, D., and Hirshfield, M. (1968). The effect of high hydrostatic pressure on the mechanical properties of the membrane of the sea urchin egg. *Biol. Bull.* **135**, 442–443.

Yaqui, R., and Shigenaka, Y. (1963). Electron microscopy of the longitudinal fibrillar bundle and the contractile fibrillar system in *Spirostomum ambiguum*. *J. Protzool.* **10**, 364–369.

Zimmerman, A. M., and Marsland, D. (1956). Induction of premature cleavage furrows in the eggs of *Arbacia punctulata*. *Biol. Bull.* **111**, 317.

Zimmerman, A. M., and Marsland, D. (1960). Experimental induction of the furrowing reaction in the eggs of *Arbacia punctulata*. *Ann. N. Y. Acad. Sci.* **90**, 480–485.

Zimmerman, A. M., and Marsland, D. (1964). Cell division: Effects of pressure on the mitotic mechanisms of marine eggs (*Arbacia punctulata*). *Exptl. Cell Res.* **35**, 293–302.

Zimmerman, A. M., Landau, J. V., and Marsland, D. (1957). Cell Division: A pressure-temperature analysis of the effects of sulfhydryl reagents on the cortical plasmagel structure and furrowing strength of dividing eggs (*Arbacia* and *Chaetopterus*). *J. Cellular Comp. Physiol.* **49**, 395–435.

AUTHOR INDEX

Numbers in italics refer to the pages on which the complete references are listed.

N

O

P

Q

R

S

T

U

V

W

SUBJECT INDEX

L

M

N

O

P